About the Authors

Titu Andreescu received his Ph.D. from the West University of Timisoara, Romania. The topic of his dissertation was "Research on Diophantine Analysis and Applications." Professor Andreescu currently teaches at The University of Texas at Dallas. He is past chairman of the USA Mathematical Olympiad, served as director of the MAA American Mathematics Competitions (1998–2003), coach of the USA International Mathematical Olympiad Team (IMO) for 10 years (1993–2002), director of the Mathematical Olympiad Summer Program (1995–2002), and leader of the USA IMO Team (1995–2002). In 2002 Titu was elected member of the IMO Advisory Board, the governing body of the world's most prestigious mathematics competition. Titu co-founded in 2006 and continues as director of the AwesomeMath Summer Program (AMSP). He received the Edyth May Sliffe Award for Distinguished High School Mathematics Teaching from the MAA in 1994 and a "Certificate of Appreciation" from the president of the MAA in 1995 for his outstanding service as coach of the Mathematical Olympiad Summer Program in preparing the US team for its perfect performance in Hong Kong at the 1994 IMO. Titu's contributions to numerous textbooks and problem books are recognized worldwide.

Dorin Andrica received his Ph.D in 1992 from "Babeş-Bolyai" University in Cluj-Napoca, Romania; his thesis treated critical points and applications to the geometry of differentiable submanifolds. Professor Andrica has been chairman of the Geometry Division at "Babeş-Bolyai" University since 1995 and head of the Department of Pure Mathematics at the same University since 2007. He has written and contributed to numerous mathematics textbooks, problem books, articles and scientific papers at various levels. Dorin has been an invited lecturer at university conferences around the world: Austria, Bulgaria, Czech Republic, Egypt, France, Germany, Greece, Italy, the Netherlands, Portugal, Serbia, Turkey, and the USA. Dorin is a member of the Romanian Committee for the Mathematics Olympiad and is a member on the editorial boards of several international journals. Also, he is well known for his conjecture about consecutive primes called "Andrica's Conjecture one of the most important open problems in Number Theory". Dorin has been a regular faculty member at the Canada–USA Mathcamps (2001–2005) and at the AwesomeMath Summer Program (AMSP) since 2006.

Titu Andreescu   Dorin Andrica

Number Theory

Structures, Examples, and Problems

Birkhäuser
Boston • Basel • Berlin

Titu Andreescu
Department of Science
  and Mathematics Education
University of Texas at Dallas
2601 N. Floyd Road, FN 33
Richardson, TX 75002, USA
titu.andreescu@utdallas.edu

Dorin Andrica
Faculty of Mathematics
  and Computer Science
"Babeş-Bolyai" University
Str. Kogalniceanu 1
3400 Cluj-Napoca, Romania
dandrica@math.ubbcluj.ro

ISBN: 978-0-8176-3245-8   e-ISBN: 978-0-8176-4645-5
DOI: 10.1007/b11856

Library of Congress Control Number: 2009921128

Mathematics Subject Classification (2000): 11A05, 11A07, 11A15, 11A25, 11A25, 11B37, 11B39, 11B65, 11D04, 11D09, 11D25

Cover Design by Joseph Sherman

Printed on acid-free paper.

www.birkhauser.com

Die ganzen Zahlen hat der liebe Gott gemacht, alles andere ist Menschenwerk.
God made the integers, all else is the work of man.

Leopold Kronecker

Contents

Preface

One of the oldest and liveliest branches of mathematics, number theory is noted for its theoretical depth and applications to other fields, including representation theory, physics, and cryptography. The forefront of number theory is replete with sophisticated and famous open problems; at its foundation, however, are basic, elementary ideas that can stimulate and challenge beginning students. This textbook takes a problem-solving approach to number theory, situating each theoretical concept within the framework of some examples or some problems for readers to solve. Starting with the essentials, the text covers divisibility, powers of integers, the floor function and fractional part, digits of numbers, basic methods of proof (extremal arguments, pigeonhole principle, induction, infinite descent, inclusion–exclusion), arithmetic functions, divisibility theorems, and Diophantine equations. Emphasis is also placed on the presentation of some special problems involving quadratic residues; Fermat, Mersenne, and perfect numbers; as well as famous sequences of integers such as Fibonacci, Lucas, and those defined by recursive relations. By thoroughly discussing interesting examples and applications and by introducing and illustrating every key idea with relevant problems of various levels of difficulty, the book motivates, engages, and challenges the reader. The exposition proceeds incrementally and intuitively, and rigorously uncovers deeper properties.

A special feature of the book is an outstanding selection of genuine Olympiad and other mathematical contest problems solved using the methods already presented. The book brings about the unique and vast experience of the authors. It captures the spirit of the mathematical literature and distills the essence of a rich problem-solving culture.

Number Theory: Structures, Examples, and Problems will appeal to senior high school and undergraduate students and their instructors, as well as to all who would like to expand their mathematical horizons. It is a source of fascinating problems for readers at all levels and widely opens the gate to further explorations in mathematics.

Titu Andreescu, University of Texas at Dallas

Dorin Andrica, "Babeş-Bolyai" University
Cluj-Napoca, Romania

March 2008

Acknowledgments

Many problems are either inspired by or adapted from various mathematical contests in different countries. We express our deepest appreciation to the original proposers of the problems. Special thanks are given to Gabriel Dospinescu (École Normale Supérieure, Paris, France) and Paul Stanford (University of Texas at Dallas) for their careful proofreading of the manuscript and for many helpful suggestions. We are deeply grateful to Richard Stong for carefully reading the manuscript and considerably improving its quality.

Notation

$\mathbb{Z}$	the set of integers
$\mathbb{Z}_n$	the set of integers modulo n
$\mathbb{N}$	the set of positive integers
$\mathbb{N}_0$	the set of nonnegative integers
$\mathbb{Q}$	the set of rational numbers
$\mathbb{Q}^+$	the set of positive rational numbers
$\mathbb{Q}^{\geq}$	the set of nonnegative rational numbers
$\mathbb{Q}^n$	the set of n-tuples of rational numbers
$\mathbb{R}$	the set of real numbers
$\mathbb{R}^+$	the set of positive real numbers
$\mathbb{R}^{\geq}$	the set of nonnegative real numbers
$\mathbb{R}^n$	the set of n-tuples of real numbers
$\mathbb{C}$	the set of complex numbers
$|A|$	the number of elements in the set A
$A \subset B$	A is a proper subset of B
$A \subseteq B$	A is a subset of B
$A \setminus B$	A without B (set difference)
$A \cap B$	the intersection of sets A and B
$A \cup B$	the union of sets A and B
$a \in A$	the element a belongs to the set A
$n \mid m$	n divides m
$\gcd(m, n)$	the greatest common divisor of m, n
$\operatorname{lcm}(m, n)$	the least common multiple of m, n
$\pi(n)$	the number of primes $\leq n$
$\tau(n)$	number of divisors of n
$\sigma(n)$	sum of all positive divisors of n
$a \equiv b \pmod{m}$	a and b are congruent modulo m
φ	Euler's totient function
$\operatorname{ord}_m(a)$	order of a modulo m
μ	Möbius function

$\overline{a_k a_{k-1} \cdots a_0}_{(b)}$	base-b representation
$S(n)$	the sum of the digits of n
$(f_1, f_2, \ldots, f_m)$	factorial base expansion
$\lfloor x \rfloor$	floor of x
$\lceil x \rceil$	ceiling of x
$\{x\}$	fractional part of x
e_p	Legendre function
$p^k \parallel n$	p^k fully divides n
f_n	Fermat number
M_n	Mersenne number
$\left(\frac{a}{p}\right)$	Legendre symbol
F_n	Fibonacci number
L_n	Lucas number
P_n	Pell number
$\binom{n}{k}$	binomial coefficient

Part I

Fundamentals

1

Divisibility

1.1 Divisibility

For integers a and b, $a \neq 0$, we say that a *divides* b if $b = ac$ for some integer c. We denote this by $a \mid b$. We also say that b is divisible by a or that b is a multiple of a.

Because $0 = a \cdot 0$, it follows that $a \mid 0$ for all integers a. We have $0 \mid 0$, since $0 = 0 \cdot 0$.

Straight from the definition we can derive the following properties:

1. If $a \mid b$, $b \neq 0$, then $|a| \leq |b|$;

2. If $a \mid b$ and $a \mid c$, then $a \mid \alpha b + \beta c$ for any integers α and β;

3. If $a \mid b$ and $a \mid b \pm c$, then $a \mid c$;

4. $a \mid a$ (reflexivity);

5. If $a \mid b$ and $b \mid c$, then $a \mid c$ (transitivity);

6. If $a \mid b$ and $b \mid a$, then $|a| = |b|$.

The following result is called the *division algorithm*, and it plays an important role:

Theorem. *For any positive integers a and b there exists a unique pair (q, r) of nonnegative integers such that*

$$b = aq + r, \quad r < a.$$

T. Andreescu and D. Andrica, *Number Theory*, DOI: 10.1007/b11856_1,
© Birkhäuser Boston, a part of Springer Science + Business Media, LLC 2009

Proof. Since $a \geq 1$, there exist positive integers n such that $na > b$ (for example, $n = b$ is one such). Let q be the least positive integer for which $(q + 1)a > b$. Then $qa \leq b$. Let $r = b - aq$. It follows that $b = aq + r$ and $0 \leq r < a$.

For the uniqueness, assume that $b = aq' + r'$, where q' and r' are also non-negative integers satisfying $0 \leq r' < a$. Then $aq + r = aq' + r'$, implying $a(q - q') = r' - r$, and so $a \mid r' - r$. Hence $|r' - r| \geq a$ or $|r' - r| = 0$. Because $0 \leq r, r' < a$ yields $|r' - r| < a$, we are left with $|r' - r| = 0$, implying $r' = r$ and, consequently, $q' = q$. $\qquad \square$

In the theorem above, when b is divided by a, q is called the *quotient* and r the *remainder*.

Remark. The division algorithm can be extended for integers as follows: For any integers a and b, $a \neq 0$, there exists a unique pair (q, r) of integers such that

$$b = aq + r, \quad 0 \leq r < |a|.$$

Example. Prove that for all positive integers n, the fraction

$$\frac{21n + 4}{14n + 3}$$

is irreducible.

(1st International Mathematical Olympiad)

Indeed, from the equality

$$2(21n + 4) - 3(14n + 3) = -1$$

it follows that $21n + 4$ and $14n + 3$ have no common divisor except for 1; hence the conclusion.

Problem 1.1.1. *Prove that for all integers n:*
   (a) $n^5 - 5n^3 + 4n$ is divisible by 120;
   (b) $n^2 + 3n + 5$ is not divisible by 121.
Solution. (a) $n^5 - 5n^3 + 4n = n(n^2 - 1)(n^2 - 4)$
$$= n(n - 1)(n + 1)(n - 2)(n + 2),$$

the product of five consecutive integers: $n - 2, n - 1, n, n + 1, n + 2$.
   If $n \in \{-2, -1, 0, 1, 2\}$ we get $n^5 - 5n^3 + 4n = 0$ and the property holds.
   If $n \geq 3$ we can write

$$n^5 - 5n^3 + 4n = 5! \binom{n + 2}{5} = 120 \binom{n + 2}{5},$$

and the conclusion follows.

If $n \leq -3$, write $n = -m$, where $m \geq 3$, and obtain

$$n^5 - 5n^3 + 4n = -120\binom{m+2}{5},$$

and we are done.

(b) Observe that

$$n^2 + 3n + 5 = (n+7)(n-4) + 33,$$

so that $11 \mid n^2+3n+5$ if and only if $11 \mid (n+7)(n-4)$. Thus, if $11 \nmid (n+7)(n-4)$ then 11 (and hence 121) does not divide $n^2 + 3n + 5$. So, assume 11 divides $(n+7)(n-4)$. Then $11 \mid n+7$ or $11 \mid n-4$; but then 11 must divide both of $n+7$ and $n-4$, since $(n+7) - (n-4) = 11$. Thus, $121 \mid (n+7)(n-4)$. However, $121 \nmid 33$. So $121 \nmid n^2 + 3n + 5 = (n+7)(n-4) + 33$. Hence, in all cases, $121 \nmid n^2 + 3n + 5$.

Problem 1.1.2. *Let $a > 2$ be an odd number and let n be a positive integer. Prove that a divides $1^{a^n} + 2^{a^n} + \cdots + (a-1)^{a^n}$.*

Solution. Define $k = a^n$ and note that k is odd. Then

$$d^k + (a-d)^k = a[d^{k-1} - d^{k-2}(a-d) + \cdots + (a-d)^{k-1}]$$

Summing up the equalities from $d = 1$ to $d = \frac{a-1}{2}$ implies that p divides $1^k + 2^k + \cdots + (a-1)^k$, as claimed.

Problem 1.1.3. *Prove that*

$$3^{4^5} + 4^{5^6}$$

is a product of two integers each of which is larger than 10^{2002}.

Solution. Write $3^{4^5} + 4^{5^6}$ as $m^4 + 4n^4$, where $n = 2^{(5^6-1)/2}$. Then the desired factorization is

$$m^4 + 4n^4 = (m^2 + 2n^2)^2 - 4m^2n^2 = (m^2 - 2mn + 2n^2)(m^2 + 2mn + 2n^2).$$

Since the smaller factor is

$$m^2 - 2mn + 2n^2 = (m-n)^2 + n^2 \geq n^2 = 2^{5^6-1} > 2^{8008} = (2^4)^{2002} > 10^{2002},$$

we are done.

Problem 1.1.4. *Find all positive integers n such that for all odd integers a, if $a^2 \leq n$ then $a \mid n$.*

Solution. Consider a fixed positive integer n. Let a be the greatest odd integer such that $a^2 < n$ and hence $n \leq (a+2)^2$. If $a \geq 7$, then $a-4, a-2, a$ are odd integers that divide n. Note that any two of these numbers are relatively prime, so $(a-4)(a-2)a$ divides n. It follows that $(a-4)(a-2)a \le (a+2)^2$, so $a^3-6a^2+8a \le a^2+4a+4$. Then $a^3-7a^2+4a \le 0$ or $a^2(a-7)+4(a-1) \le 0$. This is false, because $a \ge 7$; hence $a = 1, 3$, or 5.

If $a = 1$, then $1^2 \le n \le 3^2$, so $n \in \{1, 2, \ldots, 8\}$.

If $a = 3$, then $3^2 \le n \le 5^2$ and $1 \cdot 3 \mid n$, so $n \in \{9, 12, 15, 18, 21, 24\}$.

If $a = 5$, then $5^2 \le n \le 7^2$ and $1 \cdot 3 \cdot 5 \mid n$, so $n \in \{30, 45\}$. Therefore $n \in \{1, 2, 3, 4, 5, 6, 7, 8, 9, 12, 15, 18, 21, 24, 30, 45\}$.

Problem 1.1.5. *Find the elements of the set*

$$S = \left\{ x \in \mathbb{Z} \ \middle| \ \frac{x^3 - 3x + 2}{2x + 1} \in \mathbb{Z} \right\}.$$

Solution. Since $\frac{x^3-3x+2}{2x+1} \in \mathbb{Z}$, then

$$\frac{8x^3 - 24x + 16}{2x + 1} = 4x^2 - 2x - 11 + \frac{27}{2x + 1} \in \mathbb{Z}.$$

It follows that $2x + 1$ divides 27, so

$$2x + 1 \in \{\pm 1, \pm 3, \pm 9, \pm 27\} \quad \text{and} \quad x \in \{-14, -5, -2, -1, 0, 1, 4, 13\},$$

since $2x + 1$ is odd, $\frac{x^3-3x+2}{2x+1} \in \mathbb{Z} \Leftrightarrow \frac{8x^3-24x+16}{2x+1} \in \mathbb{Z}$, so all these are solutions.

Problem 1.1.6. *Find all positive integers n for which the number obtained by erasing the last digit is a divisor of n.*

Solution. Let b be the last digit of the number n and let a be the number obtained from n by erasing the last digit b. Then $n = 10a + b$. Since a is a divisor of n, we infer that a divides b. Any number n that ends in 0 is therefore a solution. If $b \ne 0$, then a is a digit and n is one of the numbers $11, 12, \ldots, 19, 22, 24, 26, 28, 33, 36, 39, 44, 48, 55, 66, 77, 88, 99$.

Problem 1.1.7. *Find the greatest positive integer x such that 2^{36+x} divides $2000!$.*

Solution. The number 23 is prime and divides every 23rd number. In all, there are $\lfloor \frac{2000}{23} \rfloor = 86$ numbers from 1 to 2000 that are divisible by 23. Among those 86 numbers, three of them, namely 23^2, $2 \cdot 23^2$, and $3 \cdot 23^2$, are divisible by 23^2. Hence $23^{89} \mid 2000!$ and $x = 89 - 6 = 83$.

Problem 1.1.8. *Find all positive integers a, b, c such that*

$$ab + bc + ac > abc.$$

Solution. Assume that $a \le b \le c$. If $a \ge 3$ then $ab + bc + ac \le 3bc \le abc$, a contradiction. Since a is an integer, all that is left is that $a = 2$ or $a = 1$.

If $a = 2$, then the inequality becomes $2b + 2c + bc > 2bc$; hence $\frac{1}{c} + \frac{1}{b} > \frac{1}{2}$.

If $b \geq 5$, then $c \geq 5$ and

$$\frac{1}{2} < \frac{1}{b} + \frac{1}{c} < \frac{1}{5} + \frac{1}{5} = \frac{2}{5},$$

which is false.

Therefore $b < 5$, that is, $b = 4$, $b = 3$, or $b = 2$.

The case $b = 4$ gives $c < 4$, which is not possible, since $b \leq c$.

If $b = 3$, then we get $c < 6$, whence $c \in \{3, 4, 5\}$. In this case we find the triples $(2, 3, 3)$, $(2, 3, 4)$, $(2, 3, 5)$.

If $b = 2$, then we find the solutions $(2, 2, c)$, where c is any positive integer.

If $a = 1$, then the solutions are $(1, b, c)$, where b and c are any positive integers.

In conclusion, the solutions are given by the triples $(2, 3, 3)$, $(2, 3, 4)$, $(2, 3, 5)$, $(2, 2, c)$, $(1, b, c)$, where b, c are arbitrary positive integers. Because of symmetry we have also to consider all permutations.

Problem 1.1.9. *Let n be a positive integer. Show that any number greater than $n^4/16$ can be written in at most one way as the product of two of its divisors having difference not exceeding n.*

(1998 St. Petersburg City Mathematical Olympiad)

First Solution. Suppose, on the contrary, that there exist $a > c \geq d > b$ with $a - b \leq n$ and $ab = cd > n^4/16$. Put $p = a+b$, $q = a-b$, $r = c+d$, $s = c-d$. Now

$$p^2 - q^2 = 4ab = 4cd = r^2 - s^2 > n^4/4.$$

Thus $p^2 - r^2 = q^2 - s^2 \leq q^2 \leq n^2$. But $r^2 > n^4/4$ (so $r > n^2/2$) and $p > r$, so

$$p^2 - r^2 > (n^2/2 + 1)^2 - (n^2/2)^2 \geq n^2 + 1,$$

a contradiction.

Second solution. Suppose $a < c \leq d < b$ with $ab = cd = N$ and $b - a \leq n$. Note that

$$(a+b)^2 - n^2 \leq (a+b)^2 - (b-a)^2 = 4ab = 4cd = (c+d)^2 - (d-c)^2 \leq (c+d)^2,$$

so that $(a + b)^2 - (c + d)^2 \leq n^2$. But since $a + b > c + d$ (since the function $f : x \to x + N/x$ decreases for $N < \sqrt{x}$, which means that $f(a) > f(c)$), we obtain

$$n^2 \geq (c + d + 1)^2 - (c + d)^2 = 2c + 2d + 1.$$

Finally, the arithmetic–geometric means (AM–GM) inequality (see the glossary) gives

$$N = cd \leq \left(\frac{c+d}{2}\right)^2 \leq \frac{(n^2 - 1)^2}{16} < \frac{n^4}{16},$$

proving the claim.

Additional Problems

Problem 1.1.10. Show that for any natural number $n \geq 2$, one can find three distinct natural numbers a, b, c between n^2 and $(n + 1)^2$ such that $a^2 + b^2$ is divisible by c.

(1998 St. Petersburg City Mathematical Olympiad)

Problem 1.1.11. Find all odd integers n greater than 1 such that for any relatively prime divisors a and b of n, the number $a + b - 1$ is also a divisor of n.

(2001 Russian Mathematical Olympiad)

Problem 1.1.12. Find all positive integers n such that $3^{n-1}+5^{n-1}$ divides 3^n+5^n.

(1996 St. Petersburg City Mathematical Olympiad)

Problem 1.1.13. Find all positive integers n such that the set

$$\{n, n + 1, n + 2, n + 3, n + 4, n + 5\}$$

can be split into two disjoint subsets such that the products of elements in these subsets are the same.

(12th International Mathematical Olympiad)

Problem 1.1.14. The positive integers $d_1, d_2, \ldots, d_n$ are distinct divisors of 1995. Prove that there exist d_i and d_j among them such that the numerator of the reduced fraction d_i/d_j is at least n.

(1995 Israeli Mathematical Olympiad)

Problem 1.1.15. Determine all pairs (a, b) of positive integers such that $ab^2 + b + 7$ divides $a^2b + a + b$.

(39th International Mathematical Olympiad)

Problem 1.1.16. Find all integers a, b, c with $1 < a < b < c$ such that $(a - 1)(b - 1)(c - 1)$ is a divisor of $abc - 1$.

(33rd International Mathematical Olympiad)

Problem 1.1.17. Find all pairs of positive integers (x, y) for which

$$\frac{x^2 + y^2}{x - y}$$

is an integer that divides 1995.

(1995 Bulgarian Mathematical Olympiad)

Problem 1.1.18. Find all positive integers (x, n) such that $x^n + 2^n + 1$ is a divisor of $x^{n+1} + 2^{n+1} + 1$.

(1998 Romanian International Mathematical Olympiad Team Selection Test)

Problem 1.1.19. Find the smallest positive integer K such that every K-element subset of $\{1, 2, \ldots, 50\}$ contains two distinct elements a, b such that $a + b$ divides ab.

(1996 Chinese Mathematical Olympiad)

1.2 Prime Numbers

The integer $p > 1$ is called a *prime* if its only divisors are 1 and p itself. Any integer $n > 1$ has at least one prime divisor. If n is a prime, then that prime divisor is n itself. If n is not a prime, then let a be its least divisor greater than 1. If a were not a prime, then $a = a_1 a_2$ with $1 < a_1 \le a_2 < a$ and $a_1 \mid n$, contradicting the minimality of a.

An integer $n > 1$ that is not a prime is called *composite*. If n is a composite integer, then it has a prime divisor p not exceeding $\sqrt{n}$. Indeed, writing again $n = ab$, with $1 < a \le b$, we see that $n \ge a^2$; hence $a \le \sqrt{n}$.

The following result has been known for more than 2000 years:

Theorem 1.2.1. (Euclid[1]) *There are infinitely many primes.*

Proof. Assume by way of contradiction that there are only a finite number of primes: $p_1 < p_1 < \cdots < p_m$. Consider the number $P = p_1 p_2 \cdots p_n + 1$.

If P is a prime, then $P > p_m$, contradicting the maximality of p_m. Hence P is composite and, consequently, it has a prime divisor $p > 1$ that is one of the primes $p_1, p_2, \ldots, p_m$, say p_k. It follows that $p_k \mid p_1 \cdots p_k \cdots p_m + 1$. This, together with $p_k \mid p_1 \cdots p_k \cdots p_m$, implies $p_k \mid 1$, a contradiction. $\square$

Remark. The largest known prime at the present time is $2^{32582657} - 1$. It was discovered in 2006 and it has 9808358 digits.

The most fundamental result in arithmetic pertains to the factorization of integers:

Theorem 1.2.2. (The prime factorization theorem) *Any integer $n > 1$ has a unique representation as a product of primes.*

Proof. The existence of such a representation can be obtained as follows: Let p_1 be a prime divisor (factor) of n. If $p_1 = n$, then $n = p_1$ is the prime factorization of n. If $p_1 < n$, then $n = p_1 r_1$, where $r_1 > 1$. If r_1 is a prime, then $n =$

---

[1] Euclid of Alexandria (ca. 325–365 B.C.E.) is one of the most prominent mathematician of antiquity, best known for his treatise on mathematics *The Elements*. The long-lasting nature of *The Elements* must make Euclid the leading mathematics teacher of all time.

p_1p_2, where $p_2 = r_1$, is the desired factorization of n. If r_1 is composite, then $r_1 = p_2r_2$, where p_2 is a prime, $r_2 > 1$ and so $n = p_1p_2r_2$. If r_2 is a prime, then $n = p_1p_2p_3$, where $r_2 = p_3$ and we are done. If r_2 is composite, then we continue this algorithm, obtaining a sequence of integers $r_1 > r_2 > \cdots \geq 1$. After a finite number of steps (see also FMID Variant 1 in Section 5.3), we reach $r_k = 1$, that is, $n = p_1p_2\cdots p_k$.

For the uniqueness, let us assume that there is at least one positive integer n with two distinct representations, i.e.,

$$n = p_1p_2\cdots p_k = q_1q_2\cdots q_h,$$

where $p_1, p_2, \ldots, p_k, q_1, q_2, \ldots, q_h$ are primes. It is clear that $k \geq 2$ and $h \geq 2$. Let n be the minimal such integer. We claim that $p_i \neq q_j$ for every $i = 1, 2, \ldots, k$, $j = 1, 2, \ldots, h$. If, for example, $p_k = q_h = p$, then $n' = n/p = p_1\cdots p_{k-1} = q_1\cdots q_{h-1}$ and $1 < n' < n$, contradicting the minimality of n. Assume without loss of generality that p_1 is the least prime factor of n in the above representations. By applying the division algorithm, it follows that

$$q_1 = p_1c_1 + r_1,$$
$$q_2 = p_1c_2 + r_2,$$
$$\ldots$$
$$q_h = p_1c_h + r_h,$$

where $1 \leq r_i < p_1, i = 1, \ldots, h$.

We have

$$n = q_1q_2\cdots q_h = (p_1c_1 + r_1)(p_1c_2 + r_2)\cdots(p_1c_h + r_h).$$

Expanding the last product, we obtain $n = Ap_1 + r_1r_2\cdots r_h$. Setting $n' = r_1r_2\cdots r_h$ we have $n = p_1p_2\cdots p_k = Ap_1 + n'$. It follows that $p_1 \mid n'$ and $n' = p_1s_1s_2\cdots s_i$, where $s_1, s_2, \ldots, s_i$ are primes.

On the other hand, using the factorization of $r_1, r_2, \ldots, r_h$ into primes, all their factors are less than $r_i < p_1$. From $n' = r_1r_2\cdots r_h$, it follows that n' has a factorization into primes of the form $n' = t_1t_2\cdots t_j$, where $t_s < p_1$, $s = 1, 2, \ldots, j$. This factorization is different from $n' = p_1s_1s_2\cdots s_i$. But $n' < n$, contradicting the minimality of n. □

From the above theorem it follows that any integer $n > 1$ can be written uniquely in the form

$$n = p_1^{\alpha_1}\cdots p_k^{\alpha_k},$$

where $p_1, \ldots, p_k$ are distinct primes and $\alpha_1, \ldots, \alpha_k$ are positive integers and $p_1 < p_2 < \cdots < p_k$. This representation is called the *canonical factorization* of n.

An immediate application of the prime factorization theorem is an alternative way of proving that there are infinitely many primes.

As in the previous proof, assume that there are only finitely many primes: $p_1 < p_2 < \cdots < p_m$. Let

$$x = \prod_{i=1}^{m} \left(1 + \frac{1}{p_i} + \cdots + \frac{1}{p_i^k} + \cdots \right) = \prod_{i=1}^{m} \frac{1}{1 - \frac{1}{p_i}}.$$

On the other hand, by expanding and by using the canonical factorization of positive integers, we obtain

$$x = 1 + \frac{1}{2} + \frac{1}{3} + \cdots$$

yielding $\prod_{i=1}^{m} \frac{p_i}{p_i - 1} = \infty$, a contradiction. We have used the well-known fact that the harmonic series

$$1 + \frac{1}{2} + \frac{1}{3} + \cdots$$

diverges and the expansion formula

$$\frac{1}{1 - x} = 1 + x + x^2 + \cdots \quad \text{(for } |x| < 1\text{)},$$

which can also be interpreted as the summation formula for the infinite geometric progression $1, x, x^2, \ldots$.

From the formula

$$\prod_{i=1}^{\infty} \frac{p_i}{p_i - 1} = \infty,$$

using the inequality $1 + t \le e^t$, $t \in \mathbb{R}$, we can easily derive

$$\sum_{i=1}^{\infty} \frac{1}{p_i} = \infty.$$

Even though there are no definitive ways to find all primes, the density of primes (that is, the average appearances of primes among integers) has been known for about 100 years. This was a remarkable result in the mathematical field of *analytic number theory* showing that

$$\lim_{n \to \infty} \frac{\pi(n)}{n / \log n} = 1,$$

where $\pi(n)$ denotes the number of primes $\leq n$. The relation above is known as the prime number theorem. It was proved by Hadamard[2] and de la Vallée Poussin[3] in 1896. An elementary but difficult proof was given by Erdős[4] and Selberg.[5]

The most important open problems in number theory involve primes. The recent book of David Wells [*Prime Numbers: The Most Mysterious Figures in Maths*, John Wiley and Sons, 2005] contains just few of them. We mention here only three such open problems:

(1) Consider the sequence $(A_n)_{n\geq 1}$, $A_n = \sqrt{p_{n+1}} - \sqrt{p_n}$, where p_n denotes the nth prime. **Andrica's conjecture** states that the following inequality holds:

$$A_n < 1,$$

for any positive integer n. Results connected to this conjecture are given in D. Andrica [*On a Conjecture in Prime Number Theory*, Proc. Algebra Symposium, "Babeş-Bolyai" University of Cluj, 2005, pp. 1–8]. A search conducted by H.J. Smith has grown past $n = 26 \cdot 10^{10}$, so it seems highly likely that the conjecture is true.

(2) If p is a prime such that $p + 2$ is also a prime, then p and $p + 2$ are called **twin primes**. It is not known whether there are infinitely many twin primes. The largest known such pair is $100314512544015 \cdot 2^{171960} \pm 1$, and it was found in 2006.

(3) The following property has been conjectured by Michael Th. Rassias, an International Mathematical Olympiad Silver Medal winner in 2003 in Tokyo: For any prime p greater than two there are two distinct primes p_1, p_2 with $p_1 < p_2$ such that

$$p = \frac{p_1 + p_2 + 1}{p_1}.$$

This is equivalent to the following statement: For any prime p greater than two there are two primes p_1, p_2 with $p_1 < p_2$ such that $(p-1)p_1$, p_2 are consecutive integers [*Octogon Mathematical Magazine*, **13** (1.B), 2005, p. 885].

For a prime p we say that p^k *fully divides* n and write $p^k \parallel n$ if k is the greatest positive integer such that $p^k \mid n$.

Problem 1.2.1. *Prove that for any integer $n > 1$ the number $n^5 + n^4 + 1$ is not a prime.*

---

[2]Jacques Salomon Hadamard (1865–1963), French mathematician whose most important result is the prime number theorem, which he proved in 1896.

[3]Charles Jean Gustave Nicolas de la Vallée Poussin (1866–1962), Belgian mathematician who proved the prime number theorem independently of Hadamard in 1896.

[4]Paul Erdős (1913–1996), one of the greatest mathematicians of the twentieth century. Erdős posed and solved problems in number theory and other areas and founded the field of discrete mathematics.

[5]Atle Selberg (1917–2007), Norwegian mathematician known for his work in analytic number theory and in the theory of automorphic forms.

Solution. We have

$$n^5 + n^4 + 1 = n^5 + n^4 + n^3 - n^3 - n^2 - n + n^2 + n + 1$$
$$= n^3(n^2 + n + 1) - n(n^2 + n + 1) + (n^2 + n + 1)$$
$$= (n^2 + n + 1)(n^3 - n + 1),$$

the product of two integers greater than 1. Hence $n^5 + n^4 + 1$ is not a prime.

Problem 1.2.2. *Find the positive integers n with exactly 12 divisors $1 = d_1 < d_2 < \cdots < d_{12} = n$ such that the divisor with index $d_4 - 1$ (that is, d_{d_4-1}) is $(d_1 + d_2 + d_4)d_8$.*

(1989 Russian Mathematical Olympiad)

Solution. Of course, there is $1 \leq i \leq 12$ such that $d_i = d_1 + d_2 + d_4$. Since $d_i > d_4$, we have $i \geq 5$. Also, observe that $d_j d_{13-j} = n$ for all j and since $d_i d_8 = d_{d_4-1} \leq n$, we must have $i \leq 5$, thus $i = 5$ and $d_1 + d_2 + d_4 = d_5$. Also, $d_{d_4-1} = d_5 d_8 = n = d_{12}$, thus $d_4 = 13$ and $d_5 = 14 + d_2$. Of course, d_2 is the smallest prime divisor of n, and since $d_4 = 13$, we can only have $d_2 \in \{2, 3, 5, 7, 11\}$. Also, since n has 12 divisors, it has at most 3 prime divisors. If $d_2 = 2$ then $d_5 = 16$ and then 4 and 8 are divisors of n smaller than $d_4 = 13$, impossible. A similar argument shows that $d_2 = 3$ and $d_5 = 17$. Since n has 12 divisors and is a multiple of $3 \cdot 13 \cdot 17$, the only possibilities are $9 \cdot 13 \cdot 17$, $3 \cdot 169 \cdot 7$ and $3 \cdot 13 \cdot 289$. One can easily check that only $9 \cdot 13 \cdot 17 = 1989$ is a solution.

Problem 1.2.3. *Find all positive integers a, b for which $a^4 + 4b^4$ is a prime.*

Solution. Observe that

$$a^4 + 4b^4 = a^4 + 4b^4 + 4a^2b^2 - 4a^2b^2$$
$$= (a^2 + 2b^2)^2 - 4a^2b^2$$
$$= (a^2 + 2b^2 + 2ab)(a^2 + 2b^2 - 2ab)$$
$$= [(a + b)^2 + b^2][(a - b)^2 + b^2].$$

Since $(a + b)^2 + b^2 > 1$, then $a^4 + 4b^4$ can be a prime number only if $(a - b)^2 + b^2 = 1$. This implies $a = b = 1$, which is the only solution of the problem.

Problem 1.2.4. *Let p, q be distinct primes. Prove that there are positive integers a, b such that the arithmetic mean of all the divisors of the number $n = p^a \cdot q^b$ is also an integer.*

(2002 Romanian Mathematical Olympiad)

Solution. The sum of all divisors of n is given by the formula

$$(1 + p + p^2 + \cdots + p^a)(1 + q + q^2 + \cdots + q^b),$$

as can easily be seen by expanding the parentheses. The number n has $(a+1) \times (b+1)$ positive divisors and their arithmetic mean is

$$M = \frac{(1 + p + p^2 + \cdots + p^a)(1 + q + q^2 + \cdots + q^b)}{(a+1)(b+1)}.$$

If p and q are both odd numbers, we can take $a = p$ and $b = q$, and it is easy to see that M is an integer.

If $p = 2$ and q odd, choose again $b = q$ and consider $a + 1 = 1 + q + q^2 + \cdots + q^{q-1}$. Then $M = 1 + 2 + 2^2 + \cdots + 2^a$, and it is an integer.

For p odd and $q = 2$, set $a = p$ and $b = p + p^2 + p^3 + \cdots + p^{p-1}$. The solution is complete.

Problem 1.2.5. *Find all primes p such that $p^2 + 11$ has exactly six different divisors (including 1 and the number itself).*

(1995 Russian Mathematical Olympiad)

Solution. For $p \neq 3$, $3 \mid p^2 - 1$, and so $3 \mid (p^2 + 11)$. Similarly, for $p \neq 2$, $4 \mid p^2 - 1$, and so $4 \mid (p^2 + 11)$. Except in these two cases, then, $12 \mid (p^2 + 11)$; since 12 itself has six divisors (1, 2, 3, 4, 6, 12) and $p^2 + 11 > 12$ for $p > 1$, $p^2 + 11$ must have more than six divisors. The only cases to check are $p = 2$ and $p = 3$. If $p = 2$, then $p^2 + 11 = 15$, which has only four divisors (1, 3, 5, 15), while if $p = 3$, then $p^2 + 11 = 20$, which indeed has six divisors (1, 2, 4, 5, 10, 20). Hence $p = 3$ is the only solution.

Problem 1.2.6. *Let a, b, c be nonzero integers, $a \neq c$, such that*

$$\frac{a}{c} = \frac{a^2 + b^2}{c^2 + b^2}.$$

Prove that $a^2 + b^2 + c^2$ cannot be a prime.

(1999 Romanian Mathematical Olympiad)

First solution. The equality $\frac{a}{c} = \frac{a^2 + b^2}{c^2 + b^2}$ is equivalent to $(a - c)(b^2 - ac) = 0$. Since $a \neq c$, it follows that $b^2 = ac$ and therefore:

$$a^2 + b^2 + c^2 = a^2 + ac + c^2 = a^2 + 2ac + c^2 - b^2$$
$$= (a + c)^2 - b^2 = (a + c - b)(a + c + b).$$

Now, clearly, $a^2 + b^2 + c^2 > 3$, so, if $a^2 + b^2 + c^2$ is a prime number, then only four cases are possible:

(1) $a + c - b = 1$ and $a + c - b = a^2 + b^2 + c^2$;

(2) $a + c + b = 1$ and $a + c + b = a^2 + b^2 + c^2$;

(3) $a + c - b = -1$ and $a + c + b = -(a^2 + b^2 + c^2)$;

(4) $a + c + b = -1$ and $a + c - b = -(a^2 + b^2 + c^2)$.

In the first two cases we are led to $a^2 + b^2 + c^2 - 2(a + c) + 1 = 0$, or $(a - 1)^2 + (c - 1)^2 + b^2 = 1$; hence $a = c = 1$.

In other cases we obtain $(a + 1)^2 + (c + 1)^2 + b^2 = 1$; hence $a = c = -1$. But $a = c$ is a contradiction.

Second solution. As in the previous solution we get $b^2 = ac$, then observe that this forces a and c to have the same sign. Hence replacing them with their negatives if necessary we may assume $a, c > 0$. Similarly, we may assume $b > 0$. Then continue with the given solution to get $a^2 + b^2 + c^2 = (a + c - b)(a + b + c)$. If this is a prime then the smaller factor $a + c - b$ must be 1. But since $a \neq c$, $a + c > 2\sqrt{ac} = 2b$ so $a + c - b > b \geq 1$, a contradiction.

Problem 1.2.7. *Show that each natural number can be written as the difference of two natural numbers having the same number of prime factors.*

(1999 Russian Mathematical Olympiad)

Solution. If n is even, then we can write it as $(2n) - (n)$. If n is odd, let d be the smallest odd prime that does not divide n. Then write $n = (dn) - ((d - 1)n)$. The number dn contains exactly one more prime factor than n. As for $(d - 1)n$, it is divisible by 2 because $d - 1$ is even. Its odd factors are less than d, so they all divide n. Therefore $(d - 1)n$ also contains exactly one more prime factor than n, and dn and $(d - 1)n$ have the same number of prime factors.

Problem 1.2.8. *Let p be a prime number. Find all $k \in \mathbb{Z}$, $k \neq 0$, such that $\sqrt{k^2 - pk}$ is a positive integer.*

(1997 Spanish Mathematical Olympiad)

Solution. We will use the following simple but important property: If ab is a square and a and b are coprime, then a and b are both squares.

The values are $k = (p + 1)^2/4$ for p odd (and none for $p = 2$).

We first rule out the case that k is divisible by p: if $k = np$, then $k^2 - pk = p^2n(n-1)$. Since n and $n - 1$ are consecutive, they are coprime, and by the lemma above they are both squares. However, the only consecutive squares are 0 and 1, so this gives $k = p$ and $\sqrt{k^2 - pk} = 0$, which is not positive.

We thus assume that k and p are coprime, in which case k and $k - p$ are coprime. Thus $k^2 - pk = k(k - p)$ is a square if and only if k and $k - p$ are squares, say $k = m^2$ and $k - p = n^2$. Then $p = m^2 - n^2 = (m + n)(m - n)$, which implies $m + n = p$, $m - n = 1$ and hence $k = (p + 1)^2/4$. Since k must be an integer, this forces p to be odd.

Problem 1.2.9. *Let $p > 5$ be a prime number and*

$$X = \{p - n^2 \mid n \in \mathbb{N}, n^2 < p\}.$$

Prove that X contains two distinct elements x, y such that $x \neq 1$ and x divides y.

(1996 Balkan Mathematical Olympiad)

Solution. Take m such that $m^2 < p < (m+1)^2$ and write $p = k + m^2$, with $1 \leq k \leq 2m$. Since $p - (m-k)^2 = k(2m - k + 1)$, we have $p - m^2 \mid p - (m-k)^2$. Thus we are done by taking $x = p - m^2 = k$ and $y = p - |m - k|^2$ unless either $k = p - m^2 = 1$ (which gives $x = 1$), $k = m$ (which gives $|m - k| = 0$), or $k = 2m$ (which gives $x = y$). The latter two cases give $p = m(m+1)$ and $p = m(m+2)$, respectively, which cannot occur, since p is prime. In the remaining case, $p = m^2 + 1$, which forces m be even. Hence $x = p - (m-1)^2 = 2m$ divides $y = p - 1 = m^2$ is the required example.

Additional Problems

Problem 1.2.10. For each integer n such that $n = p_1 p_2 p_3 p_4$, where p_1, p_2, p_3, p_4 are distinct primes, let

$$d_1 = 1 < d_2 < d_3 < \cdots < d_{16} = n$$

be the sixteen positive integers that divide n. Prove that if $n < 1995$, then $d_9 - d_8 \neq 22$.

(1995 Irish Mathematical Olympiad)

Problem 1.2.11. Prove that there are infinitely many positive integers a such that the sequence $(z_n)_{n \geq 1}$, $z_n = n^4 + a$, does not contain any prime number.

(11th International Mathematical Olympiad)

Problem 1.2.12. Let p, q, r be distinct prime numbers and let A be the set

$$A = \{p^a q^b r^c : 0 \leq a, b, c \leq 5\}.$$

Find the smallest integer n such that every n-element subset of A contains two distinct elements x, y such that x divides y.

(1997 Romanian Mathematical Olympiad)

Problem 1.2.13. Prove Bonse's inequality:

$$p_1 p_2 \cdots p_n > p_{n+1}^2$$

for $n \geq 4$, where $p_1 = 2$, $p_2 = 3, \ldots$ is the increasing sequence of prime numbers.

Problem 1.2.14. Show that there exists a set A of positive integers with the following property: for any infinite set S of primes, there exist two positive integers $m \in A$ and $n \notin A$ each of which is a product of k distinct elements of S for some $k \geq 2$.

(35th International Mathematical Olympiad)

Problem 1.2.15. Let n be an integer, $n \geq 2$. Show that if $k^2 + k + n$ is a prime number for every integer k, $0 \leq k \leq \sqrt{n/3}$, then $k^2 + k + n$ is a prime number for every k, $0 \leq k \leq n - 2$.

(28th International Mathematical Olympiad)

Problem 1.2.16. A sequence $q_1, q_2, \ldots$ of primes satisfies the following condition: for $n \geq 3$, q_n is the greatest prime divisor of $q_{n-1} + q_{n-2} + 2000$. Prove that the sequence is bounded.

(2000 Polish Mathematical Olympiad)

Problem 1.2.17. Let $a > b > c > d$ be positive integers and suppose

$$ac + bd = (b + d + a - c)(b + d - a + c).$$

Prove that $ab + cd$ is not prime.

(42nd International Mathematical Olympiad)

Problem 1.2.18. Find the least odd positive integer n such that for each prime p, $\frac{n^2-1}{4} + np^4 + p^8$ is divisible by at least four primes.

(Mathematical Reflections)

1.3 The Greatest Common Divisor and the Least Common Multiple

For a positive integer k we denote by D_k the set of all its positive divisors. It is clear that D_k is a finite set. For positive integers m, n the maximal element in the set $D_m \cap D_n$ is called the *greatest common divisor* of m and n and is denoted by $\gcd(m, n)$.

Another characterization of $\gcd(m, n)$ is given by the property $d \mid \gcd(m, n)$ if and only if $d \mid m$ and $d \mid n$. Note that $\gcd(0, n) = n$.

If $D_m \cap D_n = \{1\}$, we have $\gcd(m, n) = 1$ and we say that m and n are *relatively prime*.

The following properties can be directly derived from the definition above.

(1) If $d = \gcd(m, n)$, $m = dm'$, $n = dn'$, then $\gcd(m', n') = 1$.

(2) If $m = dm'$, $n = dn'$, and $\gcd(m', n') = 1$, then $d = \gcd(m, n)$.

(3) If d' is a common divisor of m and n, then d' divides $\gcd(m, n)$.

This property says, in particular, that $\gcd(m, n)$ is the largest common divisor of m and n, as the name implies.

(4) If $m = p_1^{\alpha_1} \cdots p_k^{\alpha_k}$ and $n = p_1^{\beta_1} \cdots p_k^{\beta_k}$, $\alpha_i, \beta_i \geq 0$, $i = 1, \ldots, k$, then

$$\gcd(m, n) = p_1^{\min(\alpha_1, \beta_1)} \cdots p_k^{\min(\alpha_k, \beta_k)}.$$

(5) If $m = nq + r$, then $\gcd(m, n) = \gcd(n, r)$.

Let us prove the last property. Set $d = \gcd(m, n)$ and $d' = \gcd(n, r)$. Because $d \mid m$ and $d \mid n$ it follows that $d \mid r$. Hence $d \mid d'$. Conversely, from $d' \mid n$ and $d' \mid r$ it follows that $d' \mid m$, so $d' \mid d$. Thus $d = d'$.

A useful algorithm for finding the greatest common divisor of two positive integers is the *Euclidean algorithm*. It consists in repeated application of the division algorithm:

$$m = nq_1 + r_1, \quad 1 \leq r_1 < n,$$
$$n = r_1 q_2 + r_2, \quad 1 \leq r_2 < r_1,$$
$$\cdots$$
$$r_{k-2} = r_{k-1} q_k + r_k, \quad 1 \leq r_k < r_{k-1},$$
$$r_{k-1} = r_k q_{k+1} + r_{k+1}, \quad r_{k+1} = 0.$$

This chain of equalities is finite because the descending sequence $n > r_1 > r_2 > \cdots > r_k \geq 0$ of integers cannot go on indefinitely.

The last nonzero remainder, r_k, is the greatest common divisor of m and n. Indeed, by applying successively property (5) above, we obtain

$$\gcd(m, n) = \gcd(n, r_1) = \gcd(r_1, r_2) = \cdots = \gcd(r_{k-1}, r_k) = r_k.$$

Proposition 1.3.1. *For positive integers m and n, there exist integers a and b such that $am + bn = \gcd(m, n)$.*

Proof. From the Euclidean algorithm it follows that

$$r_1 = m - nq_1, \quad r_2 = -mq_2 + n(1 + q_1 q_2), \quad \ldots.$$

In general, $r_i = m\alpha_i + n\beta_i$, $i = 1, \ldots, k$. Because $r_{i+1} = r_{i-1} - r_i q_{i+1}$, it follows that

$$\begin{cases} \alpha_{i+1} = \alpha_{i-1} - q_{i+1}\alpha_i, \\ \beta_{i+1} = \beta_{i-1} - q_{i+1}\beta_i, \end{cases}$$

$i = 2, \ldots, k - 1$. Finally, we obtain $\gcd(m, n) = r_k = \alpha_k m + \beta_k n$. $\qquad\square$

Proposition 1.3.2. (Euclid's lemma). *Let p be a prime and $a, b \in \mathbb{Z}$. If $p \mid ab$, then $p \mid a$ or $p \mid b$.*

Proof. Suppose that $p \nmid a$. Since $\gcd(p, a) \mid p$, we have $\gcd(p, a) = 1$ or $\gcd(p, a) = p$; but the latter implies $p \mid a$, contradicting our assumption, thus $\gcd(p, a) = 1$. Let $r, s \in \mathbb{Z}$ be such that $rp + sa = 1$. Then $rpb + sab = b$ and so $p \mid b$. □

More generally, if a prime p divides a product of integers $a_1 \cdots a_n$ then $p \mid a_j$ for some j. This can be proved by induction on the number n.

We can define the greatest common divisor of several positive integers m_1, $m_2, \ldots, m_s$ by considering

$$d_1 = \gcd(m_1, m_2), \ d_2 = \gcd(d_1, m_3), \ldots, d_{s-1} = \gcd(d_{s-2}, m_s).$$

The integer $d = d_{s-1}$ is called the greatest common divisor of $m_1, \ldots, m_s$ and denoted by $\gcd(m_1, \ldots, m_s)$. The following properties can be easily verified:

(i) $\gcd(\gcd(m, n), p) = \gcd(m, \gcd(n, p))$, proving that $\gcd(m, n, p)$ is well defined.

(ii) If $d \mid m_i, i = 1, \ldots, s$, then $d \mid \gcd(m_1, \ldots, m_s)$.

(iii) If $m_i = p_1^{\alpha_{1i}} \cdots p_k^{\alpha_{ki}}, i = 1, \ldots, s$, then

$$\gcd(m_1, \ldots, m_s) = p_1^{\min(\alpha_{11}, \ldots, \alpha_{1k})} \cdots p_k^{\min(\alpha_{1s}, \ldots, \alpha_{ks})}.$$

For a positive integer k we denote by M_k the set of all multiples of k. Unlike the set D_k defined earlier in this section, M_k is an infinite set.

For positive integers s and t, the minimal element of the set $M_s \cap M_t$ is called the *least common multiple* of s and t and is denoted by $\mathrm{lcm}(s, t)$.

The following properties are easily obtained from the definition above:

(1') If $m = \mathrm{lcm}(s, t), m = ss' = tt'$, then $\gcd(s', t') = 1$.

(2') If m' is a common multiple of s and t and $m' = ss' = tt'$, $\gcd(s', t') = 1$, then $m' = m$.

(3') If m' is a common multiple of s and t, then $m \mid m'$.

(4') If $s = p_1^{\alpha_1} \cdots p_k^{\alpha_k}$ and $t = p_1^{\beta_1} \cdots p_k^{\beta_k}, \alpha_i, b_i \geq 0, i = 1, \ldots, k$, then

$$\mathrm{lcm}(s, t) = p_1^{\max(\alpha_1, \beta_1)} \cdots p_k^{\max(\alpha_k, \beta_k)}.$$

The following property establishes an important connection between greatest common divisor and least common multiple:

Proposition 1.3.3. *For any positive integers m, n the following relation holds:*

$$mn = \gcd(m, n) \cdot \mathrm{lcm}(m, n).$$

Proof. Let $m = p_1^{\alpha_1} \cdots p_k^{\alpha_k}$, $n = p_1^{\beta_1} \cdots p_k^{\beta_k}$, $\alpha_i, \beta_i \geq 0$, $i = 1, \ldots, k$. From properties (4) and (4') we have

$$\gcd(m, n) \cdot \mathrm{lcm}(m, n) = p_1^{\min(\alpha_1, \beta_1) + \max(\alpha_1, \beta_1)} \cdots p_k^{\min(\alpha_k, \beta_k) + \max(\alpha_k, \beta_k)}$$

$$e = p_1^{\alpha_1 + \beta_1} \cdots p_k^{\alpha_k + \beta_k} = mn. \qquad \square$$

It is also not difficult to see that if $m \mid s$ and $n \mid s$, then $\mathrm{lcm}(m, n) \mid s$. Another useful property is the following.

Proposition 1.3.4. *Let n, a, b be positive integers. Then*

$$\gcd(n^a - 1, n^b - 1) = n^{\gcd(a,b)} - 1.$$

Proof. Without loss of generality, we assume that $a \geq b$. Then

$$\gcd(n^a - 1, n^b - 1) = \gcd(n^a - 1 - n^{a-b}(n^b - 1), n^b - 1) = \gcd(n^{a-b} - 1, n^b - 1).$$

Recall the process of finding $\gcd(a, b) = \gcd(a - b, b)$ given by the Euclidean algorithm. We see that the process of computing $\gcd(n^a - 1, n^b - 1)$ is the same as the process of computing $\gcd(a, b)$ as the exponents, from which the conclusion follows. $\qquad \square$

Problem 1.3.1. *Prove that for odd n and odd integers $a_1, a_2, \ldots, a_n$, the greatest common divisor of numbers $a_1, a_2, \ldots, a_n$ is equal to the greatest common divisor of $\frac{a_1 + a_2}{2}, \frac{a_2 + a_3}{2}, \ldots, \frac{a_n + a_1}{2}$.*

Solution. Let

$$a = \gcd(a_1, a_2, \ldots, a_n) \quad \text{and} \quad b = \gcd\left(\frac{a_1 + a_2}{2}, \frac{a_2 + a_3}{2}, \ldots, \frac{a_n + a_1}{2}\right).$$

Then $a_k = b_k a$, for some integers b_k, $k = 1, 2, \ldots, n$. It follows that

$$\frac{a_k + a_{k+1}}{2} = \frac{b_k + b_{k+1}}{2} a, \tag{1}$$

where $a_{n+1} = a_1$ and $b_{n+1} = b_1$. Since a_k are odd numbers, b_k are also odd, so $\frac{b_k + b_{k+1}}{2}$ are integers.

From relation (1) it follows that a divides $\frac{a_k + a_{k+1}}{2}$ for all $k \in \{1, 2, \ldots, n\}$ so a divides b.

On the other hand, $\frac{a_k + a_{k+1}}{2} = \beta_k b$, for some integers β_k. Then

$$2b \mid a_k + a_{k+1} \tag{2}$$

for all $k \in \{1, 2, \ldots, n\}$. Summing up from $k = 1$ to $k = n$ yields

$$2b \mid 2(a_1 + a_2 + \cdots + a_n),$$

hence

$$b \mid a_1 + a_2 + \cdots + a_n. \tag{3}$$

Summing up for $k = 1, 3, \ldots, n - 2$ implies

$$2b \mid a_1 + a_2 + \cdots + a_{n-1}$$

and furthermore

$$b \mid a_1 + a_2 + \cdots + a_{n-1}. \tag{4}$$

From (3) and (4) we get that b divides a_n; then using relation (2) we obtain $b \mid a_k$ for all k. Hence $b \mid a$ and the proof is complete.

Problem 1.3.2. *Prove that for all nonnegative integers a, b, c, d such that a and b are relatively prime, the system*

$$\begin{aligned} ax - yz - c &= 0, \\ bx - yt + d &= 0, \end{aligned}$$

has infinitely many solutions in nonnegative integers.

Solution. We start with a useful lemma that is a corollary to Proposition 1.3.1.

Lemma. *If a and b are relatively prime positive integers, then there are positive integers u and v such that*

$$au - bv = 1.$$

Proof. Here we give a different argument as in the proof of Proposition 1.3.1. Consider the numbers

$$1 \cdot a, \ 2 \cdot a, \ldots, (b - 1) \cdot a. \tag{1}$$

When divided by b, the remainders of these numbers are distinct. Indeed, otherwise we would have $k_1 \neq k_2 \in \{1, 2, \ldots, b - 1\}$ such that

$$k_1 a = p_1 b + r, \quad k_2 a = p_2 b + r,$$

for some integers p_1, p_2. Hence

$$(k_1 - k_2)a = (p_1 - p_2)b.$$

Since a and b are relatively prime, it follows that b divides $k_1 - k_2$, which is false because $1 \leq |k_1 - k_2| < b$.

On the other hand, none of the numbers listed in (1) is divisible by b. Indeed, if so, then there is $k \in \{1, 2, \ldots, n-1\}$ such that

$$k \cdot a = p \cdot b \text{ for some integer } p.$$

Let d be the greatest common divisor of k and p. Then $k = k_1 d$, $p = p_1 d$, for some integers p_1, k_1 with $\gcd(p_1, k_1) = 1$. Hence $k_1 a = p_1 b$, and since $\gcd(a, b) = 1$, we have $k_1 = b$, $p_1 = a$. This is false, because $k_1 < b$.

It follows that one of the numbers from (1) has the remainder 1 when divided by b, so there is $u \in \{1, 2, \ldots, b-1\}$ such that $au = bv + 1$ and the lemma is proved. In fact, there are infinitely many positive integers u and v satisfying this property, that is, $u = u_0 + kb$, $v = v_0 + ka$, where u_0 and v_0 satisfy $au_0 - bv_0 = 1$.

We now prove that the system

$$\begin{cases} ax - yz - c = 0, \\ bx - yt + d = 0, \end{cases}$$

with a, b, c, d nonnegative integers and $\gcd(a, b) = 1$, has infinitely many solutions in nonnegative integers.

Because $\gcd(a, b) = 1$, using the lemma, we see that there are infinitely many positive integers u and v such that $au - bv = 1$. Hence

$$x = cu + dv, \quad y = ad + bc, \quad z = v, \quad t = u,$$

are solutions to the system.

Problem 1.3.3. *Find all the pairs of integers (m, n) such that the numbers $A = n^2 + 2mn + 3m^2 + 2$, $B = 2n^2 + 3mn + m^2 + 2$, $C = 3n^2 + mn + 2m^2 + 1$ have a common divisor greater than 1.*

Solution. A common divisor of A, B, and C is also a divisor for $D = 2A - B$, $E = 3A - C$, $F = 5E - 7D$, $G = 5D - E$, $H = 18A - 2F - 3E$, $I = nG - mF$, and $126 = 18nI - 5H + 11F = 2 \cdot 3^2 \cdot 7$. Since $A + B + C = 6(m^2 + mn + n^2) + 5$, it follows that 2 and 3 do not divide A, B, and C. Therefore $d = 7$. We get that (m, n) is equal to $(7a + 2, 7b + 3)$ or $(7c + 5, 7d + 4)$.

Problem 1.3.4. *Let n be an even positive integer and let a, b be positive coprime integers. Find a and b if $a + b$ divides $a^n + b^n$.*

(2002 Romanian Mathematical Olympiad)

Solution. Since n is even, we have

$$a^n - b^n = (a^2 - b^2)(a^{n-2} + a^{n-4}b^2 + \cdots + b^{n-2}).$$

Since $a + b$ is a divisor of $a^2 - b^2$, it follows that $a + b$ is a divisor of $a^n - b^n$. In turn, $a + b$ divides $2a^n = (a^n + b^n) + (a^n - b^n)$, and $2b^n = (a^n + b^n) - (a^n - b^n)$. But a and b are coprime numbers, and so $\gcd(2a^n, 2b^n) = 2$. Therefore $a + b$ is a divisor of 2; hence $a = b = 1$.

Problem 1.3.5. *M is the set of all values of the greatest common divisor d of the numbers* $A = 2n + 3m + 13$, $B = 3n + 5m + 1$, $C = 6n + 8m - 1$, *where m and n are positive integers. Prove that M is the set of all divisors of an integer k.*

Solution. If d is a common divisor of the numbers A, B, and C, then d divides $E = 3A - C = m + 40$, $F = 2B - C = 2m + 3$, and $G = 2E - F = 77$.

Hence d must be a divisor of 77. To complete the problem we need only show that every divisor of 77 occurs as d for some m and n. Taking $m = n = 1$ gives $(A, B, C) = (18, 9, 13)$ and hence $d = 1$. If $d = 7$, then $7 \mid 2m + 3$. The smallest solution is $m = 2$. Taking $m = 2$ and $n = 1$ gives $(A, B, C) = (21, 14, 21)$ and $d = 7$, as desired. If $d = 11$, then $11 \mid 2m + 3$, which has smallest solution $m = 4$. Taking $m = n = 4$ gives $(A, B, C) = (33, 33, 55)$ and $d = 11$, as desired. If $d = 77$, then $77 \mid 2m + 3$. Taking $m = 37$ and $n = 15$ gives $(A, B, C) = (154, 231, 385)$ and $d = 77$.

Problem 1.3.6. *Let a, b, and c be integers. Prove that*

$$\frac{\gcd(a, b)\gcd(b, c)\gcd(c, a)}{\gcd(a, b, c)^2} \quad and \quad \frac{\operatorname{lcm}(a, b)\operatorname{lcm}(b, c)\operatorname{lcm}(c, a)}{\operatorname{lcm}(a, b, c)^2}$$

are equal integers.

Solution. Let $a = p_1^{\alpha_1} \cdots p_n^{\alpha_n}$, $b = p_1^{\beta_1} \cdots p_n^{\beta_n}$, and $c = p_1^{\gamma_1} \cdots p_n^{\gamma_n}$, where $p_1, \ldots, p_n$ are distinct primes, and $\alpha_1, \ldots, \alpha_n, \beta_1, \ldots, \beta_n, \gamma_1, \ldots, \gamma_n$ are nonnegative integers. Then

$$\frac{\gcd(a, b)\gcd(b, c)\gcd(c, a)}{\gcd(a, b, c)^2} = \frac{\prod_{i=1}^{n} p_i^{\min\{\alpha_i, \beta_i\}} \prod_{i=1}^{n} p_i^{\min\{\beta_i, \gamma_i\}} \prod_{i=1}^{n} p_i^{\min\{\gamma_i, \alpha_i\}}}{\prod_{i=1}^{n} p_i^{2\min\{\alpha_i, \beta_i, \gamma_i\}}}$$

$$= \prod_{i=1}^{n} p_i^{\min\{\alpha_i, \beta_i\} + \min\{\beta_i, \gamma_i\} + \min\{\gamma_i, \alpha_i\} - 2\min\{\alpha_i, \beta_i, \gamma_i\}}$$

and

$$\frac{\operatorname{lcm}(a, b)\operatorname{lcm}(b, c)\operatorname{lcm}(c, a)}{\operatorname{lcm}(a, b, c)^2} = \frac{\prod_{i=1}^{n} p_i^{\max\{\alpha_i, \beta_i\}} \prod_{i=1}^{n} p_i^{\max\{\beta_i, \gamma_i\}} \prod_{i=1}^{n} p_i^{\max\{\gamma_i, \alpha_i\}}}{\prod_{i=1}^{n} p_i^{2\max\{\alpha_i, \beta_i, \gamma_i\}}}$$

$$= \prod_{i=1}^{n} p_i^{\max\{\alpha_i, \beta_i\} + \max\{\beta_i, \gamma_i\} + \max\{\gamma_i, \alpha_i\} - 2\max\{\alpha_i, \beta_i, \gamma_i\}}.$$

It suffices to show that for each nonnegative numbers α, β, and γ

$$\min\{\alpha, \beta\} + \min\{\beta, \gamma\} + \min\{\gamma, \alpha\} - 2\min\{\alpha, \beta, \gamma\}$$
$$= \max\{\alpha, \beta\} + \max\{\beta, \gamma\} + \max\{\gamma, \alpha\} - 2\max\{\alpha, \beta, \gamma\}.$$

By symmetry, we may assume that $\alpha \leq \beta \leq \gamma$. It is not difficult to deduce that both sides are equal to β, completing our proof.

Problem 1.3.7. *Let $m \geq 2$ be an integer. A positive integer n is called m-good if for every positive integer a relatively prime to n, one has $n \mid a^m - 1$.*

Show that every m-good number is at most $4m(2^m - 1)$.

(2004 Romanian International Mathematical Olympiad Team Selection Test)

Solution. If m is odd, then $n \mid (n-1)^m - 1$ implies $n \mid 2$; hence $n \leq 2$.

Take now $m = 2^t q, t \geq 1, q$ odd. If $n = 2^u(2v+1)$ is m-good, then $(2v+1) \mid (2v-1)^m - 1$; hence $(2v+1) \mid 2^m - 1$. Also, if $a = 8v + 5$, then $\gcd(a, n) = 1$, so

$$2^u \mid (a^q)^{2^t} - 1 = (a^q - 1)(a^q + 1)(a^{2q} + 1)\cdots(a^{2^{t-1}q} + 1).$$

Since q is odd, $a^q = 8V + 5$ for some integer V. Hence the first term in the product is divisible by 2^2, but not 2^3. Similarly, the other terms are divisible by 2, but not 2^2. Hence the product is divisible by 2^{t+1} and no higher power of 2. Therefore $u \leq t + 2$, whence $n \leq 4 \cdot 2^t(2v+1) \leq 4m(2^m - 1)$.

Remark. The estimate is optimal only for $m = 2, m = 4$.

Problem 1.3.8. *Find all triples of positive integers (a, b, c) such that $a^3 + b^3 + c^3$ is divisible by $a^2 b$, $b^2 c$, and $c^2 a$.*

(2001 Bulgarian Mathematical Olympiad)

Solution. Answer: triples of the form (k, k, k) or $(k, 2k, 3k)$ or their permutations.

Let g be the positive greatest common divisor of a and b. Then g^3 divides $a^2 b$, so g^3 divides $a^3 + b^3 + c^3$, and g divides c. Thus, the gcd of any two of a, b, c is the gcd of all three.

Let $(l, m, n) = (a/g, b/g, c/g)$. Then (l, m, n) is a triple satisfying the conditions of the problem, and l, m, n are pairwise relatively prime. Because l^2, m^2, and n^2 all divide $l^3 + m^3 + n^3$, we have

$$l^2 m^2 n^2 \mid (l^3 + m^3 + n^3).$$

We will prove that (l, m, n) is either $(1, 1, 1)$ or a permutation of $(1, 2, 3)$. Assume without loss of generality that $l \geq m \geq n$. We have

$$3l^3 \geq l^3 + m^3 + n^3 \geq l^2 m^2 n^2,$$

and therefore $l \geq m^2 n^2/3$. Because $l^2 \mid (m^3 + n^3)$, we also have

$$2m^3 \geq m^3 + n^3 \geq l^2 \geq m^4 n^4/9.$$

If $n \geq 2$, then $m \leq 2 \cdot 9/2^4 < 2 \leq n$, which contradicts the assumption that $m \geq n$. Therefore, n must be 1. If $m = n = 1$, then $l^2 \mid m^3 + n^3 = 2$ so $l = 1$ and we get the unique solution $(1, 1, 1)$.

If $m \geq 2$, then $l > m$, because l and m are relatively prime, so

$$2l^3 > l^3 + m^3 + 1 \geq l^2 m^2,$$

and $l > m^2/2$, so

$$m^3 + 1 \geq l^2 > m^4/4,$$

and $m \leq 4$. If $m = 2$, then $l^2 \mid m^3 + 1 = 9$ and $l \geq m$; hence $l = 3$ and we get the solution $(3, 2, 1)$. If $m = 3$ or 4, then $l^2 \mid m^3 + 1$ and $l \geq m$ gives a contradiction.

Additional Problems

Problem 1.3.9. A sequence $a_1, a_2, \ldots$ of natural numbers satisfies

$$\gcd(a_i, a_j) = \gcd(i, j) \text{ for all } i \neq j.$$

Prove that $a_i = i$ for all i.

(1995 Russian Mathematical Olympiad)

Problem 1.3.10. The natural numbers a and b are such that

$$\frac{a+1}{b} + \frac{b+1}{a}$$

is an integer. Show that the greatest common divisor of a and b is not greater than $\sqrt{a+b}$.

(1996 Spanish Mathematical Olympiad)

Problem 1.3.11. The positive integers m, n, m, n are written on a blackboard. A generalized Euclidean algorithm is applied to this quadruple as follows: if the numbers x, y, u, v appear on the board and $x > y$, then $x - y, y, u + v, v$ are written instead; otherwise $x, y - x, u, v + u$ are written instead. The algorithm stops when the numbers in the first pair become equal (they will equal the greatest common divisor of m and n). Prove that the arithmetic mean of the numbers in the second pair at that moment equals the least common multiple of m and n.

(1996 St. Petersburg City Mathematical Olympiad)

Problem 1.3.12. How many pairs (x, y) of positive integers with $x \leq y$ satisfy $\gcd(x, y) = 5!$ and $\text{lcm}(x, y) = 50!$?

(1997 Canadian Mathematical Olympiad)

Problem 1.3.13. Several positive integers are written on a blackboard. One can erase any two distinct integers and write their greatest common divisor and least common multiple instead. Prove that eventually the numbers will stop changing.

(1996 St. Petersburg City Mathematical Olympiad)

Problem 1.3.14. (a) For which positive integers n do there exist positive integers x, y such that
$$\text{lcm}(x, y) = n!, \quad \gcd(x, y) = 1998?$$

(b) For which n is the number of such pairs x, y with $x \leq y$ less than 1998?

(1998 Hungarian Mathematical Olympiad)

Problem 1.3.15. Determine all integers k for which there exists a function $f : \mathbb{N} \to \mathbb{Z}$ such that
(a) $f(1997) = 1998$;
(b) for all $a, b \in \mathbb{N}$, $f(ab) = f(a) + f(b) + kf(\gcd(a, b))$.

(1997 Taiwanese Mathematical Olympiad)

Problem 1.3.16. Find all triples (x, y, n) of positive integers such that

$$\gcd(x, n + 1) = 1 \text{ and } x^n + 1 = y^{n+1}.$$

(1998 Indian Mathematical Olympiad)

Problem 1.3.17. Find all triples (m, n, l) of positive integers such that

$$m + n = \gcd(m, n)^2, \quad m + l = \gcd(m, l)^2, \quad n + l = \gcd(n, l)^2.$$

(1997 Russian Mathematical Olympiad)

Problem 1.3.18. Let a, b be positive integers such that $\gcd(a, b) = 1$. Find all pairs (m, n) of positive integers such that $a^m + b^m$ divides $a^n + b^n$.

(Mathematical Reflections)

1.4   Odd and Even

The set $\mathbb{Z}$ of integers can be partitioned into two subsets, the set of odd integers and the set of even integers: $\{\pm 1, \pm 3, \pm 5, \ldots\}$ and $\{0, \pm 2, \pm 4, \ldots\}$, respectively. Although the concepts of odd and even integers appear straightforward, they come in handy in various number theory problems. Here are some basic ideas:

(1) an odd number is of the form $2k + 1$, for some integer k;
(2) an even number is of the form $2m$, for some integer m;
(3) the sum of two odd numbers is an even number;
(4) the sum of two even numbers is an even number;
(5) the sum of an odd number and an even number is an odd number;
(6) the product of two odd numbers is an odd number;
(7) a product of integers is even if and only if at least one of its factors is even.

Problem 1.4.1. *Let m and n be integers $m \geq 1$ and $n > 1$. Prove that m^n is the sum of m odd consecutive integers.*

Solution. The equality

$$m^n = (2k + 1) + (2k + 3) + \cdots + (2k + 2m - 1)$$

is equivalent to

$$m^n = 2km + (1 + 3 + \cdots + 2m - 1)$$

or $m^n = 2km + m^2$. It follows that $k = m(m^{n-2} - 1)/2$, which is an integer because m and $m^{n-2} - 1$ have different parities.

Problem 1.4.2. *Let n be a positive integer. Find the sum of all even numbers between $n^2 - n + 1$ and $n^2 + n + 1$.*

Solution. We have $n^2 - n + 1 = n(n - 1) + 1$ and $n^2 + n + 1 = n(n + 1) + 1$, both odd numbers. It follows that the least even number to be considered is $n^2 - n + 2$ and the greatest is $n^2 + n$. The desired sum is

$$(n^2 - n + 2) + (n^2 - n + 4) + \cdots + (n^2 + n - 2) + (n^2 + n)$$
$$= (n^2 - n) + 2 + (n^2 - n) + 4 + \cdots + (n^2 - n) + 2n - 2 + (n^2 - n) + 2n$$
$$= n(n^2 - n) + 2(1 + 2 + \cdots + n) = n^3 - n^2 + n^2 + n = n^3 + n.$$

Problem 1.4.3. *Let n be a positive integer and let $\varepsilon_1, \varepsilon_2, \ldots, \varepsilon_n \in \{-1, 1\}$ be such that $\varepsilon_1 \varepsilon_2 + \varepsilon_2 \varepsilon_3 + \cdots + \varepsilon_n \varepsilon_1 = 0$. Prove that n is divisible by 4.*

(Kvant)

Solution. The sum $\varepsilon_1 \varepsilon_2 + \varepsilon_2 \varepsilon_3 + \cdots + \varepsilon_n \varepsilon_1$ has n terms equal to 1 or -1, so n is even, say $n = 2k$. It is clear that k of the terms in the sequence $\varepsilon_1 \varepsilon_2, \varepsilon_2 \varepsilon_3, \ldots, \varepsilon_n \varepsilon_1$

are equal to 1 and k of them are equal to -1. On the other hand, the product of the terms in the sum is

$$(\varepsilon_1 \varepsilon_2)(\varepsilon_2 \varepsilon_3) \cdots (\varepsilon_n \varepsilon_1) = \varepsilon_1^2 \varepsilon_2^2 \ldots \varepsilon_n^2 = 1;$$

hence $(+1)^k(-1)^k = 1$. That is, k is even and the conclusion follows.

Note that the result of this problem is sharp.

For any integer $n = 4m$ there exist $\varepsilon_1, \varepsilon_2, \ldots, \varepsilon_n$ such that

$$\varepsilon_1 \varepsilon_2 + \varepsilon_2 \varepsilon_3 + \cdots + \varepsilon_n \varepsilon_1 = 0;$$

for example,

$$\varepsilon_1 = \varepsilon_4 = \varepsilon_5 = \varepsilon_8 = \cdots = \varepsilon_{4m-3} = \varepsilon_{4m} = +1,$$
$$\varepsilon_2 = \varepsilon_3 = \varepsilon_6 = \varepsilon_7 = \cdots = \varepsilon_{4m-2} = \varepsilon_{4m-1} = -1.$$

Problem 1.4.4. *A table of numbers with m rows and n columns has all entries -1 or 1 such that for each row and each column the product of entries is -1. Prove that m and n have the same parity.*

Solution. We compute the product P of the $m \cdot n$ entries in two ways, by rows and by columns, respectively:

$$P = \underbrace{(-1)(-1) \cdots (-1)}_{m \text{ times}} = (-1)^m = (-1)^n = \underbrace{(-1)(-1) \cdots (-1)}_{n \text{ times}}.$$

The conclusion now follows.

We show such a table for $m = 3$ and $n = 5$:

$$
\begin{array}{rrrrr}
-1 & 1 & 1 & -1 & -1 \\
1 & 1 & -1 & 1 & 1 \\
1 & -1 & 1 & 1 & 1
\end{array}
$$

Remark. If m and n have the same parity, then the number of tables with the above property is $2^{(m-1)(n-1)}$.

Additional Problems

Problem 1.4.5. We are given three integers a, b, c such that $a, b, c, a + b - c$, $a + c - b$, $b + c - a$, and $a + b + c$ are seven distinct primes. Let d be the difference between the largest and smallest of these seven primes. Suppose that $800 \in \{a + b, b + c, c + a\}$. Determine the maximum possible value of d.

Problem 1.4.6. Let n be an integer ≥ 1996. Determine the number of functions $f : \{1, 2, \ldots, n\} \rightarrow \{1995, 1996\}$ that satisfy the condition that $f(1) + f(2) + \cdots + f(1996)$ is odd.

(1996 Greek Mathematical Olympiad)

Problem 1.4.7. Is it possible to place 1995 different natural numbers around a circle so that in each pair of these numbers, the ratio of the larger to the smaller is a prime?

(1995 Russian Mathematical Olympiad)

Problem 1.4.8. Let a, b, c, d be odd integers such that $0 < a < b < c < d$ and $ad = bc$. Prove that if $a + d = 2^k$ and $b + c = 2^m$ for some integers k and m, then $a = 1$.

(25th International Mathematical Olympiad)

1.5 Modular Arithmetic

Let a, b, n be integers, with $n \neq 0$. We say that a and b are *congruent modulo n* if $n \mid a - b$. We denote this by $a \equiv b \pmod{n}$. The relation "$\equiv$" on the set $\mathbb{Z}$ of integers is called the *congruence relation*. If m does not divide $a - b$, then we say that integers a and b are not congruent modulo n and we write $a \not\equiv b \pmod{n}$. It is clear that if a is divided by b with remainder r, then a is congruent to r modulo b. In this case r is called the *residue of a modulo b*. The following properties can be directly derived:

(1) $a \equiv a \pmod{n}$ (reflexivity).

(2) If $a \equiv b \pmod{n}$ and $b \equiv c \pmod{n}$, then $a \equiv c \pmod{n}$ (transitivity).

(3) If $a \equiv b \pmod{n}$, then $b \equiv a \pmod{n}$ (symmetry).

(4) If $a \equiv b \pmod{n}$ and $c \equiv d \pmod{n}$, then $a + c \equiv b + d \pmod{n}$ and $a - c \equiv b - d \pmod{n}$.

(5) If $a \equiv b \pmod{n}$, then for any integer k, $ka \equiv kb \pmod{n}$.

(6) If $a \equiv b \pmod{n}$ and $c \equiv d \pmod{n}$, then $ac \equiv bd \pmod{n}$.

(7) If $a_i \equiv b_i \pmod{n}, i = 1, \ldots, k$, then $a_1 + \cdots + a_k \equiv b_1 + \cdots + b_k \pmod{n}$ and $a_1 \cdots a_k \equiv b_1 \cdots b_k \pmod{n}$. In particular, if $a \equiv b \pmod{n}$, then for any positive integer k, $a^k \equiv b^k \pmod{n}$.

(8) We have $a \equiv b \pmod{m_i}, i = 1, \ldots, k$, if and only if

$$a \equiv b \pmod{\operatorname{lcm}(m_1, \ldots, m_k)}.$$

In particular, if $m_1, \ldots, m_k$ are pairwise relatively prime, then $a \equiv b \pmod{m_i}$, $i = 1, \ldots, k$, if and only if $a \equiv b \pmod{m_1, \ldots, m_k}$.

Let us prove the last property. From $a \equiv b \pmod{m_i}, i = 1, \ldots, k$, it follows that $m_i \mid a - b, i = 1, \ldots, k$. Hence $a - b$ is a common multiple of $m_1, \ldots, m_k$, and so $\mathrm{lcm}(m_1, \ldots, m_k) \mid a - b$. That is, $a \equiv b \pmod{\mathrm{lcm}(m_1, \ldots, m_k)}$. Conversely, from $a \equiv b \pmod{\mathrm{lcm}(m_1, \ldots, m_k)}$, and the fact that each m_i divides $\mathrm{lcm}(m_1, \ldots, m_k)$ we obtain $a \equiv b \pmod{m_i}, i = 1, \ldots, k$.

Theorem 1.5.1. *Let a, b, n be integers, $n \neq 0$, such that $a = nq_1 + r_1$, $b = nq_2 + r_2$, $0 \le r_1, r_2 < |n|$. Then $a \equiv b \pmod n$ if and only if $r_1 = r_2$.*

Proof. Because $a - b = n(q_1 - q_2) + (r_1 - r_2)$, it follows that $n \mid a - b$ if and only if $n \mid r_1 - r_2$. Taking into account that $|r_1 - r_2| < |n|$, we have $n \mid r_1 - r_2$ if and only if $r_1 = r_2$. $\qquad\square$

Problem 1.5.1. *For all the positive integers $k \le 1999$, let $S_1(k)$ be the sum of all the remainders of the numbers $1, 2, \ldots, k$ when divided by 4, and let $S_2(k)$ be the sum of all the remainders of the numbers $k+1, k+2, \ldots, 2000$ when divided by 3. Prove that there is a unique positive integer $m \le 1999$ so that $S_1(m) = S_2(m)$.*

(1999 Romanian Mathematical Olympiad)

Solution. Let $A_k = \{1, 2, 3, \ldots, k\}$ and $B_k = \{k+1, k+2, \ldots, 2000\}$. From the division of integers we have

$$k = 4q_1 + r_1, \text{ with } r_1 \in \{0, 1, 2, 3\}. \tag{1}$$

If $s_1(k)$ is the sum of the remainders after division by 4 of the last r_1 elements of A_k, then let

$$S_1(k) = 6q_1 + s_1(k), \text{ with } 0 \le s_1(k) \le 6. \tag{2}$$

If $r_1 = 0$, then set $s_1(k) = 0$.

Using again the division of integers, there exist integers q_2, r_2 such that

$$2000 - k = 3q_2 + r_2, \text{ with } r_2 \in \{0, 1, 2\}. \tag{3}$$

If $s_2(k)$ is the sum of the remainders on division by 3 of the last r_2 elements of B_k, then let

$$s_2(k) = 3q_2 + s_2(k), \text{ with } 0 \le s_2(k) \le 3. \tag{4}$$

Again we set $S_2(k) = 0$ if $r_2 = 0$.

Since $S_1(k) = S_2(k)$, $s_2(k) - s_1(k) = 3(2q_1 - q_2)$, so $3 \mid 2q_1 - q_2 \mid = |s_2(k) - s_1(k)| \le 6$, and $|2q_1 - q_2| \le 2$. In other words, $|2q_1 - q_2| \in \{0, 1, 2\}$.

If $2q_1 = q_2$, then (1) and (3) imply $2000 - (r_1 + r_2) = 10q_1$; hence $10 \mid (r_1 + r_2)$. Then $r_1 = r_2 = 0$ and $q_1 = 200$. From (1) it follows that $k = 800$, and from (2) and (4) we have $S_1(800) = S_2(800) = 1200$.

Furthermore, $S_1(k) \le S_1(k+1)$, and $S_2(k) \ge S_2(k+1)$ for all $k \in \{1, 2, \ldots, 1998\}$. Since $S_1(799) = S_1(800)$ and $S_2(799) = S_2(800) + 2 > S_1(800)$, we deduce that $S_1(k) < S_2(k)$ for all $k \in \{1, 2, \ldots, 799\}$. Since $S_1(801) = S_1(800) + 1 > S_2(800) \geq S_2(801)$, we derive that $S_1(k) > S_2(k)$ for all $k \in \{801, 802, \ldots, 1999\}$. Consequently, $S_1(m) = S_2(m)$ if and only if $m = 800$.

Problem 1.5.2. *Let n be a positive integer. Show that if a and b are integers greater than 1 such that $2^n - 1 = ab$, then $ab - (a - b) - 1$ can be written as $k \cdot 2^{2m}$ for some odd integer k and some positive integer m.*

(2001 Balkan Mathematical Olympiad)

Solution. Note that $ab - (a - b) - 1 = (a + 1)(b - 1)$. We shall show that the highest powers of 2 dividing $(a + 1)$ and $(b - 1)$ are the same. Let 2^s and 2^t be the highest powers of 2 dividing $(a + 1)$ and $(b - 1)$, respectively. Because $a + 1, b - 1 \leq ab + 1 = 2^n$, we have $s, t \leq n$.

Note that 2^s divides $2^n = ab + 1$ and $a + 1$, so that

$$ab \equiv a \equiv -1 \pmod{2^s}.$$

Hence, $b \equiv 1 \pmod{2^s}$, or $2^s \mid b - 1$, so that $s \leq t$.

Similarly, $ab \equiv -b \equiv -1 \pmod{2^t}$, so $a \equiv -1 \pmod{2^t}$, and $2^t \mid a + 1$. Thus, $t \leq s$.

Therefore, $s = t$, the highest power of 2 dividing $(a + 1)(b - 1)$ is $2s$, and $ab - (a - b) - 1 = k \cdot 2^{2s}$ for some odd k.

Problem 1.5.3. *Find all nonnegative integers m such that $(2^{2m+1})^2 + 1$ is divisible by at most two different primes.*

(2002 Baltic Mathematics Competition)

Solution. We claim $m = 0, 1, 2$ are the only such integers. It is easy to check that these values of m satisfy the requirement. Suppose some $m \geq 3$ works. Write

$$(2^{2m+1})^2 + 1 = (2^{2m+1} + 1)^2 - 2 \cdot 2^{2m+1}$$
$$= (2^{2m+1} + 2^{m+1} + 1)(2^{2m+1} - 2^{m+1} + 1).$$

The two factors are both odd, and their difference is 2^{m+2}; hence, they are relatively prime. It follows that each is a prime power. We also know that $(2^{2m+1})^2 = 4^{2m+1} \equiv -1 \pmod 5$, so one of the factors $2^{2m+1} \pm 2^{m+1} + 1$ must be a power of 5. Let $2^{2m+1} + 2^{m+1}s + 1 = 5^k$, where $s = \pm 1$ is the appropriate sign.

Taking the above equation modulo 8, and using the assumption $m \geq 3$, we obtain $5^k \equiv 1 \pmod 8$, so that k is even. Writing $k = 2l$, we have

$$2^{m+1}(2^m + s) = (5^l - 1)(5^l + 1).$$

The factor $5^l + 1$ is congruent to 2 (mod 4), so $5^l - 1 = 2^m a$ for some odd integer a. But if $a = 1$, then

$$2 = (5^l + 1) - (5^l - 1) = 2(2^m + s) - 2^m = 2^m + 2s \geq 2^3 - 2,$$

a contradiction, whereas if $a \geq 3$, then $5^l - 1 \geq 3 \cdot 2^m$ while $5^l + 1 \leq 2(2^m + s)$, another contradiction.

Problem 1.5.4. *Find an integer n with* $100 \leq n \leq 1997$ *such that n divides* $2^n + 2$.

(1997 Asian Pacific Mathematics Olympiad)

Solution. Note that 2 divides $2^n + 2$ for all n. Also, 11 divides $2^n + 2$ if and only if $n \equiv 6 \pmod{10}$, and 43 divides $2^n + 2$ if and only if $n \equiv 8 \pmod{14}$. Since $n = 946 = 2 \cdot 11 \cdot 43$ satisfies both congruences, n divides $2^n + 2$.

Remark. Actually, one can prove that there are infinitely many n such that $n \mid 2^n + 2$. Also, any such n is even, since by a theorem of W. Sierpiński[6] we cannot have $n \mid 2^{n-1} + 1$ unless $n = 1$ (see also Problem 7.1.16).

Problem 1.5.5. *The number* $99\ldots99$ *(with 1997 nines) is written on a black-board. Each minute, one number written on the blackboard is factored into two factors and erased, each factor is (independently) increased or diminished by 2, and the resulting two numbers are written. Is it possible that at some point all of the numbers on the blackboard equal 9?*

(1997 St. Petersburg City Mathematical Olympiad)

Solution. The answer is No. Indeed, note that $99\ldots99$ (with 1997 nines) $= 10^{1997} - 1$ is congruent to 3 modulo 4. If $99\ldots99$ (with 1997 nines) $= ab$, for some integers a and b, then for example a is congruent to 1 modulo 4 and b is congruent to 3 modulo 4. Changing a and b by 2 in either direction we find numbers congruent to 3, respectively to 1 modulo 4. Hence at any point we get numbers of different residues modulo 4, so these numbers cannot be equal.

Problem 1.5.6. *Find the smallest positive integer that can be written both as (i) a sum of* 2002 *positive integers (not necessarily distinct), each of which has the same sum of digits and (ii) as a sum of* 2003 *positive integers (not necessarily distinct) each of which has the same sum of digits.*

(2002 Russian Mathematical Olympiad)

Solution. The answer is 10010. First observe that this is indeed a solution: $10010 = 2002 \cdot 5 = 1781 \cdot 4 + 222 \cdot 13$, so we may express 10010 as the sum of 2002 fives or of 1781 fours and 222 thirteens, where $1781 + 222 = 2003$. To prove minimality, observe that a number is congruent modulo 9 to the sum of its digits, so two positive integers with the same digit sum have the same remainders modulo 9. Let k_1 be the digit sum of the 2002 numbers and k_2 the digit sum of the 2003 numbers. Then $4k_1 \equiv 2002k_1 \equiv 2003k_2 \equiv 5k_2 \pmod 9$. If $k_1 \geq 5$, the sum of the 2002 numbers is at least 10010; if $k_2 \geq 5$, the sum of the 2003 numbers is greater than 10010. Since $k_1 + k_2 \equiv 0 \pmod 9$, we have $k_1 + k_2 \geq 9$. Hence either $k_1 \geq 5$ or $k_2 \geq 5$, and the minimal integer is 10010.

---

[6]Wacław Sierpiński (1882–1969), Polish mathematician with fundamental work in the area of set theory, point set topology, and number theory.

Additional Problems

Problem 1.5.7. Find all integers $n > 1$ such that any prime divisor of $n^6 - 1$ is a divisor of $(n^3 - 1)(n^2 - 1)$.

(2002 Baltic Mathematics Competition)

Problem 1.5.8. Let $f(n)$ be the number of permutations $a_1, \ldots, a_n$ of the integers $1, \ldots, n$ such that

(i) $a_1 = 1$;

(ii) $|a_i - a_{i+1}| \leq 2$, $i = 1, \ldots, n - 1$.

Determine whether $f(1996)$ is divisible by 3.

(1996 Canadian Mathematical Olympiad)

Problem 1.5.9. For natural numbers m, n, show that $2^n - 1$ is divisible by $(2^m - 1)^2$ if and only if n is divisible by $m(2^m - 1)$.

(1997 Russian Mathematical Olympiad)

Problem 1.5.10. Suppose that n is a positive integer and let

$$d_1 < d_2 < d_3 < d_4$$

be the four smallest positive integer divisors of n. Find all integers n such that

$$n = d_1^2 + d_2^2 + d_3^2 + d_4^2.$$

(1999 Iranian Mathematical Olympiad)

Problem 1.5.11. Let p be an odd prime. For each $i = 1, 2, \ldots, p - 1$ denote by r_i the remainder when i^p is divided by p^2. Evaluate the sum

$$r_1 + r_2 + \cdots + r_{p-1}.$$

(Kvant)

Problem 1.5.12. Find the number of integers x with $|x| \leq 1997$ such that 1997 divides $x^2 + (x + 1)^2$.

(1998 Indian Mathematical Olympiad)

Problem 1.5.13. Find the greatest common divisor of the numbers

$$A_n = 2^{3n} + 3^{6n+2} + 5^{6n+2}$$

when $n = 0, 1, \ldots, 1999$.

(1999 Junior Balkan Mathematical Olympiad)

1.6   Chinese Remainder Theorem

There are many situations in which one wishes to solve a congruence equation $f(x) \equiv a \pmod{N}$ for some large N. Suppose we factor N as $N = p_1^{k_1} p_2^{k_2} \cdots$ $p_r^{k_r}$. Then any solution must also have $f(x) \equiv a \pmod{p_i^{k_i}}$ for all i. Conversely, if $f(x) \equiv a \pmod{p_u^{k_i}}$, then all the $p_i^{k_i}$ divide $f(x) - a$. Hence their least common multiple N divides $f(x) - a$. Thus one congruence mod a very large N is equivalent to lots of congruences mod its prime power factors. These congruences are often much easier to solve, either because $p_i^{k_i}$ is much smaller than N, or because of special facts about primes such as Fermat's little theorem. Thus it might be easier to solve $f(x) \equiv a \pmod{p_i^{k_i}}$ for all i. However, this solution will often give us a list of congruences that x must satisfy, say

$$x \equiv c_1 \pmod{m_1}, \quad \ldots, \quad x \equiv c_n \pmod{m_n}.$$

This leaves us with the problem whether such a system has a solution and how to find the solutions. In solving systems of this form an important part is played by the following very important result:

Theorem 1.6.1. (Chinese remainder theorem) *Let $m_1, \cdots, m_n$ be positive integers different from 1 and pairwise relatively prime. Then for any nonzero integers $a_1, \ldots, a_r$ the system of linear congruences*

$$x \equiv a_1 \pmod{m_1}, \quad \ldots, \quad x \equiv a_r \pmod{m_r}$$

has solutions, and any two such solutions are congruent modulo $m = m_1 \cdots m_r$.
Proof. It is clear that $\gcd\left(\frac{m}{m_j}, m_j\right) = 1$, $j = 1, \ldots, r$. Applying Proposition 1.3.1, it follows that there is an integer b_j such that

$$\frac{m}{m_j} b_j \equiv 1 \pmod{m_j}, \quad j = 1, \ldots, r.$$

Then

$$\frac{m}{m_j} b_j a_j \equiv a_j \pmod{m_j}, \quad j = 1, \ldots, r.$$

Now consider the integer

$$x_0 = \sum_{j=1}^{r} \frac{m}{m_j} b_j a_j.$$

We have

$$x_0 \equiv \left(\sum_{j=1}^{r} \frac{m}{m_j} b_j a_j \right) \pmod{m_i} \equiv \frac{m}{m_i} b_i a_i \pmod{m_i}$$

$$\equiv a_i \pmod{m_i}, \quad i = 1, \ldots, r,$$

that is, x_0 is a solution to the system of linear congruences.

If x_1 is another solution, then $x_1 \equiv x_0 \pmod{m_i}$, $i = 1, \ldots, r$. Applying property (8) in Section 1.5, the conclusion follows. $\square$

Example. Let us find the solutions to the system of linear congruences

$$x \equiv 2 \pmod 3, \quad x \equiv 1 \pmod 4, \quad x \equiv 3 \pmod 5.$$

We proceed as in the proof of the theorem. Because in this case $m = 3 \cdot 4 \cdot 5 = 60$, we have to find a solution to each of the congruences

$$\frac{60}{3}b_1 \equiv 1 \pmod 3, \quad \frac{60}{4}b_2 \equiv 1 \pmod 4, \quad \frac{60}{5}b_3 \equiv 1 \pmod 5.$$

This is equivalent to finding solutions to the congruences

$$2b_1 \equiv 1 \pmod 3, \quad 3b_2 \equiv 1 \pmod 4, \quad 2b_3 \equiv 1 \pmod 5.$$

We obtain $b_1 = 2, b_2 = 3, b_3 = 3$. Then

$$x_0 = 20 \cdot 2 \cdot 2 + 15 \cdot 3 \cdot 1 + 12 \cdot 3 \cdot 3 = 233.$$

Taking into account that all solutions are congruent modulo 60, it follows that it suffices to take $x_0 = 53$. All solutions are given by $x = 53 + 60k$, $k \in \mathbb{Z}$.

Problem 1.6.1. *We call a lattice point X in the plane visible from the origin O if the segment $\overline{OX}$ does not contain any other lattice points besides O and X. Show that for any positive integer n, there exists a square of n^2 lattice points (with sides parallel to the coordinate axes) such that none of the lattice points inside the square is visible from the origin.*

(2002 Taiwanese Mathematical Olympiad)

Solution. Suppose that the lower-left lattice point of such a square has coordinates (x_1, y_1). We shall show that it is possible to select (x_1, y_1) such that the square of lattice points with (x_1, y_1) at its corner and n points on a side contains only invisible points. This can be accomplished by ensuring that each point has both coordinates divisible by some prime number; this would imply that by dividing both coordinates by this prime, we could find another lattice point that is between the origin and this point.

In fact, we note that a lattice point $X = (x, y)$ is visible from the origin if and only if $\gcd(x, y) = 1$.

Select n^2 distinct prime numbers and call them $p_{i,j}$, $1 \leq 1, j \leq n$. Now find x_1 satisfying the following congruences:

$$x_1 \equiv 0 \pmod{p_{1,1}p_{1,2}\cdots p_{1,n}},$$
$$x_1 + 1 \equiv 0 \pmod{p_{2,1}p_{2,2}\cdots p_{2,n}},$$
$$\cdots$$
$$x_1 + n - 1 \equiv 0 \pmod{p_{n,1}p_{n,2}\cdots p_{n,n}}.$$

Likewise select y_1 satisfying

$$y_1 \equiv 0 \quad (\text{mod } p_{1,1}p_{2,1} \cdots p_{n,1}),$$
$$y_1 + 1 \equiv 0 \quad (\text{mod } p_{1,2}p_{2,2} \cdots p_{n,2}),$$
$$\cdots$$
$$y_1 + n - 1 \equiv 0 \quad (\text{mod } p_{1,n}p_{2,n} \cdots p_{n,n}).$$

Both values must exist by the Chinese remainder theorem. Thus we have proved that it is possible to determine a position for (x_1, y_1) such that every point in the square of n^2 lattice points with (x_1, y_1) at its lower left corner is associated with some prime by which both of its coordinates are divisible; thus no points in this square are visible from the origin.

Problem 1.6.2. *Show that there exists an increasing sequence $\{a_n\}_{n=1}^{\infty}$ of natural numbers such that for any $k \geq 0$, the sequence $\{k + a_n\}$ contains only finitely many primes.*

(1997 Czech and Slovak Mathematical Olympiad)

Solution. Let p_k be the kth prime number, $k \geq 1$. Set $a_1 = 2$. For $n \geq 1$, let a_{n+1} be the least integer greater than a_n that is congruent to $-k$ modulo p_{k+1} for all $k \leq n$. Such an integer exists by the Chinese remainder theorem. Thus, for all $k \geq 0$, $k + a_n \equiv 0 \ (\text{mod } p_{k+1})$ for $n \geq k + 1$. Then at most $k + 1$ values in the sequence $\{k + a_n\}$ can be prime; from the $(k + 2)$th term onward, the values are nontrivial multiples of p_{k+1} and must be composite. This completes the proof.

Additional Problems

Problem 1.6.3. Let $P(x)$ be a polynomial with integer coefficients. Suppose that the integers $a_1, a_2, \ldots, a_n$ have the following property: For any integer x there exists an $i \in \{1, 2, \ldots, n\}$ such that $P(x)$ is divisible by a_i. Prove that there is an $i_0 \in \{1, 2, \ldots, n\}$ such that a_{i_0} divides $P(x)$ for every integer x.

(St. Petersburg City Mathematical Olympiad)

Problem 1.6.4. For any set of positive integers $\{a_1, a_2, \ldots, a_n\}$ there exists a positive integer b such that the set $\{ba_1, ba_2, \ldots, ba_n\}$ consists of perfect powers.

1.7 Numerical Systems

1.7.1 Representation of Integers in an Arbitrary Base

The fundamental result in this subsection is given by the following theorem:

Theorem 1.7.1. *Let b be an integer greater than 1. For any integer $n \geq 1$ there is a unique system $(k, a_0, a_1, \ldots, a_k)$ of integers such that $0 \leq a_i \leq b - 1$, $i = 0, 1, \ldots, k$, $a_k \neq 0$, and*

$$n = a_k b^k + a_{k-1} b^{k-1} + \cdots + a_1 b + a_0. \tag{1}$$

Proof. For the existence, we repeatedly apply the division algorithm:

$$n = q_1 b + r_1, \quad 0 \leq r_1 \leq b - 1,$$
$$q_1 = q_2 b + r_2, \quad 0 \leq r_2 \leq b - 1,$$
$$\cdots$$
$$q_{k-1} = q_k b + r_k, \quad 0 \leq r_k \leq b - 1,$$

where q_k is the last nonzero quotient.

Let

$$a_0 = r_1 = n - q_1 b, \quad a_1 = q_1 - q_2 b, \quad \ldots, \quad a_{k-1} = q_{k-1} - q_k b, \quad a_k = q_k.$$

Then

$$\sum_{i=0}^{k} a_i b^i = \sum_{i=0}^{k-1} (q_i - q_{i+1} b) b^i + q_k b^k = q_0 + \sum_{i=1}^{k} q_i b^i - \sum_{i=1}^{k} q_i b^i = q_0 = n.$$

For the uniqueness, assume that $n = c_0 + c_1 b + \cdots + c_h b^h$ is another such representation.

If $h \neq k$, for example $h > k$, then $n \geq b^h \geq b^{k+1}$. But

$$n = a_0 + a_1 b + \cdots + a_k b^k \leq (b-1)(1 + b + \cdots + b^k) = b^{k+1} - 1 < b^{k+1},$$

a contradiction.

If $h = k$, then

$$a_0 + a_1 b + \cdots + a_k b^k = c_0 + c_1 b + \cdots + c_k b^k,$$

and so $b \mid a_0 - c_0$. On the other hand, $|a_0 - c_0| < b$; hence $a_0 = c_0$, Therefore

$$a_1 + a_2 b + \cdots + a_k b^{k-1} = c_1 + c_2 b + \cdots + c_k b^{k-1}.$$

Repeating the procedure above, it follows that $a_1 = c_1$, $a_2 = c_2, \ldots$, $a_k = c_k$. $\qquad\square$

Relation (1) is called *the base-b representation* of n and is denoted by

$$n = \overline{a_k a_{k-1} \cdots a_0}_{(b)}$$

The usual *decimal representation* corresponds to $b = 10$.

Examples. (1) $4567 = 4 \cdot 10^3 + 5 \cdot 10^2 + 6 \cdot 10 + 7 = \overline{4567}_{(10)}$.

(2) Let us write $\overline{1010011}_{(2)}$ in base 10. We have

$$\overline{1010011}_{(2)} = 1 \cdot 2^6 + 0 \cdot 2^5 + 1 \cdot 2^4 + 0 \cdot 2^3 + 0 \cdot 2^2 + 1 \cdot 2 + 1 = 64 + 16 + 2 + 1 = 83.$$

(3) Let us write 1211 in base 3. As above, dividing by 3 successively, the remainders give the digits of the base-3 representation, beginning with the last. The first digit is the last nonzero quotient. We can arrange the computations as follows:

$$
\begin{array}{rrrrrr}
1211| & 3 & & & & \\
1209\,403| & 3 & & & & \\
2 & 402\,134| & 3 & & & \\
& 1 & 132\,44| & 3 & & \\
& & 2 & 42\,14|3 & & \\
& & & 2 & 12\,4|3 & \\
& & & & 2 & 3\,1 \\
& & & & & 1
\end{array}
$$

Hence $1211 = \overline{1122212}_{(3)}$.

1.7.2   Divisibility Criteria in the Decimal System

We will prove some divisibility criteria for integers in decimal representation. In this subsection, we will write $n = \overline{a_h a_{h-1} \cdots a_0}$ with the understanding that we operate in base 10.

Criterion 1. (a) *The integer* $n = \overline{a_h a_{h-1} \cdots a_0}$ *is divisible by 3 if and only if the sum* $s(n)$ *of its digits is divisible by 3.*

(b) *The integer* $n = \overline{a_h a_{h-1} \cdots a_0}$ *is divisible by 9 if and only if* $s(n)$ *is divisible by 9.*

Proof. We have $10^k \equiv 1 \pmod 9$, since $10 \equiv 1 \pmod 9$; hence

$$n = \sum_{k=0}^{h} a_k 10^k \equiv \sum_{k=0}^{h} a_k \equiv s(n) \pmod 9.$$

Both conclusions follow.   □

Criterion 2. *The integer* $n = \overline{a_h a_{h-1} \cdots a_0}$ *is divisible by 11 if and only if* $a_0 - a_1 + \cdots + (-1)^h a_h$ *is divisible by 11.*

Proof. We have $10^k \equiv (11 - 1)^k \equiv (-1)^k \pmod{11}$; hence

$$n = \sum_{k=0}^{h} a_k 10^k \equiv \sum_{k=0}^{h} (-1)^k a_k \pmod{11},$$

and the conclusion follows.   □

Criterion 3. *The integer* $n = \overline{a_h a_{h-1} \cdots a_0}$ *is divisible by* 7, 11, *or* 13 *if and only if* $\overline{a_h a_{h-1} \cdots a_3} - \overline{a_2 a_1 a_0}$ *has this property.*

Proof. We have

$$n = \overline{a_2 a_1 a_0} + (1001 - 1)\overline{a_h a_{h-1} \cdots a_3}$$
$$= (7 \cdot 11 \cdot 13)\overline{a_h a_{h-1} \cdots a_3} - (\overline{a_h a_{h-1} \cdots a_3} - \overline{a_2 a_1 a_0}),$$

hence the desired conclusion. □

Criterion 4. *The integer* $n = \overline{a_h a_{h-1} \cdots a_0}$ *is divisible by* 27 *or* 37 *if and only if* $\overline{a_h a_{h-1} \cdots a_3} + \overline{a_2 a_1 a_0}$ *has this property.*

Proof. We have

$$n = \overline{a_2 a_1 a_0} + (999 + 1)\overline{a_h a_{h-1} \cdots a_3}$$
$$= (27 \cdot 37)\overline{a_h a_{h-1} \cdots a_3} + (\overline{a_h a_{h-1} \cdots a_3} + \overline{a_2 a_1 a_0}),$$

and the conclusion follows. □

Examples. (1) The integer 123456789 is divisible by 9 because the sum of its digits $1 + 2 + \cdots + 9 = 45$ has this property (Criterion 1(b)).

(2) The integer $\underbrace{20 \ldots 04}_{2004}$ is not a perfect square because the sum of its digits is 6, a multiple of 3 but not of 9; hence the integer itself has these properties (Criteria 1(a) and 1(b)).

(3) All integers of the form $\overline{abcdef}$, where $a + c + e = 8$ and $b + d + f = 19$, are divisible by 99, because $a + b + c + d + f = 8 + 19$, a multiple of 9, and $f - e + d - c + b - a = 19 - 8$, a multiple of 11, and the conclusion follows from Criteria 1(b) and 2.

(4) For any nonzero digit a, the integer $\overline{a1234567}$ is not divisible by 37. Indeed, applying Criterion 4, we have $\overline{a1234} + 567 = \overline{a1801}$ and $\overline{a1} + 801 = \overline{8a2} = 800 + 10a + 2 = 37 \cdot 21 + 10a + 25$. The integer $10a + 25 = 5(2a + 5)$ is not divisible by 37 because $7 \le 2a + 5 \le 23$.

Problem 1.7.1. *Find all integers written as* $\overline{abcd}$ *in decimal representation and* $\overline{dcba}$ *in base* 7.

Solution. We have

$$\overline{abcd}_{(10)} = \overline{dcba}_{(7)} \Leftrightarrow 999a + 93b = 39c + 342d \Leftrightarrow 333a + 31b = 13c + 114d;$$

hence $b \equiv c \pmod{3}$. Since $b, c \in \{0, 1, 2, 3, 4, 5, 6\}$, the possibilities are:

(i) $b = c$;

(ii) $b = c + 3$;

(iii) $b + 3 = c$.

Also, we note that $13c + 114d \le 762 < 3 \cdot 333$; hence $a \le 2$.

In the first case we must have $a = 2$, $d = 3d'$, $37 + b = 19d'$, $d' = 2$. Hence $a = 2$, $d = 6$, $b = 1$, $c = 1$, and the number $\overline{abcd}$ is 2116.

In the other cases $a = 1$. Considering $a = 1$, we obtain no solutions.

Problem 1.7.2. *Prove that every integer $k > 1$ has a multiple less than k^4 whose decimal expansion has at most four distinct digits.*

(1996 German Mathematical Olympiad)

Solution. Let n be the integer such that $2^{n-1} \le k < 2^n$. For $n \le 6$ the result is immediate, so assume $n > 6$.

Let S be the set of nonnegative integers less than 10^n whose decimal digits are all 0 or 1. Since $|S| = 2^n > k$, we can find two elements $a < b$ of S that are congruent modulo k, and $b - a$ has only the digits 8, 9, 0, 1 in its decimal representation. On the other hand,

$$b - a \le 1 + 10 + \cdots + 10^{n-1} < 10^n < 16^{n-1} \le k^4;$$

hence $b - a$ is the desired multiple.

Problem 1.7.3. *A positive integer is written on a board. We repeatedly erase its unit digit and add 5 times that digit to what remains. Starting with 7^{1998}, can we ever end up at 1998^7?*

(1998 Russian Mathematical Olympiad)

Solution. The answer is no. Let a_n be the nth number written on the board; let u_n be the unit digit and $a_n = 10t_n + u_n$. We have

$$a_{n+1} = t_n + 5u_n \equiv 50t_n + 5u_n = 5(10t_n + u_n) = 5a_n \quad (\mathrm{mod}\ 7).$$

Since $a_1 = 7^{1998} \equiv 0 \not\equiv 1998^7 \pmod 7$, we can never obtain 1998^7 from 7^{1998}.

Problem 1.7.4. *Find all the three-digit numbers $\overline{abc}$ such that the 6003-digit number $\overline{abcabc \ldots abc}$ is divisible by 91 ($\overline{abc}$ occurs 2001 times).*

Solution. The number is equal to

$$\overline{abc}(1 + 10^3 + 10^6 + \cdots + 10^{6000}).$$

Since 91 is a divisor of $1001 = 1 + 10^3$ and the sum $S = 1 + 10^3 + 10^6 + \cdots + 10^{6000}$ has 2001 terms, it follows that 91 and $(1 + 10^3) + 10^6(1 + 10^3) + \cdots + 10^{6 \cdot 999}(1 + 10^3) + 10^{6000}$ are relatively prime. Thus $\overline{abc}$ is divisible by 91. The numbers are

$$182, \ 273, \ 364, \ 455, \ 546, \ 637, \ 728, \ 819, \ 910.$$

Problem 1.7.5. *Let n be an integer greater than* 10 *such that each of its digits belongs to the set* $S = \{1, 3, 7, 9\}$. *Prove that n has some prime divisor greater than or equal to* 11.

(1999 Iberoamerican Mathematical Olympiad)

Solution. Note that any product of any two numbers from $\{1, 3, 7, 9\}$ taken modulo 20 is still in $\{1, 3, 7, 9\}$. Therefore any finite product of such numbers is still in this set. Specifically, any number of the form $3^j 7^k$ is congruent to 1, 3, 7, or 9 (mod 20).

Now if all the digits of $n \geq 10$ are in S, then its tens digit is odd and we cannot have $n \equiv 1, 3, 7$, or 9 (mod 20). Thus, n cannot be of the form $3^j 7^k$. Nor can n be divisible by 2 or 5 (otherwise, its last digit would not be 1, 3, 7, or 9). Hence n must be divisible by some prime greater than or equal to 11, as desired.

Problem 1.7.6. *Find all natural numbers with the property that when the first digit is moved to the end, the resulting number is* $3\frac{1}{2}$ *times the original one.*

(1997 South African Mathematical Olympiad)

Solution. Such numbers are those of the form

$$153846153846153846 \ldots 153846.$$

Obviously, since the number has the same number of digits when multiplied by 3.5, it must begin with either 1 or 2.

Case 1. The number is of the form $10^N + A$, $A < 10^N$. So $7/2 \times (10^N + A) = 10A + 1$ is equivalent to $A = (7 \times 10^N - 2)/13$. The powers of 10 repeat with a period of 6 mod 13 (10, 9, 12, 3, 4, 1), so A will be an integer iff $n \equiv 5$ (mod 6). This gives the family of solutions above.

Case 2. The number is of the form $2 \times 10^N + A$, $A < 10^N$. Then, as before, $A = (14 \times 10^N - 4)/13$. But since $A < 10^N$, this implies $10^N < 4$, which is impossible.

Problem 1.7.7. *Any positive integer m can be written uniquely in base 3 as a string of 0's, 1's, and 2's (not beginning with a zero). For example,*

$$98 = 81 + 9 + 2 \times 3 + 2 \times 1 = (10122)_3.$$

Let c(m) denote the sum of the cubes of the digits of the base-3 form of m; thus, for instance,

$$c(98) = 1^3 + 0^3 + 1^3 + 2^3 + 2^3 = 18.$$

Let n be any fixed positive integer. Define the sequence $\{u_r\}$ as

$$u_1 = n, \text{ and } u_r = c(u_{r-1}) \text{ for } r \geq 2.$$

Show that there is a positive integer r such that $u_r = 1, 2$, or 17.

(1999 United Kingdom Mathematical Olympiad)

Solution. If m has $d \geq 5$ digits then we have $m \geq 3^{d-1} = (80+1)^{(d-1)/4} \geq 80 \cdot \frac{d-1}{4} + 1 > 8d$ by Bernoulli's inequality. Thus $m > c(m)$.

If $m > 32$ has four digits in base 3, then $c(m) \leq 2^3 + 2^3 + 2^3 + 2^3 = 32 < m$. On the other hand, if $27 \leq m \leq 32$, then m starts with the digits 10 in base 3 and $c(m) < 1^3 + 0^3 + 2^3 + 2^3 = 17 < m$.

Therefore $0 < c(m) < m$ for all $m \geq 27$. Hence, eventually, we have $u_s < 27$. Because u_s has at most three digits, u_{s+1} can equal only 8, 16, 24, 1, 9, 17, 2, 10, or 3. If it equals 1, 2, or 17 we are already done; if it equals 3 or 9 then $u_{s+2} = 1$. Otherwise, a simple check shows that u_r will eventually equal 2:

$$\left. \begin{array}{l} 8 = (22)_3 \\ 24 = (220)_3 \end{array} \right\} \rightarrow 16 = (121)_3 \rightarrow 10 = (101)_3 \rightarrow 2.$$

Problem 1.7.8. *Do there exist n-digit numbers M and N such that all of the digits of M are even, all of the digits of N are odd, each digit from 0 to 9 occurs exactly once among M and N, and N divides M?*

(1998 Russian Mathematical Olympiad)

Solution. The answer is no. We proceed by indirect proof. Suppose that such M and N exist, and let $a = M/N$. Then $M \equiv 0 + 2 + 4 + 6 + 8 \equiv 2 \pmod 9$ and $N \equiv 1 + 3 + 5 + 7 + 9 \equiv 7 \pmod 9$; they are both relatively prime to 9. Now $a \equiv M/N \equiv 8 \pmod 9$, and so $a \geq 8$. But $N \geq 13579$, so $M = aN \geq 8(13579) > 99999$, a contradiction.

Problem 1.7.9. *Let $k \geq 1$ be an integer. Show that there are exactly 3^{k-1} positive integers n with the following properties:*

(a) *The decimal representation of n consists of exactly k digits.*

(b) *All digits of n are odd.*

(c) *The number n is divisible by 5.*

(d) *The number $m = n/5$ has k (decimal) digits.*

(1996 Austrian–Polish Mathematics Competition)

Solution. The multiplication in each place must produce an even carry digit, since these will be added to 5 in the next place and an odd digit must result. Hence all of the digits of m must be 1, 5 or 9, and the first digit must be 1, since m and n have the same number of decimal digits. Hence there are 3^{k-1} choices for m and hence for n.

Problem 1.7.10. *Can the number obtained by writing the numbers from 1 to n in order ($n > 1$) be the same when read left to right and right to left?*

(1996 Russian Mathematical Olympiad)

Solution. It is not possible. Suppose $10^p \le n < 10^{p+1}$. If N does not end with 321, then it will not be palindromic, so we may assume $p \ge 2$. The longest run of consecutive zeros in N will have length p. These runs occur exactly where we write down the multiples of 10^p. Suppose further $k \cdot 10^p \le n < (k+1) \cdot 10^p$ for a digit k. Then there are exactly k runs of p consecutive zeros, and the kth is bracketed by $\dots 9k00 \dots 0k0 \dots$. Thus none of the runs of p zeros can be sent to itself or another run by reversing the order of the digits.

Problem 1.7.11. *Three boxes with at least one marble in each are given. In a step we choose two of the boxes, doubling the number of marbles in one of the boxes by taking the required number of marbles from the other box. Is it always possible to empty one of the boxes after a finite number of steps?*

(1999 Slovenian Mathematical Olympiad)

Solution. Without loss of generality suppose that the number of marbles in the boxes are a, b, and c with $a \le b \le c$. Write $b = qa + r$ where $0 \le r < a$ and $q \ge 1$. Then express q in binary:

$$q = m_0 + 2m_1 + \cdots + 2^k m_k,$$

where each $m_i \in \{0, 1\}$ and $m_k = 1$. Now for each $i = 0, 1, \dots, k$, add $2^i a$ marbles to the first box: if $m_i = 1$ take these marbles from the second box; otherwise, take them from this third box. In this way we take at most $(2^k - 1)a < qa \le b \le c$ marbles from the third box and exactly qa marbles from the second box altogether.

In the second box there are now $r < a$ marbles left. Thus the box with the least number of marbles now contains fewer than a marbles. Then by repeating the described procedure, we will eventually empty one of the boxes.

Additional Problems

Problem 1.7.12. The natural number A has the following property: the sum of the integers from 1 to A, inclusive, has decimal expansion equal to that of A followed by three digits. Find A.

(1999 Russian Mathematical Olympiad)

Problem 1.7.13. A positive integer is said to be *balanced* if the number of its decimal digits equals the number of its distinct prime factors. For instance, 15 is balanced, while 49 is not. Prove that there are only finitely many balanced numbers.

(1999 Italian Mathematical Olympiad)

Problem 1.7.14. Let $p \geq 5$ be a prime and choose $k \in \{0, \ldots, p - 1\}$. Find the maximum length of an arithmetic progression, none of whose elements contain the digit k when written in base p.

(1997 Romanian Mathematical Olympiad)

Problem 1.7.15. How many 10-digit numbers divisible by 66667 are there whose decimal representation contains only the digits 3, 4, 5, and 6?

(1999 St. Petersburg City Mathematical Olympiad)

Problem 1.7.16. Call positive integers *similar* if they are written using the same digits. For example, for the digits 1, 1, 2, the similar numbers are 112, 121, and 211. Prove that there exist three similar 1995-digit numbers containing no zero digit such that the sum of two them equals the third.

(1995 Russian Mathematical Olympiad)

Problem 1.7.17. Let k and n be positive integers such that

$$(n + 2)^{n+2}, \quad (n + 4)^{n+4}, \quad (n + 6)^{n+6}, \quad \ldots, \quad (n + 2k)^{n+2k}$$

end in the same digit in decimal representation. At most how large is k?

(1995 Hungarian Mathematical Olympiad)

Problem 1.7.18. Let

$$\prod_{n=1}^{1996}(1 + nx^{3^n}) = 1 + a_1 x^{k_1} + a_2 x^{k_2} + \cdots + a_m x^{k_m},$$

where $a_1, a_2, \ldots, a_m$ are nonzero and $k_1 < k_2 < \cdots < k_m$. Find a_{1996}.

(1996 Turkish Mathematical Olympiad)

Problem 1.7.19. For any positive integer k, let $f(k)$ be the number of elements in the set $\{k + 1, k + 2, \ldots, 2k\}$ whose base-2 representation has precisely three 1's.

(a) Prove that, for each positive integer m, there exists at least one positive integer k, such that $f(k) = m$.

(b) Determine all positive integers m for which there exists exactly one k with $f(k) = m$.

(35th International Mathematical Olympiad)

Problem 1.7.20. For each positive integer n, let $S(n)$ be the sum of digits in the decimal representation of n. Any positive integer obtained by removing several (at least one) digits from the right-hand end of the decimal representation of n is called a stump of n. Let $T(n)$ be the sum of all stumps of n. Prove that $n = S(n) + 9T(n)$.

(2001 Asian Pacific Mathematical Olympiad)

Problem 1.7.21. Let p be a prime number and m a positive integer. Show that there exists a positive integer n such that there exist m consecutive zeros in the decimal representation of p^n.

(2001 Japanese Mathematical Olympiad)

Problem 1.7.22. Knowing that 2^{29} is a 9-digit number whose digits are distinct, without computing the actual number determine which of the ten digits is missing. Justify your answer.

Problem 1.7.23. It is well known that the divisibility tests for division by 3 and 9 do not depend on the order of the decimal digits. Prove that 3 and 9 are the only positive integers with this property. More exactly, if an integer $d > 1$ has the property that $d \mid n$ implies $d \mid n_1$, where n_1 is obtained from n through an arbitrary permutation of its digits, then $d = 3$ or $d = 9$.

2

Powers of Integers

An integer n is a *perfect square* if $n = m^2$ for some integer m. Taking into account the prime factorization, if $m = p_1^{\alpha_1} \cdots p_k^{\alpha_k}$, then $n = p_1^{2\alpha_1} \cdots p_k^{2\alpha_k}$. That is, n is a perfect square if and only if all exponents in its prime factorization are even.

An integer n is a *perfect power* if $n = m^s$ for some integers m and s, $s \geq 2$. Similarly, n is an sth perfect power if and only if all exponents in its prime factorization are divisible by s.

We say that the integer n is *square-free* if for any prime divisor p, p^2 does not divide n. Similarly, we can define the sth power-free integers.

These preliminary considerations seem trivial, but as you will see shortly, they have significant rich applications in solving various problems.

2.1   Perfect Squares

Problem 2.1.1. *Find all nonnegative integers n such that there are integers a and b with the property*
$$n^2 = a + b \text{ and } n^3 = a^2 + b^2.$$

(2004 Romanian Mathematical Olympiad)

Solution. From the inequality $2(a^2 + b^2) \geq (a + b)^2$ we get $2n^3 \geq n^4$, that is, $n \leq 2$. Thus:

for $n = 0$, we choose $a = b = 0$,
for $n = 1$, we take $a = 1$, $b = 0$, and
for $n = 2$, we may take $a = b = 2$.

Problem 2.1.2. *Find all integers n such that n − 50 and n + 50 are both perfect squares.*

T. Andreescu and D. Andrica, *Number Theory*, DOI: 10.1007/b11856_2,
© Birkhäuser Boston, a part of Springer Science + Business Media, LLC 2009

Solution. Let $n - 50 = a^2$ and $n + 50 = b^2$. Then $b^2 - a^2 = 100$, so $(b-a)(b+a) = 2^2 \cdot 5^2$. Because $b-a$ and $b+a$ are of the same parity, we have the following possibilities: $b - a = 2$, $b + a = 50$, yielding $b = 26$, $a = 24$, and $b - a = 10$, $b + a = 10$ with $a = 0$, $b = 10$. Hence the integers with this property are $n = 626$ and $n = 50$.

Problem 2.1.3. *Let $n \geq 3$ be a positive integer. Show that it is possible to eliminate at most two numbers among the elements of the set $\{1, 2, \ldots, n\}$ such that the sum of the remaining numbers is a perfect square.*

(2003 Romanian Mathematical Olympiad)

Solution. Let $m = \lfloor \sqrt{n(n+1)/2} \rfloor$. From $m^2 \leq n(n+1)/2 < (m+1)^2$ we obtain

$$\frac{n(n+1)}{2} - m^2 < (m+1)^2 - m^2 = 2m + 1.$$

Therefore, we have

$$\frac{n(n+1)}{2} - m^2 \leq 2m \leq \sqrt{2n^2 + 2n} \leq 2n - 1.$$

Since any number k, $k \leq 2n - 1$, can be obtained by adding at most two numbers from $\{1, 2, \ldots, n\}$, we obtain the result.

Problem 2.1.4. *Let k be a positive integer and $a = 3k^2 + 3k + 1$.*

(i) *Show that $2a$ and a^2 are sums of three perfect squares.*

(ii) *Show that if a is a divisor of a positive integer b, and b is a sum of three perfect squares, then any power b^n is a sum of three perfect squares.*

(2003 Romanian Mathematical Olympiad)

Solution. (i) $2a = 6k^2 + 6k + 2 = (2k+1)^2 + (k+1)^2 + k^2$ and $a^2 = 9k^4 + 18k^3 + 15k^2 + 6k + 1 = (k^2 + k)^2 + (2k^2 + 3k + 1)^2 + k^2(2k+1)^2 = a_1^2 + a_2^2 + a_3^2$. (ii) Let $b = ca$. Then $b = b_1^2 + b_2^2 + b_3^2$ and $b^2 = c^2 a^2 = c^2(a_1^2 + a_2^2 + a_3^2)$. To end the proof, we proceed as follows: for $n = 2p + 1$ we have $b^{2p+1} = (b^p)^2(b_1^2 + b_2^2 + b_3^2)$, and for $n = 2p + 2$, $b^n = (b^p)^2 b^2 = (b^p)^2 c^2(a_1^2 + a_2^2 + a_3^2)$.

Problem 2.1.5. *(a) Let k be an integer number. Prove that the number*

$$(2k + 1)^3 - (2k - 1)^3$$

is the sum of three squares. (b) Let n be a positive number. Prove that the number $(2n + 1)^3 - 2$ can be represented as the sum of $3n - 1$ squares greater than 1.

(2000 Romanian Mathematical Olympiad)

Solution. (a) It is easy to check that

$$(2k+1)^3 - (2k-1)^3 = (4k)^2 + (2k+1)^2 + (2k-1)^2.$$

(b) Observe that

$$(2n+1)^3 - 1 = (2n+1)^3 - (2n-1)^3 + (2n-1)^3 - (2n-3)^3 + \cdots + 3^3 - 1^3.$$

Each of the n differences in the right-hand side can be written as a sum of three squares greater than 1, except for the last one:

$$3^3 - 1^3 = 4^2 + 3^2 + 1^2.$$

It follows that

$$(2n+1)^3 - 2 = 3^2 + 4^2 + \sum_{k=2}^{n} \left[(4k)^2 + (2k+1)^2 + (2k-1)^2 \right]$$

as desired.

Problem 2.1.6. *Prove that for any positive integer n the number*

$$\frac{\left(17 + 12\sqrt{2}\right)^n - \left(17 - 12\sqrt{2}\right)^n}{4\sqrt{2}}$$

is an integer but not a perfect square.

Solution. Note that $17 + 12\sqrt{2} = \left(\sqrt{2}+1\right)^4$ and $17 - 12\sqrt{2} = \left(\sqrt{2}-1\right)^4$, so

$$\frac{\left(17 + 12\sqrt{2}\right)^n - \left(17 - 12\sqrt{2}\right)^n}{4\sqrt{2}} = \frac{\left(\sqrt{2}+1\right)^{4n} - \left(\sqrt{2}-1\right)^{4n}}{4\sqrt{2}}$$

$$= \frac{\left(\sqrt{2}+1\right)^{2n} + \left(\sqrt{2}-1\right)^{2n}}{2} \cdot \frac{\left(\sqrt{2}+1\right)^{2n} - \left(\sqrt{2}-1\right)^{2n}}{2\sqrt{2}}.$$

Define

$$A = \frac{\left(\sqrt{2}+1\right)^{2n} + \left(\sqrt{2}-1\right)^{2n}}{2} \quad \text{and} \quad B = \frac{\left(\sqrt{2}+1\right)^{2n} - \left(\sqrt{2}-1\right)^{2n}}{2\sqrt{2}}.$$

Using the binomial expansion formula we obtain positive integers x and y such that

$$\left(\sqrt{2}+1\right)^{2n} = x + y\sqrt{2}, \quad \left(\sqrt{2}-1\right)^{2n} = x - y\sqrt{2}.$$

Then

$$x = \frac{\left(\sqrt{2}+1\right)^{2n} + \left(\sqrt{2}-1\right)^{2n}}{2} = A$$

and

$$y = \frac{\left(\sqrt{2}+1\right)^{2n} - \left(\sqrt{2}-1\right)^{2n}}{2\sqrt{2}} = B,$$

and so AB is as integer, as claimed. Observe that

$$A^2 - 2B^2 = (A + \sqrt{2}B)(A - \sqrt{2}B) = (\sqrt{2}+1)^{2n}(\sqrt{2}-1)^{2n} = 1,$$

so A and B are relatively prime. It is sufficient to prove that at least one of them is not a perfect square. We have

$$A = \frac{\left(\sqrt{2}+1\right)^{2n} + \left(\sqrt{2}-1\right)^{2n}}{2} = \left[\frac{\left(\sqrt{2}+1\right)^{n} + \left(\sqrt{2}-1\right)^{n}}{\sqrt{2}}\right]^2 - 1 \quad (1)$$

and

$$A = \frac{\left(\sqrt{2}+1\right)^{2n} + \left(\sqrt{2}-1\right)^{2n}}{2} = \left[\frac{\left(\sqrt{2}+1\right)^{n} - \left(\sqrt{2}-1\right)^{n}}{\sqrt{2}}\right]^2 + 1. \quad (2)$$

Since one of the numbers

$$\frac{\left(\sqrt{2}+1\right)^{n} + \left(\sqrt{2}-1\right)^{n}}{\sqrt{2}}, \quad \frac{\left(\sqrt{2}+1\right)^{n} - \left(\sqrt{2}-1\right)^{n}}{\sqrt{2}}$$

is an integer, depending on the parity of n, from the relations (1) and (2) we derive that A is not a square. This completes the proof.

Problem 2.1.7. *The integers a and b have the property that for every nonnegative integer n, the number $2^n a + b$ is a perfect square. Show that $a = 0$.*

(2001 Polish Mathematical Olympiad)

Solution. If $a \neq 0$ and $b = 0$, then at least one of $2^1 a + b$ and $2^2 a + b$ is not a perfect square, a contradiction. If $a \neq 0$ and $b \neq 0$, then each $(x_n, y_n) = (2\sqrt{2^n a + b}, \sqrt{2^{n+2} a + b})$ satisfies

$$(x_n + y_n)(x_n - y_n) = 3b.$$

Hence, $x + n + y_n$ divides $3b$ for each n. But this is impossible because $3b \neq 0$ but $|x_n + y_n| > |3b|$ for large enough n. Therefore, $a = 0$.

Remark. We invite the courageous reader to prove that if $f \in \mathbb{Z}[X]$ is a polynomial and $f(2^n)$ is a perfect square for all n, then there is $g \in \mathbb{Z}[X]$ such that $f = g^2$.

Problem 2.1.8. *Prove that the number*

$$\underbrace{11\ldots11}_{1997}\underbrace{22\ldots22}_{1998}5$$

is a perfect square.

First solution.

$$N = \underbrace{11\ldots11}_{1997}\cdot 10^{1999} + \underbrace{22\ldots22}_{1998}\cdot 10 + 5$$

$$= \frac{1}{9}(10^{1997}-1)\cdot 10^{1999} + \frac{2}{9}(10^{1998}-1)\cdot 10 + 5$$

$$= \frac{1}{9}(10^{3996}+2\cdot 5\cdot 10^{1998}+25) = \left[\tfrac{1}{3}(10^{1998}+5)\right]^2$$

$$= \left(\frac{\overbrace{1\,00\ldots00}^{1997}5}{3}\right)^2 = \underbrace{33\ldots33}_{1997}5^2.$$

Second solution. Note that

$$9N = \underbrace{1\,00\ldots00}_{1996}\underbrace{1\,00\ldots00}_{1997}25 = 10^{3996}+10^{1999}+25 = (10^{1998}+5)^2;$$

hence N is a square.

Problem 2.1.9. *Find all positive integers n, $n \geq 1$, such that $n^2 + 3^n$ is a perfect square.*

Solution. Let m be a positive integer such that

$$m^2 = n^2 + 3^n.$$

Since $(m-n)(m+n) = 3^n$, there is $k \geq 0$ such that $m-n = 3^k$ and $m+n = 3^{n-k}$. From $m-n < m+n$ follows $k < n-k$, and so $n-2k \geq 1$. If $n-2k = 1$, then $2n = (m+n)-(m-n) = 3^{n-k}-3^k = 3^k(3^{n-2k}-1) = 3^k(3^1-1) = 2\cdot 3^k$, so $n = 3^k = 2k+1$. We have $3^m = (1+2)^m = 1+2m+2^2\binom{m}{2}+\cdots > 2m+1$. Therefore $k = 0$ or $k = 1$, and consequently $n = 1$ or $n = 3$. If $n-2k > 1$, then $n-2k \geq 2$ and $k \leq n-k-2$. It follows that $3^k \leq 3^{n-k-2}$, and consequently

$$2n = 3^{n-k}-3^k \geq 3^{n-k}-3^{n-k-2} = 3^{n-k-2}(3^2-1) = 8\cdot 3^{n-k-2}$$

$$\geq 8[1+2(n-k-2)] = 16n-16k-24,$$

which implies $8k+12 \geq 7n$. On the other hand, $n \geq 2k+2$; hence $7n \geq 14k+14$, contradiction. In conclusion, the only possible values for n are 1 and 3.

Problem 2.1.10. *Find the number of five-digit perfect squares having the last two digits equal.*

Solution. Suppose $n = \overline{abcdd}$ is a perfect square. Then $n = 100\overline{abc} + 11d = 4m + 3d$ for some m. Since all squares have the form $4m$ or $4m + 1$ and $d \in \{0, 1, 4, 5, 6, 9\}$ as the last digit of a square, it follows that $d = 0$ or $d = 4$. If $d = 0$, then $n = 100\overline{abc}$ is a square if $\overline{abc}$ is a square. Hence $\overline{abc} \in \{10^2, 11^2, \dots, 31^2\}$, so there are 22 numbers. If $d = 4$, then $100\overline{abc} + 44 = n = k^2$ implies $k = 2p$ and $\overline{abc} = \frac{p^2-11}{25}$. (1) If $p = 5x$, then $\overline{abc}$ is not an integer, false. (2) If $p = 5x + 1$, then $\overline{abc} = \frac{25x^2+10x-1}{25} = x^2 + \frac{2(x-1)}{5} \Rightarrow x \in \{11, 16, 21, 26, 31\}$, so there are 5 solutions. (3) If $p = 5x + 2$, then $\overline{abc} = x^2 + \frac{20x-7}{25} \notin \mathbb{N}$, false. (4) If $p = 5x + 3$, then $\overline{abc} = x^2 + \frac{30x-2}{25} \notin \mathbb{N}$, false. (5) If $p = 5x + 4$ then $\overline{abc} = x^2 + \frac{8x+1}{5}$; hence $x = 5m + 3$ for some $m \Rightarrow x \in \{13, 18, 23, 28\}$, so there are four solutions. Finally, there are $22 + 5 + 4 = 31$ squares.

Problem 2.1.11. *The last four digits of a perfect square are equal. Prove they are all zero.*

(2002 Romanian Team Selection Test for JBMO)

Solution. Denote by k^2 the perfect square and by a the digit that appears in the last four positions. It easily follows that a is one of the numbers 0, 1, 4, 5, 6, 9. Thus $k^2 \equiv a \cdot 1111 \pmod{16}$. (1) If $a = 0$, we are done. (2) Suppose that $a \in \{1, 5, 9\}$. Since $k^2 \equiv 0 \pmod 8$, $k^2 \equiv 1 \pmod 8$ or $k^2 \equiv 4 \pmod 8$ and $1111 \equiv 7 \pmod 8$, we obtain $1111 \equiv 7 \pmod 8$, $5 \cdot 1111 \equiv 3 \pmod 8$, and $9 \cdot 1111 \equiv 7 \pmod 8$. Thus the congruence $k^2 \equiv a \cdot 1111 \pmod{16}$ cannot hold. (3) Suppose $a \in \{4, 6\}$. Since $1111 \equiv 7 \pmod{16}$, $4 \cdot 1111 \equiv 12 \pmod{16}$, and $6 \cdot 1111 \equiv 10 \pmod{16}$, we conclude that in this case the congruence $k^2 \equiv a \cdot 1111 \pmod{16}$ cannot hold. Thus $a = 0$.

Remark. $38^2 = 1444$ ends in three equal digits, so the problem is sharp.

Problem 2.1.12. *Let $1 < n_1 < n_2 < \cdots < n_k < \cdots$ be a sequence of integers such that no two are consecutive. Prove that for all positive integers m between $n_1 + n_2 + \cdots + n_m$ and $n_2 + n_2 + \cdots + n_{m+1}$ there is a perfect square.*

Solution. It is easy to prove that between numbers $a > b \geq 0$ such that $\sqrt{a} - \sqrt{b} > 1$ there is a perfect square: take for example $([\sqrt{b}] + 1)^2$. It suffices to prove that

$$\sqrt{n_1 + \cdots + n_{m+1}} - \sqrt{n_1 + \cdots + n_m} > 1, \quad m \geq 1.$$

This is equivalent to

$$n_1 + \cdots + n_m + n_{m+1} > (1 + \sqrt{n_1 + n_2 + \cdots + n_m})^2,$$

and then

$$n_{m+1} > 1 + 2\sqrt{n_1 + n_2 + \cdots + n_m}, \quad m \geq 1.$$

We induct on m. For $m = 1$ we have to prove that $n_2 > 1 + 2\sqrt{n_1}$. Indeed, $n_2 > n_1 + 2 = 1 + (1 + n_1) > 1 + 2\sqrt{n_1}$. Assume that the claim holds for some $m \geq 1$. Then

$$n_{m+1} - 1 > 2\sqrt{n_1 + \cdots + n_m}$$

so $(n_{m+1} - 1)^2 > 4(n_1 + \cdots + n_m)$ hence

$$(n_{m+1} + 1)^2 > 4(n_1 + \cdots + n_{m+1}).$$

This implies

$$n_{m+1} + 1 > 2\sqrt{n_1 + \cdots + n_{m+1}},$$

and since $n_{m+2} - n_{m+1} \geq 2$, it follows that

$$n_{m+2} > 1 + 2\sqrt{n_1 + \cdots + n_{m+1}},$$

as desired.

Problem 2.1.13. *Find all integers x, y, z such that $4^x + 4^y + 4^z$ is a square.*

Solution. It is clear that there are no solutions with $x < 0$. Without loss of generality assume that $x \leq y \leq z$ and let $4^x + 4^y + 4^z = u^2$. Then $2^{2x}(1 + 4^{y-x} + 4^{z-x}) = u^2$. We have two situations.

Case 1. $1 + 4^{y-x} + 4^{z-x}$ is odd, i.e., $1 + 4^{y-x} + 4^{z-x} = (2a + 1)^2$. It follows that

$$4^{y-x-1} + 4^{z-x-1} = a(a + 1),$$

and then

$$4^{y-x-1}(1 + 4^{z-y}) = a(a + 1).$$

We consider two cases. (1) The number a is even. Then $a + 1$ is odd, so $4^{y-x-1} = a$ and $1 + 4^{z-y} = a + 1$. It follows that $4^{y-x-1} = 4^{z-y}$; hence $y - x - 1 = z - y$. Thus $z = 2y - x - 1$ and

$$4^x + 4^y + 4^z = 4^x + 4^y + 4^{2y-x-1} = (2^x + 2^{2y-x-1})^2.$$

(2) The number a is odd. Then $a + 1$ is even, so $a = 4^{z-y} + 1$, $a + 1 = 4^{y-x-1}$ and $4^{y-x-1} - 4^{z-y} = 2$. It follows that $2^{2y-2x-3} = 2^{2x-2y-1} + 1$, which is impossible, since $2x - 2y - 1 \neq 0$.

Case 2. $1 + 4^{y-x} + 4^{z-x}$ is even; thus $y = x$ or $z = x$. Anyway, we must have $y = x$, and then $2 + 4^{z-x}$ is a square, which is impossible, since it is congruent to $2 \pmod 4$ or congruent to $3 \pmod 4$.

Additional Problems

Problem 2.1.14. Let x, y, z be positive integers such that

$$\frac{1}{x} - \frac{1}{y} = \frac{1}{z}.$$

Let h be the greatest common divisor of x, y, z. Prove that $hxyz$ and $h(y - x)$ are perfect squares.

(1998 United Kingdom Mathematical Olympiad)

Problem 2.1.15. Let b an integer greater than 5. For each positive integer n, consider the number

$$x_n = \underbrace{11\ldots1}_{n-1}\underbrace{22\ldots2}_{n}5,$$

written in base b. Prove that the following condition holds if and only if $b = 10$: There exists a positive integer M such that for every integer n greater than M, the number x_n is a perfect square.

(44th International Mathematical Olympiad Shortlist)

Problem 2.1.16. Do there exist three natural numbers greater than 1 such that the square of each, minus one, is divisible by each of the others?

(1996 Russian Mathematical Olympiad)

Problem 2.1.17. (a) Find the first positive integer whose square ends in three 4's. (b) Find all positive integers whose squares end in three 4's. (c) Show that no perfect square ends with four 4's.

(1995 United Kingdom Mathematical Olympiad)

Problem 2.1.18. Let $\overline{abc}$ be a prime. Prove that $b^2 - 4ac$ cannot be a perfect square.

(Mathematical Reflections)

Problem 2.1.19. For each positive integer n, denote by $s(n)$ the greatest integer such that for all positive integer $k \le s(n)$, n^2 can be expressed as a sum of squares of k positive integers. (a) Prove that $s(n) \le n^2 - 14$ for all $n \ge 4$. (b) Find a number n such that $s(n) = n^2 - 14$. (c) Prove that there exist infinitely many positive integers n such that

$$s(n) = n^2 - 14.$$

(33rd International Mathematical Olympiad)

Problem 2.1.20. Let A be the set of positive integers representable in the form $a^2 + 2b^2$ for integers a, b with $b \neq 0$. Show that if $p^2 \in A$ for a prime p, then $p \in A$.

(1997 Romanian International Mathematical Olympiad Team Selection Test)

Problem 2.1.21. Is it possible to find 100 positive integers not exceeding 25000 such that all pairwise sums of them are different?

(42nd International Mathematical Olympiad Shortlist)

Problem 2.1.22. Do there exist 10 distinct integers, the sum of any 9 of which is a perfect square?

(1999 Russian Mathematical Olympiad)

Problem 2.1.23. Let n be a positive integer such that n is a divisor of the sum

$$1 + \sum_{i=1}^{n-1} i^{n-1}.$$

Prove that n is square-free.

(1995 Indian Mathematical Olympiad)

Problem 2.1.24. Let n, p be integers such that $n > 1$ and p is a prime. If $n \mid (p - 1)$ and $p \mid (n^3 - 1)$, show that $4p - 3$ is a perfect square.

(2002 Czech–Polish–Slovak Mathematical Competition)

Problem 2.1.25. Show that for any positive integer $n > 10000$, there exists a positive integer m that is a sum of two squares and such that $0 < m - n < 3\sqrt[4]{n}$.

(Russian Mathematical Olympiad)

Problem 2.1.26. Show that a positive integer m is a perfect square if and only if for each positive integer n, at least one of the differences

$$(m + 1)^2 - m, \quad (m + 2)^2 - m, \quad \ldots, \quad (m + n)^2 - m$$

is divisible by n.

(2002 Czech and Slovak Mathematical Olympiad)

2.2 Perfect Cubes

Problem 2.2.1. *Prove that if n is a perfect cube, then $n^2 + 3n + 3$ cannot be a perfect cube.*

Solution. If $n = 0$, then we get 3 and the property is true. Suppose by way of contradiction that $n^2 + 3n + 3$ is a cube for some $n \neq 0$. Hence $n(n^2 + 3n + 3)$ is a cube. Note that

$$n(n^2 + 3n + 3) = n^3 + 3n^2 + 3n = (n + 1)^3 - 1,$$

and since $(n + 1)^3 - 1$ is not a cube when $n \neq 0$, we obtain a contradiction.

Problem 2.2.2. *Let m be a given positive integer. Find a positive integer n such that $m + n + 1$ is a perfect square and $mn + 1$ is a perfect cube.*

Solution. Choosing $n = m^2 + 3m + 3$, we have

$$m + n + 1 = m^2 + 4m + 4 = (m + 2)^2$$

and

$$mn + 1 = m^3 + 3m^2 + 3m + 1 = (m + 1)^3.$$

Problem 2.2.3. *Which are there more of among the natural numbers from 1 to 1000000, inclusive: numbers that can be represented as the sum of a perfect square and a (positive) perfect cube, or numbers that cannot be?*

(1996 Russian Mathematical Olympiad)

Solution. There are more numbers not of this form. Let $n = k^2 + m^3$, where $k, m, n \in \mathbb{N}$ and $n \leq 1000000$. Clearly $k \leq 1000$ and $m \leq 100$. Therefore there cannot be more numbers in the desired form than the 100000 pairs (k, m).

Problem 2.2.4. *Show that no integer of the form $\overline{xyxy}$ in base 10 can be the cube of an integer. Also find the smallest base $b > 1$ in which there is a perfect cube of the form $xyxy$.*

(1998 Irish Mathematical Olympiad)

Solution. If the 4-digit number $\overline{xyxy} = 101 \times \overline{xy}$ is a cube, then $101 \mid \overline{xy}$, which is a contradiction. Convert $\overline{xyxy} = 101 \times \overline{xy}$ from base b to base 10. We obtain $\overline{xyxy} = (b^2 + 1) \times (bx + y)$ with $x, y < b$ and $b^2 + 1 > bx + y$. Thus for $\overline{xyxy}$ to be a cube, $b^2 + 1$ must be divisible by a perfect square. We can check easily that $b = 7$ is the smallest such number, with $b^2 + 1 = 50$. The smallest cube divisible by 50 is 1000, which is $\overline{2626}$ is base 7.

Additional Problems

Problem 2.2.5. Find all the positive perfect cubes that are not divisible by 10 such that the number obtained by erasing the last three digits is also a perfect cube.

Problem 2.2.6. Find all positive integers n less than 1999 such that n^2 is equal to the cube of the sum of n's digits.

(1999 Iberoamerican Mathematical Olympiad)

Problem 2.2.7. Prove that for any nonnegative integer n the number

$$A = 2^n + 3^n + 5^n + 6^n$$

is not a perfect cube.

Problem 2.2.8. Prove that every integer is a sum of five cubes.

Problem 2.2.9. Show that every rational number can be written as a sum of three cubes.

2.3   *k*th Powers of Integers, *k* at least 4

Problem 2.3.1. *Given* 81 *natural numbers whose prime divisors belong to the set* $\{2, 3, 5\}$, *prove that there exist four numbers whose product is the fourth power of an integer.*

(1996 Greek Mathematical Olympiad)

Solution. It suffices to take 25 such numbers. To each number, associate the triple (x_2, x_3, x_5) recording the parity of the exponents of 2, 3, and 5 in its prime factorization. Two numbers have the same triple if and only if their product is a perfect square. As long as there are 9 numbers left, we can select two whose product is a square; in so doing, we obtain 9 such pairs. Repeating the process with the square roots of the products of the pairs, we obtain four numbers whose product is a fourth power.

Problem 2.3.2. *Find all collections of* 100 *positive integers such that the sum of the fourth powers of every four of the integers is divisible by the product of the four numbers.*

(1997 St. Petersburg City Mathematical Olympiad)

Solution. Such sets must be $n, n, \ldots, n$ or $3n, n, n, \ldots, n$ for some integer n. Without loss of generality, we assume that the numbers do not have a common factor. If u, v, w, x, y are five of the numbers, then uvw divides $u^4 + v^4 + w^4 + x^4$ and $u^4 + v^4 + w^4 + y^4$, and so divides $x^4 - y^4$. Likewise, $v^4 \equiv w^4 \equiv x^4$ (mod u), and from above, $3v^4 \equiv 0$ (mod u). If u has a prime divisor not equal to 3, we conclude that every other integer is divisible by the same prime, contrary to assumption. Likewise, if u is divisible by 9, then every other integer is divisible by 3. Thus all of the numbers equal 1 or 3. Moreover, if one number is 3, the others are all congruent modulo 3, so are all 3 (contrary to assumption) or 1. This completes the proof.

Problem 2.3.3. *Let M be a set of 1985 distinct positive integers, none of which has a prime divisor greater than 26. Prove that M contains at least one subset of four distinct elements whose product is the fourth power of an integer.*

(26th International Mathematical Olympiad)

Solution. There are nine prime numbers less than 26: $p_1 = 2$, $p_2 = 3, \ldots, p_9 = 23$. Any element x of M has a representation $x = \prod_{i=1}^{9} p_i^{a_i}$, $a_i \geq 0$. If $x, y \in M$ and $y = \prod_{i=1}^{9} p_i^{b_i}$, the product $xy = \prod_{i=1}^{9} p_i^{a_i+b_i}$ is a perfect square if and only if $a_i + b_i \equiv 0 \pmod{2}$. Equivalently, $a_i \equiv b_i \pmod{2}$ for all $i = 1, 2, \ldots, 9$. Because there are $2^9 = 512$ elements in $(\mathbb{Z}/2\mathbb{Z})^9$, any subset of M having at least 513 elements contains two elements x, y such that xy is a perfect square. Starting from M and eliminating such pairs, one obtains $\frac{1}{2}(1985 - 513) = 736 > 513$ distinct two-element subsets of M having a square as the product of elements. Reasoning as above, we find among these squares at least one pair (in fact many pairs) whose product is a fourth power.

Problem 2.3.4. *Let A be a subset of $\{0, 1, \ldots, 1997\}$ containing more than 1000 elements. Prove that A contains either a power of 2, or two distinct integers whose sum is a power of 2.*

(1997 Irish Mathematical Olympiad)

Solution. Suppose A did not satisfy the conclusion. Then A would contain at most half of the integers from 51 to 1997, since they can be divided into pairs whose sum is 2048 (with 1024 left over); likewise, A contains at most half of the integers from 14 to 50, at most half of the integers from 3 to 13, and possibly 0, for a total of

$$973 + 18 + 5 + 1 = 997$$

integers.

Problem 2.3.5. *Show that in the arithmetic progression with first term 1 and difference 729, there are infinitely many powers of 10.*

(1996 Russian Mathematical Olympiad)

First solution. We will show that for all natural numbers n, $10^{81n} - 1$ is divisible by 729. In fact,

$$10^{81n} - 1 = (10^{81})^n - 1^n = (10^{81} - 1) \cdot A,$$

and

$$10^{81} - 1 = \underbrace{9\ldots 9}_{81}$$

$$= \underbrace{(9\ldots 9)}_{9} \cdots \underbrace{(10\ldots 01}_{8}\, \underbrace{10\ldots 01}_{8} \ldots \underbrace{10\ldots 01)}_{8}$$

$$= 9\underbrace{(1\ldots 1)}_{9} \cdots \underbrace{(10\ldots 01}_{8}\, \underbrace{10\ldots 01}_{8} \ldots \underbrace{10\ldots 01)}_{8}.$$

The second and third factors have nine digits equal to 1 and the root of digits (if any) 0, so the sum of the digits is 9, and each is a multiple of 9. Hence $10^{81} - 1$ is divisible by $9^3 = 729$, as is $10^{81n} - 1$ for any n.

Second solution. In order to prove that $10^{81} - 1$ is divisible by 9^3, just write

$$10^{81} - 1 = (9+1)^{81} - 1 = k \cdot 9^3 + \binom{81}{2} 9^2 + \binom{81}{1} \cdot 9$$

$$= k \cdot 9^3 + 81 \cdot 40 \cdot 9^2 + 81 \cdot 9 = (k + 361) \cdot 9^3.$$

Remark. An alternative solution uses Euler's theorem (see Section 7.2). We have $10^{\varphi(729)} \equiv 1 \pmod{729}$; thus $10^{n\varphi(729)}$ is in this progression for any positive integer n.

Additional Problems

Problem 2.3.6. Let p be a prime number and a, n positive integers. Prove that if

$$2^p + 3^p = a^n,$$

then $n = 1$.

(1996 Irish Mathematical Olympiad)

Problem 2.3.7. Let x, y, p, n, k be natural numbers such that

$$x^n + y^n = p^k.$$

Prove that if $n > 1$ is odd and p is an odd prime, then n is a power of p.

(1996 Russian Mathematical Olympiad)

Problem 2.3.8. Prove that a product of three consecutive integers cannot be a power of an integer.

Problem 2.3.9. Show that there exists an infinite set A of positive integers such that for any finite nonempty subset $B \subset A$, $\sum_{x \in B} x$ is not a perfect power.

(Kvant)

Problem 2.3.10. Prove that there is no infinite arithmetic progression consisting only of powers ≥ 2.

3

Floor Function and Fractional Part

3.1 General Problems

For a real number x there is a unique integer n such that $n \le x < n + 1$.

We say that n is the *greatest integer less than or equal to x* or the *floor* of x. We write $n = \lfloor x \rfloor$. The difference $x - \lfloor x \rfloor$ is called the *fractional part* of x and is denoted by $\{x\}$.

The integer $-\lfloor -x \rfloor$ is called the *ceiling* of x and is denoted by $\lceil x \rceil$.

Examples. (1) $\lfloor 2.1 \rfloor = 2$, $\{2.1\} = 0.1$, and $\lceil 2.1 \rceil = 3$.

(2) $\lfloor -3.9 \rfloor = -4$, $\{-3.9\} = 0.1$, and $\lceil -3.9 \rceil = -3$.

The following properties are useful:

(1) If a and b are integers, $b > 0$, and q is the quotient when a is divided by b, then $q = \lfloor \frac{a}{b} \rfloor$.

(2) For any real number x and any integer n, $\lfloor x + n \rfloor = \lfloor x \rfloor + n$ and $\lceil x + n \rceil = \lceil x \rceil + n$.

(3) For any positive real number x and any positive integer n, the number of positive multiples of n not exceeding x is $\lfloor \frac{x}{n} \rfloor$.

(4) For any real number x and any positive integer n, $\lfloor \frac{\lfloor x \rfloor}{n} \rfloor = \lfloor \frac{x}{n} \rfloor$.

(5) For any real numbers x and y,

$$\lfloor x + y \rfloor - 1 \le \lfloor x \rfloor + \lfloor y \rfloor \le \lfloor x + y \rfloor.$$

We will prove the last three properties. For (3), consider all multiples

$$1 \cdot n, \quad 2 \cdot n, \quad \ldots, \quad k \cdot n,$$

where $k \cdot n \le x < (k + 1)n$. That is, $k \le \frac{x}{n} < k + 1$ and the conclusion follows. For (4) write $\lfloor x \rfloor = m$ and $\{x\} = \alpha$. From the division algorithm and

T. Andreescu and D. Andrica, *Number Theory*, DOI: 10.1007/b11856_3,
© Birkhäuser Boston, a part of Springer Science + Business Media, LLC 2009

property (1) above it follows that $m = n \left\lfloor \frac{m}{n} \right\rfloor + r$, where $0 \le r \le n - 1$. We obtain $0 \le r + \alpha \le n - 1 + \alpha < n$, that is, $\left\lfloor \frac{r+\alpha}{n} \right\rfloor = 0$ and

$$\left\lfloor \frac{x}{n} \right\rfloor = \left\lfloor \frac{m+\alpha}{n} \right\rfloor = \left\lfloor \left\lfloor \frac{m}{n} \right\rfloor + \frac{r+\alpha}{n} \right\rfloor = \left\lfloor \frac{m}{n} \right\rfloor + \left\lfloor \frac{r+\alpha}{n} \right\rfloor = \left\lfloor \frac{m}{n} \right\rfloor = \left\lfloor \frac{\lfloor x \rfloor}{n} \right\rfloor.$$

Remark. An easier proof of (4) may be this:

Since $\left\lfloor \frac{x}{n} \right\rfloor \le \frac{x}{n} < \left\lfloor \frac{x}{n} \right\rfloor + 1$, we can write $x = n \left\lfloor \frac{x}{n} \right\rfloor + s$ with $0 \le x < n$. By (2), we have $\lfloor x \rfloor = n \left\lfloor \frac{x}{n} \right\rfloor + \lfloor s \rfloor$, so

$$\frac{\lfloor x \rfloor}{n} = \left\lfloor \frac{x}{n} \right\rfloor + \frac{\lfloor s \rfloor}{n}.$$

Applying (2) again gives

$$\left\lfloor \frac{\lfloor x \rfloor}{n} \right\rfloor = \left\lfloor \frac{x}{n} \right\rfloor + \left\lfloor \frac{\lfloor x \rfloor}{n} \right\rfloor.$$

Since $0 \le \lfloor x \rfloor \le s < n$, $0 \le \frac{\lfloor s \rfloor}{n} < 1$. Hence the last term is zero and we get (4).

For (5) just set $\lfloor x \rfloor = m$, $\{x\} = \alpha$, and $\lfloor y \rfloor = n$, $\{y\} = \beta$ and reduce the inequalities to $\lfloor \alpha + \beta \rfloor - 1 \le 0 \le \lfloor \alpha + \beta \rfloor$. Because $\lfloor \alpha + \beta \rfloor$ is 0 or 1, we are done.

The following properties connecting the floor and the ceiling of x are obvious:

(6) For any real number x, $\lceil x \rceil - \lfloor x \rfloor \le 1$.

Problem 3.1.1. *Find all positive integers n such that $\lfloor \sqrt[n]{111} \rfloor$ divides 111.*

Solution. The positive divisors of 111 are 1, 3, 37, 111. So we have the following cases:

(1) $\lfloor \sqrt[n]{111} \rfloor = 1$ or $1 \le 111 < 2^n$; hence $n \ge 7$.

(2) $\lfloor \sqrt[n]{111} \rfloor = 3$, or $3^n \le 111 < 4^n$, so $n = 4$.

(3) $\lfloor \sqrt[n]{111} \rfloor = 37$, or $37^n \le 111 < 38^n$, impossible.

(4) $\lfloor \sqrt[n]{111} \rfloor = 111$, or $111^n \le 111 < 112^n$, and so $n = 1$.

Therefore $n = 1$, $n = 4$, or $n \ge 7$.

Problem 3.1.2. *Solve in $\mathbb{R}$ the equation*

$$\lfloor x \lfloor x \rfloor \rfloor = 1.$$

Solution. By definition,

$$\lfloor x \lfloor x \rfloor \rfloor = 1$$

implies

$$1 \le x \lfloor x \rfloor < 2.$$

We consider the following cases:

(a) $x \in (-\infty, -1)$. Then $\lfloor x \rfloor \leq -2$ and $x \lfloor x \rfloor > 2$, a contradiction.

(b) $x = -1 \Rightarrow \lfloor x \rfloor = -1$. Then $x \lfloor x \rfloor = (-1) \cdot (-1) = 1$ and $\lfloor x \lfloor x \rfloor \rfloor = 1$, so $x = -1$ is a solution.

(c) $x \in (-1, 0)$. We have $\lfloor x \rfloor = -1$ and $x \lfloor x \rfloor = -x < 1$, false.

(d) If $x \in [0, 1)$, then $\lfloor x \rfloor = 0$ and $x \lfloor x \rfloor = 0 < 1$, so we have no solution in this case.

(e) For $x \in [1, 2)$ we obtain $\lfloor x \rfloor = 1$ and $x \lfloor x \rfloor = x$, as needed.

(f) Finally, for $x \geq 2$ we have $\lfloor x \rfloor \geq 2$ and $x \lfloor x \rfloor \geq 2x \geq 4$, a contradiction with (1).

Consequently, $x \in \{-1\} \cup [1, 2)$.

Problem 3.1.3. *Prove that for any integer n one can find integers a and b such that*

$$n = \lfloor a\sqrt{2} \rfloor + \lfloor b\sqrt{3} \rfloor.$$

Solution. For any integer n, one can find an integer b such that

$$b - 1 < \frac{n - \sqrt{2}}{\sqrt{3}} < b.$$

Because $b - \frac{2}{\sqrt{3}} < b - 1$ we obtain

$$\sqrt{2} + b\sqrt{3} - 2 < n \leq \sqrt{2} + b\sqrt{3}.$$

Using property (5) we have to consider the following cases:

(1) If $n = \lfloor \sqrt{2} \rfloor + \lfloor b\sqrt{3} \rfloor$, we are done.

(2) If $n = \lfloor \sqrt{2} \rfloor + \lfloor b\sqrt{3} \rfloor + 1$, then $n = \lfloor 2\sqrt{2} \rfloor + \lfloor b\sqrt{3} \rfloor$.

(3) If $n = \lfloor \sqrt{2} \rfloor + \lfloor b\sqrt{3} \rfloor - 1$, then $n = \lfloor 0\sqrt{2} \rfloor + \lfloor b\sqrt{3} \rfloor$.

Problem 3.1.4. *Find all real numbers $x > 1$, such that $\sqrt[n]{\lfloor x^n \rfloor}$ is an integer for all positive integers n, $n \geq 2$.*

(2004 Romanian Regional Mathematical Contest)

Solution. Put $\sqrt[n]{\lfloor x^n \rfloor} = a_n$. Then $\lfloor x^n \rfloor = a_n^n$ and $a_n^n \leq x^n < a_n^n + 1$. Taking roots, one obtains $a_n \leq x < \sqrt[n]{a_n^n + 1}$. This shows that $\lfloor x \rfloor = a_n$.

We will show that positive integers x, $x \geq 2$, satisfy the condition and that they are the only solutions. Assume, by way of contradiction, that there is a solution x that is not a nonnegative integer. Put $x = a + \alpha$, $a \in \mathbb{Z}$, $a \geq 1$, $0 < \alpha < 1$.

It follows that $a^n < (a + \alpha)^n < a^n + 1$, and therefore,

$$1 < \left(1 + \frac{\alpha}{a}\right)^n < 1 + \frac{1}{a^n} \leq 2.$$

On the other hand, by the Bernoulli inequality,

$$\left(1 + \frac{\alpha}{a}\right)^n \geq 1 + n\frac{\alpha}{a} > 2,$$

for sufficiently large n, a contradiction.

Problem 3.1.5. *Let p be a prime and let α be a positive real number such that $p\alpha^2 < \frac{1}{4}$. Prove that*

$$\left\lfloor n\sqrt{p} - \frac{\alpha}{n} \right\rfloor = \left\lfloor n\sqrt{p} + \frac{\alpha}{n} \right\rfloor$$

for all integers $n \geq \left\lfloor \alpha/\sqrt{1 - 2\alpha\sqrt{p}} \right\rfloor + 1$.

Solution. It suffices to prove that there are no integers in the interval $\left(n\sqrt{p} - \frac{\alpha}{n}, n\sqrt{p} + \frac{\alpha}{n}\right]$ for $n > \alpha/\sqrt{1 - 2\alpha\sqrt{p}}$.

Assume by way of contradiction that there is integer k such that

$$n\sqrt{p} - \frac{\alpha}{n} < k \leq n\sqrt{p} + \frac{\alpha}{n}.$$

Hence

$$n^2 p + \frac{\alpha^2}{n^2} - 2\alpha\sqrt{p} < k^2 \leq n^2 p + \frac{\alpha^2}{n^2} + 2\alpha\sqrt{p}.$$

Observe that $\frac{\alpha^2}{n^2} - 2\alpha\sqrt{p} > -1$. If $n > \alpha/\sqrt{1 - 2\alpha\sqrt{p}}$, then $\frac{\alpha^2}{n^2} + 2\alpha\sqrt{p} < 1$,
so

$$n^2 p - 1 < k^2 < n^2 p + 1.$$

It follows that $k^2 = pn^2$ or $\sqrt{p} = k/n$, which is false, since p is prime.

Problem 3.1.6. *Find the number of different terms of the finite sequence $\left\lfloor \frac{k^2}{1998} \right\rfloor$, where $k = 1, 2, \ldots, 1997$.*

(1998 Balkan Mathematical Olympiad)

Solution. Note that

$$\left\lfloor \frac{998^2}{1998} \right\rfloor = 498 < 499 = \left\lfloor \frac{999^2}{1998} \right\rfloor,$$

so we can compute the total number of distinct terms by considering $k = 1, \ldots,$ 998 and $k = 999, \ldots, 1997$ independently. Observe that for $k = 1, \ldots, 997$,

$$\frac{(k+1)^2}{1998} - \frac{k^2}{1998} = \frac{2k+1}{1998} < 1,$$

so for $k = 1, \ldots, 998$, each of the numbers

$$\left\lfloor \frac{1^2}{1998} \right\rfloor = 0, \quad 1, \quad \ldots, \quad 498 = \left\lfloor \frac{998^2}{1998} \right\rfloor$$

appears at least once in the sequence $\lfloor k^2/1998 \rfloor$, for a total of 499 distinct terms. For $k = 999, \ldots, 1996$, we have

$$\frac{(k+1)^2}{1998} - \frac{k^2}{1998} = \frac{2k+1}{1998} > 1,$$

so the numbers $\lfloor k^2/1998 \rfloor$ ($k = 999, \ldots, 1997$) are all distinct, giving $1997 - 999 + 1 = 999$ more terms. Thus the total number of distinct terms is 1498.

Problem 3.1.7. *Determine the number of real solutions a of the equation*

$$\left\lfloor \frac{a}{2} \right\rfloor + \left\lfloor \frac{a}{3} \right\rfloor + \left\lfloor \frac{a}{5} \right\rfloor = a.$$

(1998 Canadian Mathematical Olympiad)

Solution. There are 30 solutions. Since $\lfloor a/2 \rfloor$, $\lfloor a/3 \rfloor$, and $\lfloor a/5 \rfloor$ are integers, so is a. Now write $a = 30p + q$ for integers p and q, $0 \le q < 30$. Then

$$\left\lfloor \frac{a}{2} \right\rfloor + \left\lfloor \frac{a}{3} \right\rfloor + \left\lfloor \frac{a}{5} \right\rfloor = a \Leftrightarrow 31p + \left\lfloor \frac{q}{2} \right\rfloor + \left\lfloor \frac{q}{3} \right\rfloor + \left\lfloor \frac{q}{5} \right\rfloor = 30p + q$$

$$\Leftrightarrow p = q - \left\lfloor \frac{q}{2} \right\rfloor - \left\lfloor \frac{q}{3} \right\rfloor - \left\lfloor \frac{q}{5} \right\rfloor.$$

Thus, for each value of q, there is exactly one value of p (and one value of a) satisfying the equation. Since q can equal any of thirty values, there are exactly 30 solutions, as claimed.

Problem 3.1.8. *Let λ be the positive root of the equation $t^2 - 1998t - 1 = 0$. Define the sequence $x_0, x_1, \ldots$ by setting*

$$x_0 = 1, \quad x_{n+1} = \lfloor \lambda x_n \rfloor, \quad n \ge 0.$$

Find the remainder when x_{1998} is divided by 1998.

(1998 Iberoamerican Mathematical Olympiad)

Solution. We have

$$1998 < \lambda = \frac{1998 + \sqrt{1998^2 + 4}}{2}$$

$$= 999 + \sqrt{999^2 + 1} < 1999,$$

$x_1 = 1998$, $x_2 = 1998^2$. Since $\lambda^2 - 1998\lambda - 1 = 0$,

$$\lambda = 1998 + \frac{1}{\lambda} \quad \text{and} \quad x\lambda = 1998x + \frac{x}{\lambda}$$

for all real numbers x. Since $x_n = \lfloor x_{n-1}\lambda \rfloor$ and x_{n-1} is an integer and λ is irrational, we have

$$x_n < x_{n-1}\lambda < x_n + 1 \text{ or } \frac{x_n}{\lambda} < x_{n-1} < \frac{x_n + 1}{\lambda}.$$

Since $\lambda > 1998$, $\lfloor x_n/\lambda \rfloor = x_{n-1} - 1$. Therefore,

$$x_{n+1} = \lfloor x_n \lambda \rfloor = \left\lfloor 1998 x_n + \frac{x_n}{\lambda} \right\rfloor = 1998 x_n + x_{n-1} - 1;$$

hence $x_{n+1} \equiv x_{n-1} - 1 \pmod{1998}$. Therefore by induction, $x_{1998} \equiv x_0 - 999 \equiv 1000 \pmod{1998}$.

Problem 3.1.9. (Hermite[1]) *Let n be a positive integer. Prove that for any real number x,*

$$\lfloor nx \rfloor = \lfloor x \rfloor + \left\lfloor x + \frac{1}{n} \right\rfloor + \cdots + \left\lfloor x + \frac{n-1}{n} \right\rfloor.$$

Solution. Let $f(x)$ be the difference between the right-hand side and the left-hand side of (1). Then

$$f\left(x + \frac{1}{n}\right) = \left\lfloor x + \frac{1}{n} \right\rfloor + \cdots + \left\lfloor x + \frac{1}{n} + \frac{n-1}{n} \right\rfloor - \left\lfloor n\left(x + \frac{1}{n}\right) \right\rfloor$$

$$= \left\lfloor x + \frac{1}{n} \right\rfloor + \cdots + \left\lfloor x + \frac{n-1}{n} \right\rfloor + \lfloor x + 1 \rfloor - \lfloor nx + 1 \rfloor,$$

and since $\lfloor x + k \rfloor = \lfloor x \rfloor + k$ for each integer k, it follows that

$$f\left(x + \frac{1}{n}\right) = f(x)$$

for all real x. Hence f is periodic with period $1/n$. Thus it suffices to study $f(x)$ for $0 \le x < 1/n$. But $f(x) = 0$ for all these values; hence $f(x) = 0$ for all real x, and the proof is complete.

Additional Problems

Problem 3.1.10. Let n be a positive integer. Find with proof a closed formula for the sum

$$\left\lfloor \frac{n+1}{2} \right\rfloor + \left\lfloor \frac{n+2}{2^2} \right\rfloor + \cdots + \left\lfloor \frac{n+2^k}{2^{k+1}} \right\rfloor + \cdots .$$

(10th International Mathematical Olympiad)

Problem 3.1.11. Compute the sum

$$\sum_{0 \le i < j \le n} \left\lfloor \frac{x+i}{j} \right\rfloor,$$

where x is a real number.

---

[1] Charles Hermite (1822–1901), French mathematician who did brilliant work in many branches of mathematics.

Problem 3.1.12. Evaluate the difference between the numbers

$$\sum_{k=0}^{2000} \left\lfloor \frac{3^k + 2000}{3^{k+1}} \right\rfloor \quad \text{and} \quad \sum_{k=0}^{2000} \left\lfloor \frac{3^k - 2000}{3^{k+1}} \right\rfloor.$$

Problem 3.1.13. (a) Prove that there are infinitely many rational positive numbers x such that

$$\{x^2\} + \{x\} = 0.99.$$

(b) Prove that there are no rational numbers $x > 0$ such that

$$\{x^2\} + \{x\} = 1.$$

(2004 Romanian Mathematical Olympiad)

Problem 3.1.14. Show that the fractional part of the number $\sqrt{4n^2 + n}$ is not greater than 0.25.

(2003 Romanian Mathematical Olympiad)

Problem 3.1.15. Prove that for every natural number n,

$$\sum_{k=1}^{n^2} \{\sqrt{k}\} \le \frac{n^2 - 1}{2}.$$

(1999 Russian Mathematical Olympiad)

Problem 3.1.16. The rational numbers $\alpha_1, \ldots, \alpha_n$ satisfy

$$\sum_{i=1}^{n} \{k\alpha_i\} < \frac{n}{2}$$

for every positive integer k.

(a) Prove that at least one of $\alpha_1, \ldots, \alpha_n$ is an integer.
(b) Do there exist $\alpha_1, \ldots, \alpha_n$ that satisfy, for every positive integer k,

$$\sum_{i=1}^{n} \{k\alpha_i\} \le \frac{n}{2},$$

such that no α_i is an integer?

(2002 Belarusian Mathematical Olympiad)

3.2   Floor Function and Integer Points

The following results are helpful in proving many relations involving the floor function.

Theorem 3.2.1. *Let a, b, c, d be nonnegative real numbers and let $f : [a, b] \to [c, d]$ be a bijective increasing function.*
Then

$$\sum_{a \le k \le b} \lfloor f(k) \rfloor + \sum_{c \le k \le d} \lfloor f^{-1}(k) \rfloor - n(G_f) = \lfloor b \rfloor \lfloor d \rfloor - \lceil a - 1 \rceil \lceil c - 1 \rceil, \quad (1)$$

where k ranges over integers and $n(G_f)$ is the number of points with integer coordinates on the graph of f.

Proof. Note that $\lceil x \rceil$ is the number of integers in the interval $[0, x)$, $\lfloor x \rfloor$ is the number of integers in the interval $(0, x]$, and hence $\lfloor x \rfloor + 1$ is the number of integers in the interval $[0, x]$.

For a bounded region M of the plane we denote by $n(M)$ the number of points with nonnegative integral coordinates in M.

Since f is increasing and bijective, it is continuous. Hence we can consider the following regions (see Figure 3.1):

$$M_1 = \{(x, y) \in \mathbb{R}^2 \mid a \le x \le b, \ 0 \le y \le f(x)\},$$
$$M_2 = \{(x, y) \in \mathbb{R}^2 \mid c \le y \le d, \ 0 \le x \le f^{-1}(y)\},$$
$$M_3 = \{(x, y) \in \mathbb{R}^2 \mid 0 \le x \le b, \ 0 \le y \le d\},$$
$$M_4 = \{(x, y) \in \mathbb{R}^2 \mid 0 \le x < a, \ 0 \le y < d\}.$$

Then using the remarks above, we compute

$$n(M_1) = \sum_{a \le k \le b} (\lfloor f(k) \rfloor + 1), \quad n(M_2) = \sum_{c \le k \le d} (\lfloor f^{-1}(k) \rfloor + 1)$$
$$n(M_3) = (\lfloor b \rfloor + 1)(\lfloor d \rfloor + 1), \quad n(M_4) = \lceil a \rceil \cdot \lceil c \rceil.$$

Since $M_1 \cup M_2 = M_3 \setminus M_4$ and $M_1 \cap M_2 = G_f$ is the graph of f, we get the identity

$$\sum_{a \le k \le b} (\lfloor f(k) \rfloor + 1) + \sum_{c \le k \le d} (\lfloor f^{-1}(k) \rfloor + 1) - n(G_f) = (\lfloor b \rfloor + 1)(\lfloor d \rfloor + 1) - \lceil a \rceil \cdot \lceil c \rceil.$$

The interval $[a, b] = [0, b] - [0, a)$ contains $\lfloor b \rfloor + 1 - \lceil a \rceil$ integers, and therefore the first sum on the right has this many terms. Similarly, the second sum has $\lfloor d \rfloor + 1 - \lceil c \rceil$ terms. Hence we get

$$\sum_{a \leq k \leq b} \lfloor f(k) \rfloor + \sum_{c \leq k \leq d} \lfloor f^{-1}(k) \rfloor - n(G_f)$$
$$= (\lfloor b \rfloor + 1)(\lfloor d \rfloor + 1) - \lceil a \rceil \cdot \lceil c \rceil - (\lfloor b \rfloor + 1 - \lceil a \rceil) - (\lfloor d \rfloor + 1 - \lceil c \rceil)$$
$$= \lfloor b \rfloor \lfloor d \rfloor - \lceil a - 1 \rceil \lceil c - 1 \rceil. \qquad \square$$

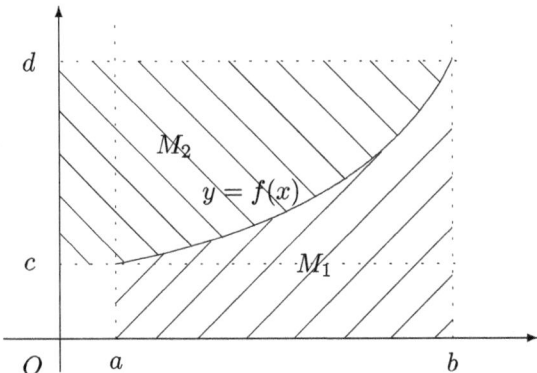

Figure 3.1.

Theorem 3.2.2. *Let m, n, s be positive integers, $m \leq n$. Then*

$$\sum_{k=1}^{s} \left\lfloor \frac{km}{n} \right\rfloor + \sum_{1 \leq k \leq \frac{ms}{n}} \left\lfloor \frac{kn}{m} \right\rfloor = s \left\lfloor \frac{ms}{n} \right\rfloor + \left\lfloor \frac{\gcd(m, n) \cdot s}{n} \right\rfloor. \qquad (2)$$

Proof. We first prove the following lemma.

Lemma. *The collection*

$$\frac{1 \cdot m}{n}, \quad \frac{2 \cdot m}{n}, \quad \dots, \quad \frac{s \cdot m}{n}$$

contains exactly $\left\lfloor \frac{\gcd(m,n) \cdot s}{n} \right\rfloor$ *integers.*

Proof of the lemma. Let d be the greatest common divisor of m and n. Hence $m = m_1 d$ and $n = n_1 d$ for some integers m_1 and n_1.

The numbers in the collection are

$$\frac{1 \cdot m_1}{n_1}, \quad \frac{2 \cdot m_1}{n_1}, \quad \dots, \quad \frac{s \cdot m_1}{n_1}$$

and since m_1, n_1 are relatively prime, there are $\lfloor s/n_1 \rfloor$ integers among them. Because $n_1 = \frac{n}{d} = \frac{n}{\gcd(m,n)}$ it follows that there are $\left\lfloor \frac{\gcd(m,n)s}{n} \right\rfloor$ integers in the collection. $\qquad \square$

In order to prove the desired result, let us consider the function $f : [1, s] \rightarrow$ $\left[\frac{m}{n}, \frac{ms}{n}\right]$, $f(x) = \frac{m}{n}x$ in Theorem 3.2.1. Using the lemma above we have $n(G_f) = \left\lfloor \frac{\gcd(m,n) \cdot s}{n} \right\rfloor$, and the conclusion follows.   □

Remark. The special case $s = n$ leads to an important result:

$$\sum_{k=1}^{n} \left\lfloor \frac{km}{n} \right\rfloor + \sum_{k=1}^{m} \left\lfloor \frac{kn}{m} \right\rfloor = mn + \gcd(m, n). \tag{3}$$

Theorem 3.2.3. *Let a, b, c, d be nonnegative real numbers and let $f : [a, b] \rightarrow [c, d]$ be a bijective, decreasing function.*
Then

$$\sum_{a \le k \le b} \lfloor f(k) \rfloor - \sum_{c \le k \le d} \lceil f^{-1}(k) \rceil = \lfloor b \rfloor \lceil c - 1 \rceil - \lfloor d \rfloor \lceil a - 1 \rceil,$$

where again k ranges over integers.

Proof. Use the notation of the proof of Theorem 3.2.1. Since f is decreasing and bijective, it is continuous and we can define the regions (see Figure 3.2)

$$N_1 = \{(x, y) \in \mathbb{R}^2 \mid a \le x \le b, \ c \le y \le f(x)\},$$
$$N_2 = \{(x, y) \in \mathbb{R}^2 \mid c \le y \le d, \ a \le x \le f^{-1}(y)\},$$
$$N_3 = \{(x, y) \in \mathbb{R}^2 \mid a \le x \le b, \ 0 \le y < c\},$$
$$N_4 = \{(x, y) \in \mathbb{R}^2 \mid a \le x \le b, \ c \le y \le d\}.$$

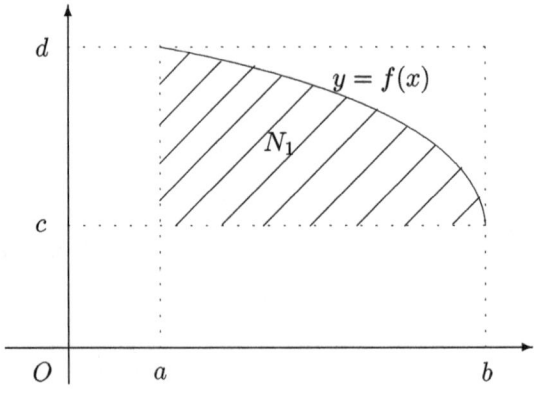

Figure 3.2.

Then $N_1 = N_2$; hence $n(N_1) = n(N_2)$ and

$$\sum_{a \leq k \leq b} (\lfloor f(k) \rfloor + 1) = n(N_1) + n(N_3),$$

$$\sum_{c \leq k \leq d} (\lfloor f^{-1}(k) \rfloor + 1) = n(N_2) + n(N_4),$$

$$n(N_3) = (\lfloor b \rfloor + 1 - \lceil a \rceil)\lceil c \rceil,$$
$$n(N_4) = (\lfloor d \rfloor + 1 - \lceil c \rceil)\lceil a \rceil.$$

It follows that

$$\sum_{a \leq k \leq b} \lfloor f(k) \rfloor - \sum_{c \leq k \leq d} \lfloor f^{-1}(k) \rfloor$$

$$= (\lfloor b \rfloor + 1 - \lceil a \rceil)\lceil c - 1 \rceil - (\lfloor d \rfloor + 1 - \lceil c \rceil)\lceil a - 1 \rceil$$
$$= \lfloor b \rfloor \lceil c - 1 \rceil - \lfloor d \rfloor \lceil a - 1 \rceil. \qquad \square$$

Remark. Combining the result in Theorem 3.2.3 and the relation (3) for the function $f : [0, n] \to [0, m]$, $f(x) = -\frac{m}{n}x + m$, $m \leq n$, yields, after some computation,

$$\sum_{k=1}^{n} \left\lfloor \frac{km}{n} \right\rfloor = \tfrac{1}{2}(mn + m - n + \gcd(m, n)). \qquad (4)$$

From the above relation we obtain

$$\gcd(m, n) = 2 \sum_{k=1}^{n-1} \left\lfloor \frac{km}{n} \right\rfloor + m + n + mn,$$

the proof of which was a 1998 Taiwanese Mathematical Olympiad problem.

From here we get

$$\sum_{k=1}^{n-1} \left\{ \frac{km}{n} \right\} = \sum_{k=1}^{n-1} \frac{km}{n} - \sum_{k=1}^{n-1} \left\lfloor \frac{km}{n} \right\rfloor$$

$$= \frac{m}{n} \cdot \frac{(n-1)n}{2} - \tfrac{1}{2}(mn - m - n + \gcd(m, n))$$

$$= \tfrac{1}{2}(n - \gcd(m, n)),$$

which was a 1995 Japanese Mathematical Olympiad problem.

Problem 3.2.1. *Express $\sum_{k=1}^{n} \lfloor \sqrt{k} \rfloor$ in terms of n and $a = \lfloor \sqrt{n} \rfloor$.*

(1997 Korean Mathematical Olympiad)

Solution. We apply Theorem 3.2.1 for the function $f : [1, n] \to [1, \sqrt{n}]$, $f(x) = \sqrt{x}$. Because $n(G_f) = \lfloor \sqrt{n} \rfloor$, we have

$$\sum_{k=1}^{n} \lfloor \sqrt{k} \rfloor + \sum_{k=1}^{\lfloor \sqrt{n} \rfloor} \lfloor k^2 \rfloor - \lfloor \sqrt{n} \rfloor = n \lfloor \sqrt{n} \rfloor,$$

hence

$$\sum_{k=1}^{n} \lfloor \sqrt{k} \rfloor = (n + 1)a - \frac{a(a + 1)(2a + 1)}{6}.$$

Problem 3.2.2. *Compute*

$$S_n = \sum_{k=1}^{\frac{n(n+1)}{2}} \left[\frac{-1 + \sqrt{1 + 8k}}{2} \right].$$

Solution. Consider the function $f : [1, n] \to \left\lfloor 1, \frac{n(n+1)}{2} \right\rfloor$,

$$f(x) = \frac{x(x + 1)}{2}.$$

The function f is increasing and bijective. Note that $n(G_f) = n$ and $f^{-1}(x) = \frac{-1+\sqrt{1+8x}}{2}$. Applying the formula in Theorem 3.2.1, we obtain

$$\sum_{k=1}^{n} \left\lfloor \frac{k(k + 1)}{2} \right\rfloor + \sum_{k=1}^{\frac{n(n+1)}{2}} \left[\frac{-1 + \sqrt{1 + 8k}}{2} \right] - n = \frac{n^2(n + 1)}{2};$$

hence

$$\sum_{k=1}^{\frac{n(n+1)}{2}} \left[\frac{-1 + \sqrt{1 + 8k}}{2} \right] = \frac{n^2(n + 1)}{2} + n - \frac{1}{2} \sum_{k=1}^{n} k(k + 1)$$

$$= \frac{n^2(n + 1)}{2} + n - \frac{n(n + 1)}{4} - \frac{n(n + 1)(2n + 1)}{12}$$

$$= \frac{n(n^2 + 2)}{3}.$$

Additional Problems

Problem 3.2.3. Prove that

$$\sum_{k=1}^{n} \left\lfloor \frac{n^2}{k^2} \right\rfloor = \sum_{k=1}^{n^2} \left\lfloor \frac{n}{\sqrt{k}} \right\rfloor$$

for all integers $n \geq 1$.

Problem 3.2.4. Let θ be a positive irrational number. Then, for any positive integer m,

$$\sum_{k=1}^{m} \lfloor k\theta \rfloor + \sum_{k=1}^{\lfloor m\theta \rfloor} \left\lfloor \frac{k}{\theta} \right\rfloor = m \lfloor m\theta \rfloor.$$

Problem 3.2.5. Let p and q be relatively prime positive integers and let m be a real number such that $1 \leq m < p$.

(1) If $s = \left\lfloor \frac{mq}{p} \right\rfloor$, then

$$\sum_{k=1}^{\lfloor m \rfloor} \left\lfloor \frac{kq}{p} \right\rfloor + \sum_{k=1}^{s} \left\lfloor \frac{kp}{q} \right\rfloor = \lfloor m \rfloor s.$$

(2) (Landau[2]) If p and q are odd, then

$$\sum_{k=1}^{\frac{p-1}{2}} \left\lfloor \frac{kq}{p} \right\rfloor + \sum_{k=1}^{\frac{q-1}{2}} \left\lfloor \frac{kp}{q} \right\rfloor = \frac{(p-1)(q-1)}{4}.$$

3.3 A Useful Result

The following theorem is also helpful in proving some relations involving the floor function.

Theorem 3.3.1. *Let p be an odd prime and let q be an integer that is not divisible by p. If $f : \mathbb{Z}_+^* \to \mathbb{R}$ is a function such that:*

(i) $\frac{f(k)}{p}$ *is not an integer, $k = 1, 2, \ldots, p-1$;*

(ii) $f(k) + f(p-k)$ *is an integer divisible by p, $k = 1, 2, \ldots, p-1$;*

then

$$\sum_{k=1}^{p-1} \left\lfloor f(k) \frac{q}{p} \right\rfloor = \frac{q}{p} \sum_{k=1}^{p-1} f(k) - \frac{p-1}{2}. \tag{1}$$

Proof. From (ii) it follows that

$$\frac{qf(k)}{p} + \frac{qf(p-k)}{p} \in \mathbb{Z}, \tag{2}$$

---

[2]Edmond Georg Hermann Landau (1877–1838), German mathematician who gave the first systematic presentation of analytic number theory and wrote an important work on the theory of analytic functions of a single variable.

and from (i) we obtain that $\frac{qf(k)}{p} \notin \mathbb{Z}$ and $\frac{qf(p-k)}{p} \notin \mathbb{Z}$, $k = 1, \ldots, p-1$; hence

$$0 < \left\{\frac{qf(k)}{p}\right\} + \left\{\frac{qf(p-k)}{p}\right\} < 2.$$

But from (2), $\left\{\frac{qf(k)}{p}\right\} + \left\{\frac{qf(p-k)}{p}\right\} \in \mathbb{Z}$; thus

$$\left\{\frac{qf(k)}{p}\right\} + \left\{\frac{qf(p-k)}{p}\right\} = 1, \quad k = 1, \ldots, p-1.$$

Summing up and dividing by 2 yields

$$\sum_{k=1}^{p-1} \left\{\frac{q}{p} f(k)\right\} = \frac{p-1}{2}.$$

It follows that

$$\sum_{k=1}^{p-1} \frac{q}{p} f(k) - \sum_{k=1}^{p-1} \left\lfloor \frac{q}{p} f(k) \right\rfloor = \frac{p-1}{2},$$

and the conclusion follows. ☐

Problem 3.3.1. *Let p and q be two relatively prime integers. The following identity holds:*

$$\sum_{k=1}^{p-1} \left\lfloor k\frac{q}{p} \right\rfloor = \frac{(p-1)(q-1)}{2} \quad (Gauss).$$

Solution. The function $f(x) = x$ satisfies both (i) and (ii) in Theorem 3.3.1; hence

$$\sum_{k=1}^{p-1} \left\lfloor k\frac{q}{p} \right\rfloor = \frac{q}{p}\frac{(p-1)p}{2} - \frac{p-1}{2},$$

and the desired relation follows.

Problem 3.3.2. *Let p be an odd prime. Prove that*

$$\sum_{k=1}^{p-1} \left\lfloor \frac{k^3}{p} \right\rfloor = \frac{(p-2)(p-1)(p+1)}{4}.$$

(2002 German Mathematical Olympiad)

Solution. The function $f(x) = x^3$ also satisfies conditions (i) and (ii); hence

$$\sum_{k=1}^{p-1} \left\lfloor k^3\frac{q}{p} \right\rfloor = \frac{q}{p} \cdot \frac{(p-1)^2 p^2}{4} - \frac{p-1}{2} = \frac{(p-1)(p^2q - pq - 2)}{4}.$$

For $q = 1$ the identity in our problem follows.

Additional Problems

Problem 3.3.3. Let p be an odd prime and let q be an integer that is not divisible by p. Shows that

$$\sum_{k=1}^{p-1} \left\lfloor (-1)^k k^2 \frac{q}{p} \right\rfloor = \frac{(p-1)(q-1)}{2}.$$

(2005 "Alexandru Myller" Romanian Regional Contest)

Problem 3.3.4. Let p be an odd prime. Show that

$$\sum_{k=1}^{p-1} \frac{k^p - k}{p} \equiv \frac{p+1}{2} \pmod{p}.$$

(2006 "Alexandru Myller" Romanian Regional Contest)

4

Digits of Numbers

4.1 The Last Digits of a Number

Let $\overline{a_n a_1 \cdots a_0}$ be the decimal representation of the positive integer N. The last digit of N is $l(N) = a_0$, and for $k \geq 2$, the last k digits of N are $l_k(N) = \overline{a_{k-1} \cdots a_0}$. These simple concepts appear in numerous situations.

It is useful to point out the last digit of k^n, where $k = 2, 3, \ldots, 9$ and $n > 0$:

$$l(2^n) = \begin{cases} 6, n \equiv 0 \pmod{4}, \\ 2, n \equiv 1 \pmod{4}, \\ 4, n \equiv 2 \pmod{4}, \\ 8, n \equiv 3 \pmod{4}, \end{cases} \quad l(3^n) = \begin{cases} 1, n \equiv 0 \pmod{4}, \\ 3, n \equiv 1 \pmod{4}, \\ 9, n \equiv 2 \pmod{4}, \\ 7, n \equiv 3 \pmod{4}, \end{cases}$$

$$l(4^n) = \begin{cases} 6, n \equiv 0 \pmod{2}, \\ 4, n \equiv 1 \pmod{2}, \end{cases} \quad l(5^n) = 5, \quad l(6^n) = 6,$$

$$l(7^n) = \begin{cases} 1, n \equiv 0 \pmod{4}, \\ 7, n \equiv 1 \pmod{4}, \\ 9, n \equiv 2 \pmod{4}, \\ 3, n \equiv 3 \pmod{4}, \end{cases} \quad l(8^n) = \begin{cases} 6, n \equiv 0 \pmod{4}, \\ 8, n \equiv 1 \pmod{4}, \\ 4, n \equiv 2 \pmod{4}, \\ 2, n \equiv 3 \pmod{4}, \end{cases}$$

$$l(9^n) = \begin{cases} 1, n \equiv 0 \pmod{2}, \\ 9, n \equiv 1 \pmod{2}. \end{cases}$$

It is clear that if $l(N) = 0$, then $l_n(N^n) = \underbrace{\overline{0 \ldots 0}}_{n \text{ times}}$, and if $l(N) = 1$, then $l(N^n) = 1$ for all $n \geq 2$.

Problem 4.1.1. *What is the final digit of $\left(\ldots \left(\left(7^7 \right)^7 \right)^7 \right) \ldots {}^7 \right)$?*
There are 1001 7's in the formula.

Solution. The final digit of a (decimal) number is its remainder modulo 10. Now

T. Andreescu and D. Andrica, *Number Theory*, DOI: 10.1007/b11856_4,
© Birkhäuser Boston, a part of Springer Science + Business Media, LLC 2009

$7^2 = 49 \equiv -1 \pmod{10}$. So $7^7 = (7^2)^3 \cdot 7 \equiv -7 \pmod{10}$, and

$$(7^7)^7 \equiv (-7)^7 \equiv -(7^7) \equiv -(-7) \equiv 7 \pmod{10}.$$

Proceeding in this way, we see that $\left((7^7)^7\right)^7 \equiv -7 \pmod{10}$, and in general,

$$\left(\dots \left(((7^7)^7)^7\right) \dots^7\right) \equiv \pm 7 \pmod{10},$$

where the sign is $+$ if altogether there is an odd number of 7's in the formula, and $-$ if there is an even number of 7's. Now, 1001 is odd. So the final digit of the given formula is 7.

Problem 4.1.2. *Prove that every positive integer has at least as many (positive) divisors whose last decimal digit is 1 or 9 as divisors whose last digit is 3 or 7.*

(1997 St. Petersburg City Mathematical Olympiad)

Solution. Let $d_1(m), d_3(m), d_7(m), d_9(m)$ be the numbers of divisors of m ending in $1, 3, 7, 9$, respectively. We prove the claim by induction on m; it holds obviously for m a prime power, and if m is composite, write $m = pq$ with p, q coprime, and note that

$$d_1(m) - d_3(m) - d_7(m) + d_9(m)$$
$$= (d_1(p) - d_3(p) - d_7(p) + d_9(p))(d_1(q) - d_3(q) - d_7(q) + d_9(q)).$$

For instance,

$$d_3(m) = d_1(p)d_3(q) + d_3(p)d_1(q) + d_7(p)d_9(q) + d_9(p)d_7(q).$$

Problem 4.1.3. *Find the least positive integer n with the following properties:*
(a) the last digit of its decimal representation is 6;
(b) by deleting the last digit 6 and replacing it in front of the remaining digits one obtains a number four times greater than the given number.

(4th International Mathematical Olympiad)

Solution. Let $n = 10^k a_k + 10^{k-1} a_{k-1} + \cdots + 10 a_1 + 6$ be the required number. Writing n in the form $n = 10N + 6$, where $10^{k-1} < N < 10^k$, the condition (b) becomes:

$$4(10N + 6) = 6 \cdot 10^k + N.$$

Thus, we obtain

$$39N = 6 \cdot 10^k - 24,$$

and equivalently

$$13N = 2(10^k - 4).$$

Thus, we obtain that $10^k \equiv 4 \pmod{13}$.
It is more convenient to write

$$(-3)^k \equiv 4 \pmod{13}.$$

From the conditions of the problem, the least k with this property is required. We have

$$(-3)^2 \equiv 9 \pmod{13}, \quad (-3)^3 \equiv -27 \pmod{13} \equiv -1 \pmod{13},$$
$$(-3)^5 \equiv (-3)^2(-3)^3 \equiv -9 \equiv 4 \pmod{13}.$$

Then, $k = 5$ is the least positive solution of the equation. Thus,

$$13N = 2 \cdot 99996 \Rightarrow N = 15384 \Rightarrow n = 153846.$$

This number satisfies (b).

Additional Problems

Problem 4.1.4. In how may zeros can the number $1^n + 2^n + 3^n + 4^n$ end for $n \in \mathbb{N}$?

(1998 St. Petersburg City Mathematical Olympiad)

Problem 4.1.5. Find the last 5 digits of the number 5^{1981}.

Problem 4.1.6. Consider all pairs (a, b) of natural numbers such that the product $a^a b^b$, written in base 10, ends with exactly 98 zeros. Find the pair (a, b) for which the product ab is smallest.

(1998 Austrian–Polish Mathematics Competition)

4.2 The Sum of the Digits of a Number

For a positive integer $N = \overline{a_n a_{n-1} \cdots a_0}$ in decimal representation we denote by $S(N)$ the sum of its digits $a_0 + \cdots + a_{n-1} + a_n$. Problems involving the function S defined above appear frequently in various contexts. We present a few basic properties.

(1) $S(N) = N - 9 \sum_{k \geq 1} \left\lfloor \frac{N}{10^k} \right\rfloor$;

(2) $9 \mid N - S(N)$;

(3) (subadditivity): $S(N_1 + N_2) \leq S(N_1) + S(N_2)$;

(4) $S(N_1 N_2) \leq \min(N_1 S(N_2), N_2 S(N_1))$;

(5) (submultiplicity): $S(N_1 N_2) \leq S(N_1) S(N_2)$.

Property (2) is in fact Property 1.7.2, Criterion 1(b).

Let us prove the last three properties. Using (1) and the inequality $\lfloor x + y \rfloor \geq \lfloor x \rfloor + \lfloor y \rfloor$, we have

$$S(N_1 + N_2) = N_1 + N_2 - 9 \sum_{k \geq 1} \left\lfloor \frac{N_1 + N_2}{10^k} \right\rfloor$$

$$\leq N_1 + N_2 - 9 \sum_{k \geq 1} \left(\left\lfloor \frac{N_1}{10^k} \right\rfloor + \left\lfloor \frac{N_2}{10^k} \right\rfloor \right)$$

$$= S(N_1) + S(N_2).$$

Because of the symmetry, in order to prove (4) it suffices to prove that $S(N_1 N_2) \leq N_1 S(N_2)$.

The last inequality follows by applying the subadditivity property repeatedly. Indeed,

$$S(2N_2) = S(N_2 + N_2) \leq S(N_2) + S(N_2) = 2S(N_2),$$

and after N_1 steps we obtain

$$S(N_1 N_2) = S(\underbrace{N_2 + N_2 + \cdots + N_2}_{N_1 \text{ times}})$$

$$\leq \underbrace{S(N_2) + S(N_2) + \cdots + S(N_2)}_{N_1 \text{ times}} = N_1 S(N_2).$$

For (5) observe that if $N_2 = \sum_{o=0}^{k} b_i \cdot 10^i$, then

$$S(N_1 N_2) = S\left(N_1 \sum_{i=0}^{h} b_i 10^i \right) = S\left(\sum_{i=0}^{h} N_1 b_i 10^i \right) \leq \sum_{i=0}^{h} S(N_1 b_i 10^i)$$

$$= \sum_{i=0}^{h} S(N_1 b_i) \leq \sum_{i=0}^{h} b_i S(N_1) = S(N_1) S(N_2).$$

Examples. (1) *In the decimal expansion of N, the digits occur in increasing order. What is $S(9N)$?*

(1999 Russian Mathematical Olympiad)

Solution. Write $N = \overline{a_k a_{k-1} \cdots a_0}$. By performing the subtraction

$$
\begin{array}{r}
a_k \; a_{k-1} \; \ldots \; a_1 \; a_0 \; 0 \\
- a_k \; \ldots \; a_2 \; a_1 \; a_0 \\
\hline
\end{array}
$$

we find that the digits of $9N = 10N - N$ are

$$a_k, \quad a_{k-1} - a_k, \quad \ldots, \quad a_1 - a_2, \quad a_0 - a_1 - 1, \quad 10 - a_0.$$

These digits sum to $10 - 1 = 9$.

(2) *Find a positive integer N such that $S(N) = 1996S(3N)$.*

(1996 Irish Mathematical Olympiad)

Solution. Consider $N = 1 \underbrace{33\ldots3}_{5986 \text{ times}} 5$. Then $3N = 4 \underbrace{00\ldots0}_{5986 \text{ times}} 5$ and

$$S(N) = 3 \cdot 5986 + 1 + 5 = 17964 = 1996 \cdot 9 = 1996S(N).$$

Problem 4.2.1. *Determine all possible values of the sum of the digits of a perfect square.*

(1995 Iberoamerican Olympiad)

Solution. The sum of the digits of a number is congruent to the number modulo 9, and so for a perfect square this must be congruent to 0, 1, 4 or 7. We show that all such numbers occur. The cases $n = 1$ and $n = 4$ are trivial, so assume $n > 4$.

If $n = 9m$, then n is the sum of the digits of $(10^m - 1)^2 = 10^m(10^m - 2) + 1$, which looks like $9\ldots980\ldots01$. If $n = 9m+1$, consider $(10^m - 2)^2 = 10^m(10^m - 4) + 4$, which looks like $9\ldots960\ldots04$. If $n = 9m + 4$, consider $(10^m - 3)^2 = 10^m(10^m - 6) + 9$, which looks like $9\ldots94\ldots09$. Finally, if $n = 9m - 2$, consider $(10^m - 5)^2 = 10^m(10^m - 10) + 25$, which looks like $9\ldots900\ldots025$.

Problem 4.2.2. *Find the number of positive 6-digit integers such that the sum of their digits is 8, and four of its digits are $1, 0, 0, 4$.*

(2004 Romanian Mathematical Olympiad)

Solution. The pair of missing digits must be 1, 2 or 0, 3.

In the first case, the first digit can be 1, 2, or 4. When 1 is the first digit, the remaining digits, 1, 2, 0, 0, 4, can be arranged in 60 ways. When 4 or 2 is the first digit, the remaining digits can be arranged in 30 ways.

In the same way, when completing with the pair $(0, 3)$, the first digit can be 1, 3, or 4. In each case, the remaining digits (three zeros and two distinct nonzero digits) can be arranged in 20 ways.

In conclusion, we have $60 + 2 \cdot 30 + 3 \cdot 20 = 180$ numbers that satisfy the given property.

Problem 4.2.3. *Find the sum of the digits of the numbers from 1 to 1,000,000.*

Solution. Write the numbers from 0 to 999,999 in a rectangular array as follows:

$$
\begin{array}{cccccc}
0 & 0 & 0 & 0 & 0 & 0 \\
0 & 0 & 0 & 0 & 0 & 1 \\
0 & 0 & 0 & 0 & 0 & 2 \\
\cdots & \cdots & \cdots & \cdots & \cdots & \cdots \\
0 & 0 & 0 & 0 & 0 & 9 \\
0 & 0 & 0 & 0 & 1 & 0 \\
0 & 0 & 0 & 0 & 1 & 1 \\
\cdots & \cdots & \cdots & \cdots & \cdots & \cdots \\
0 & 0 & 0 & 0 & 1 & 9 \\
0 & 0 & 0 & 0 & 2 & 0 \\
\cdots & \cdots & \cdots & \cdots & \cdots & \cdots \\
9 & 9 & 9 & 9 & 9 & 9
\end{array}
$$

There are 1,000,000 six-digit numbers; hence 6,000,000 digits are used. In each column every digit is equally represented, since in the units column each digit appears from 0 to 9, in the tens column each digit appears successively in blocks of 10, and so on. Thus each digit appears 600,000 times, so the required sum is

$$600,000 \cdot 45 + 1 = 27,000,001$$

(do not forget to count 1 from 1,000,000).

Problem 4.2.4. *Find every positive integer n that is equal to the sum of its digits added to the product of its digits.*

Solution. Let $\overline{a_1 a_2 \cdots a_n}$, $a_1 \neq 0$, and $a_2, \ldots, a_n \in \{0, 1, \ldots, 9\}$ be a number such that

$$\overline{a_1 a_2 \cdots a_n} = a_1 + a_2 + \cdots + a_n + a_1 a_2 \cdots a_n.$$

The relation is equivalent to

$$a_1(10^{n-1} - 1) + a_2(10^{n-2} - 1) + \cdots + 9a_{n-1} = a_1 a_2 \cdots a_n$$

and

$$a_2(10^{n-2} - 1) + \cdots + 9a_{n-1} = a_1(a_2 a_3 \cdots a_n - \underbrace{99 \ldots 9}_{n-1 \text{ digits}}).$$

The left-hand side of the equality is nonnegative, while the right-hand side is nonpositive; hence both are equal to zero. The left-hand side is zero if $n = 0$ or

$$a_2 = a_3 = \cdots = a_{n-1} = 0.$$

For $a_2 = a_3 = \cdots = a_{n-1} = 0$, the left-hand side does not equal zero; hence $n = 2$. Then $a_1(a_2 - 9) = 0$, so $a_2 = 0$ and $a_1 \in \{1, 2, \ldots, 9\}$. The numbers are 19, 29, 39, 49, 59, 69, 79, 89, 99.

Problem 4.2.5. *What is the smallest multiple of* 99 *whose digits sum to* 99 *and that begins and ends with* 97?

(1997 Rio Platense Mathematical Olympiad)

Solution. We refer to the digits of the number besides the two 97's as interior digits; the sum of these digits is $99 - 2(9 + 7) = 67$. Since each digit is at most 9, there are at least eight such digits.

Note that the sum of digits being 99 forces the number to be divisible by 9; thus it suffices to ensure that the number be divisible by 11, which is to say, the alternating sum of digits must be divisible by 11.

Suppose the number has exactly eight interior digits. If a is the sum of the odd interior places and b the sum of the even places, we have $a + b = 67$ and $a - b \equiv -4 \pmod{11}$. Since $a - b$ must also be odd, we have $a - b \geq 7$ or $a - b \leq -15$, and so either $a \geq 37$ or $b \geq 41$, contradicting the fact that a and b are each the sum of four digits.

Now suppose the number has nine interior digits. In this case, $a - b \equiv 0 \pmod{11}$, so $a - b \geq 11$ or $a - b \leq -11$. In the latter case, $b \geq 39$, again a contradiction, but in the former case, we have $a \geq 39$, which is possible because a is now the sum of five digits. To minimize the original number, we take the odd digits to be 3, 9, 9, 9, 9 and the even digits to be 1, 9, 9, 9, making the minimal number 9731999999997.

Problem 4.2.6. *Find all positive integers n such that there are nonnegative integers a and b with*
$$S(a) = S(b) = S(a + b) = n.$$

(1999 Romanian Selection Test for JBMO)

Solution. We prove that the required numbers are all multiples of 9.

(a) Let n be an integer such that there are positive integers a and b such that
$$S(a) = S(b) = S(a + b).$$

We prove that $9 \mid n$.

We have the property
$$9 \mid k - S(k). \tag{1}$$

Using the relation (1) we obtain
$$9 \mid a - S(a), \tag{2}$$
$$9 \mid b - S(b), \tag{3}$$

and
$$9 \mid (a + b) - S(a + b). \tag{4}$$

From (2) and (3) it follows that

$$9 \mid a + b - (S(a) + S(b)); \tag{5}$$

hence

$$9 \mid S(a) + S(b) - S(a + b) = n + n - n = n, \tag{6}$$

as desired.

(b) Conversely, we prove that if $n = 9p$ is a multiple of 9, then integers $a, b > 0$ with $S(a) = S(b) = S(a + b)$ can be found. Indeed, set $a = \underbrace{531531\ldots531}_{3p \text{ digits}}$

and $b = \underbrace{171171\ldots171}_{3p \text{ digits}}$. Then $a + b = \underbrace{702702\ldots702}_{3p \text{ digits}}$ and

$$S(a) = S(b) = S(a + b) = 9p = n,$$

as claimed.

Additional Problems

Problem 4.2.7. Show that there exist infinitely many natural numbers n such that $S(3^n) \geq S(3^{n+1})$.

(1997 Russian Mathematical Olympiad)

Problem 4.2.8. Do there exist three natural numbers a, b, c such that $S(a + b) < 5$, $S(b + c) < 5$, $S(c + a) < 5$, but $S(a + b + c) > 50$?

(1998 Russian Mathematical Olympiad)

Problem 4.2.9. Prove that there exist distinct positive integers $\{n_i\}_{1 \leq i \leq 50}$ such that

$$n_1 + S(n_1) = n_2 + S(n_2) = \cdots = n_{50} + S(n_{50}).$$

(1999 Polish Mathematical Olympiad)

Problem 4.2.10. The sum of the decimal digits of the natural number n is 100, and that of $44n$ is 800. What is the sum of the digits of $3n$?

(1999 Russian Mathematical Olympiad)

Problem 4.2.11. Consider all numbers of the form $3n^2 + n + 1$, where n is a positive integer.

(a) How small can the sum of the digits (in base 10) of such a number be?

(b) Can such a number have the sum of its digits (in base 10) equal to 1999?

(1999 United Kingdom Mathematical Olympiad)

Problem 4.2.12. Consider the set A of all positive integers n with the following properties: the decimal expansion contains no 0, and the sum of the (decimal) digits of n divides n.

(a) Prove that there exist infinitely many elements in A with the following property: the digits that appear in the decimal expansion of A appear the same number of times.

(b) Show that for each positive integer k, there exists an element in A with exactly k digits.

(2001 Austrian–Polish Mathematics Competition)

4.3 Other Problems Involving Digits

Problem 4.3.1. *Prove that there are at least* 666 *positive composite numbers with* 2006 *digits, having a digit equal to* 7 *and all the rest equal to* 1.

Solution. The given numbers are

$$n_k = 111\ldots17\underbrace{11\ldots1}_{k \text{ digits}} = \underbrace{111\ldots1}_{2006 \text{ digits}} + 6\underbrace{000\ldots0}_{k \text{ digits}}$$

$$= \frac{1}{9}(10^{2006} - 1) + 6 \cdot 10^k, \quad k = \overline{0, 1, \ldots, 2005}.$$

It is obvious that none of these numbers is a multiple of 2, 3, 5, or 11, since 11 divides $\underbrace{111\ldots1}_{2006 \text{ digits}}$, but not $6 \cdot 10^k$.

So we are led to the idea of counting multiples of 7 and 13. We have $9n_k = 100 \cdot 1000^{668} - 1 + 54 \cdot 10^k \equiv 2 \cdot (-1)^{668} - 1 + (-2) \cdot 10^k \equiv 1 - 2 \cdot 10^k \pmod 7$; hence $7 \mid n_k$ if $10^k \equiv 3^k \equiv 4 \pmod 7$. This happens for $k = 4, 10, 16, \ldots, 2002$, so there are 334 multiples of 7. Furthermore, $9n_k \equiv 7 \cdot (-1)^{668} - 1 + 2 \cdot 10^k = 6 + 2 \cdot 10^k \pmod{13}$; hence $13 \mid n_k$ if $10^k \equiv 10 \pmod{13}$. This happens for $k = 1, 7, 13, 19, \ldots, 2005$, so there are 335 multiples of 13. In all we have found 669 nonprime numbers.

Problem 4.3.2. *Let* $a_1, a_2, \ldots, a_{10^6}$ *be integers between* 1 *and* 9, *inclusive. Prove that at most* 100 *of the numbers* $\overline{a_1 a_2 \cdots a_k}$ $(1 \le k \le 10^6)$ *are perfect squares.*

(2001 Russian Mathematical Olympiad)

Solution. For each positive integer x, let $d(x)$ be the number of decimal digits in x.

Lemma. *Suppose that* $y > x$ *are perfect squares such that* $y = 10^{2b}x + c$ *for some positive integers* b, c *with* $c < 10^{2b}$. *Then*

$$d(y) - 1 \ge 2(d(x) - 1).$$

Proof. Because $y > 10^{2b}x$, we have $\sqrt{y} > 10^b\sqrt{x}$. Because $\sqrt{y}$ and $10^b\sqrt{x}$ are both integers, $\sqrt{y} \geq 10^b\sqrt{x} + 1$, so that $10^{2b}x + c = y \geq 10^{2b}x + 2 \cdot 10^b\sqrt{x} + 1$. Thus, $c \geq 2 \cdot 10^b\sqrt{x} + 1$.

Also, $10^{2b} > c$ by assumption, implying that

$$10^{2b} > c \geq 2 \cdot 10^b\sqrt{x} + 1.$$

Hence, $10^b > 2\sqrt{x}$. It follows that

$$y > 10^{2b}x > 4x^2.$$

Therefore,

$$d(y) \geq 2d(x) - 1,$$

as desired.

We claim that there are at most 20 perfect squares $\overline{a_1a_2\cdots a_k}$ with an even (resp. odd) number of digits. Let $s_1 < s_2 < \cdots < s_n$ be these perfect squares. Clearly $d(s_n) \leq 10^6$. We now prove that if $n > 1$, then $d(s_n) \geq 1 + 2^{n-1}$.

Because $s_1, s_2, \ldots, s_n$ all have an even (resp. odd) number of digits, for each $i = 1, 2, \ldots, n-1$, we can write $s_{i+1} = 10^{2b}s_i + c$ for some integers $b > 0$ and $0 \leq c < 10^{2b}$. Because no a_i equals 0, we further know that $0 < c$. Hence, by our lemma,

$$d(s_{i+1}) - 1 \geq 2(d(s_i) - 1)$$

for each $i = 1, 2, \ldots, n-1$. Because $d(s_2) - 1 \geq 2$, we thus have $d(s_n) - 1 \geq 2^{n-1}$, as desired.

Thus, if $n > 1$,

$$1 + 2^{n-1} \leq d(s_n) \leq 10^6,$$

and

$$n \leq \left\lfloor \frac{\log(10^6 - 1)}{\log 2} \right\rfloor + 1 = 20.$$

Hence, there are at most 20 perfect squares $\overline{a_1a_2\cdots a_k}$ with an even (resp. odd) number of digits.

Therefore, there are at most $40 < 100$ perfect squares $\overline{a_1a_2\cdots a_k}$.

Additional Problems

Problem 4.3.3. A *wobbly* number is a positive integer whose digits in base 10 are alternately nonzero and zero, the units digit being nonzero. Determine all positive integers that do not divide any wobbly number.

(35th International Mathematical Olympiad Shortlist)

Problem 4.3.4. A positive integer is called *monotonic* if its digits in base 10, read from left right, are in nondecreasing order. Prove that for each $n \in \mathbb{N}$, there exists an n-digit monotonic number that is a perfect square.

(2000 Belarusian Mathematical Olympiad)

5

Basic Principles in Number Theory

5.1 Two Simple Principles

5.1.1 Extremal Arguments

In many problems it is useful to consider the least or the greatest element with a certain property. Very often such a choice leads to the construction of other elements or to a contradiction.

Problem 5.1.1. *Show that there exist infinitely many positive integers n such that the largest prime divisor of $n^4 + 1$ is greater than $2n$.*

(2001 St. Petersburg City Mathematical Olympiad)

Solution. First we prove the following result.

Lemma. *There are infinitely many numbers that are prime divisors of $m^4 + 1$ for some m.*

Proof. Suppose that there are only finite number of such primes. Let $p_1, p_2, \ldots, p_k$ be all of them. Let p be any prime divisor of $(p_1 p_2 \cdots p_k)^4 + 1$. This number cannot equal any p_i, a contradiction to our assumption, which proves the lemma. □

Let $\mathcal{P}$ be the set of all numbers that are prime divisors of $m^4 + 1$ for some m. Pick any p from $\mathcal{P}$ and m from $\mathbb{Z}$, such that p divides $m^4 + 1$. Let r be the residue of m modulo p. We have $r < p$, $p \mid r^4 + 1$, and $p \mid (p - r)^4 + 1$. Let n be the minimum of r and $p - r$. It follows that $n < p/2$ and $p > 2n$ and of course $p \mid n^4 + 1$. Thus we have found for each $p \in \mathcal{P}$ a good number n_p. Since $n_p \geq \sqrt[4]{p} - 1$, and $\mathcal{P}$ is infinite, the set $\{n_p : p \in \mathcal{P}\}$ is also infinite.

Remark. Essentially the same proof shows that for any polynomial $P(x)$ with integer coefficients, there are infinitely many primes that divide $P(n)$ for some integer n.

T. Andreescu and D. Andrica, *Number Theory*, DOI: 10.1007/b11856_5,
© Birkhäuser Boston, a part of Springer Science + Business Media, LLC 2009

Problem 5.1.2. *Let $a_1, a_2, \ldots$ be a strictly increasing sequence of positive integers such that $\gcd(a_m, a_n) = a_{\gcd(m,n)}$ for all positive integers m and n. There exists a least positive integer k for which there exist positive integers $r < k$ and $s > k$ such that $a_k^2 = a_r a_s$. Prove that r divides k and that k divides s.*

(2001 Indian Mathematical Olympiad)

Solution. We begin by proving a lemma.

Lemma. *If positive integers a, b, c satisfy $b^2 = ac$, then*

$$\gcd(a, b)^2 = \gcd(a, c) \cdot a.$$

Proof. Consider any prime p. Let e be the highest exponent such that p^e divides b, and let e_1 and e_2 be the corresponding highest exponents for a and c, respectively. Because $b^2 = ac$, we have $2e = e_1 + e_2$. If $e_1 \geq e$, then the highest powers of p that divide $\gcd(a, b)$, $\gcd(a, c)$, and a are e, e_2, and e_1, respectively. Otherwise, these highest powers are all e_1. Therefore, in both cases, the exponent of p on the left side of the desired equation is the same as the exponent of p on the right side. The desired result follows. ☐

Applying the lemma to the given equation $a_k^2 = a_r a_s$, we have

$$\gcd(a_r, a_k)^2 = \gcd(a_r, a_s) a_r.$$

It now follows from the given equation that

$$a_{\gcd(r,k)}^2 = a_{\gcd(r,s)} a_r.$$

Assume, for sake of contradiction, that $\gcd(r, k) < r$, so that $a_{\gcd(r,k)} < a_r$. Then from the above equation, it follows that $a_{\gcd(r,k)} > a_{\gcd(r,s)}$, so that $\gcd(r, k) > \gcd(r, s)$. But then we have that $(k_0, r_0, s_0) = (\gcd(r, k), \gcd(r, s), r)$ satisfies $a_{k_0}^2 = a_{r_0} a_{s_0}$ with $r_0 < k_0 < s_0$ and $k_0 < r < k$, contradicting the minimality of k.

Thus, we must have $\gcd(r, k) = r$, implying that $r \mid k$. Then

$$\gcd(a_r, a_k) = a_{\gcd(r,k)} = a_r,$$

so $a_r \mid a_k$. Thus $a_s = a_k \frac{a_k}{a_r}$ is an integer multiple of a_k, and

$$a_{\gcd(k,s)} = \gcd(a_k, a_s) = a_k.$$

Because $a_1, a_2, \ldots$ is increasing, it follows that $\gcd(k, s) = k$. Therefore, $k \mid s$, completing the proof.

Problem 5.1.3. *Determine all pairs (n, p) of positive integers such that p is a prime, $n \leq 2p$ and $(p-1)^n + 1$ is divisible by n^{p-1}.*

(40th International Mathematical Olympiad)

Solution. All pairs $(1, p)$, where p is a prime number, satisfy the conditions. When $p = 2$, it follows that $n = 2$, and thus the pair $(2, 2)$ is also a solution of the problem. Thus, we may suppose $p \geq 3$ and let n be such that $n \leq 2p$ and n^{p-1} divides $(p-1)^n + 1$. Since $(p-1)^n + 1$ is odd number, it follows that $n < 2p$. We shall prove that $n = p$.

Let q be a minimal prime divisor of n. Since $q \mid n$ and $n^{p-1} \mid (p-1)^n + 1$, it follows that $(p-1)^n \equiv -1 \pmod{q}$. Since n and $q-1$ are relatively prime numbers, we may write $an + b(q-1) = 1$ for some integers a and b.

We have

$$p-1 \equiv (p-1)^{an+b(q-1)} \equiv (p-1)^{na}(p-1)^{(q-1)b} \equiv (-1)^a 1^b \equiv -1 \pmod{q},$$

because a must be odd. This shows that $q \mid p$, and therefore $q = p$. Since $n < 2p$, by the consideration of q, we have $n = p$.

Let consider in these conditions the original divisibility

$$p^{p-1} \mid (p-1)^p + 1 = p^p - \binom{p}{1}p^{p-1} + \binom{p}{2}p^{p-2} - \cdots + \binom{p}{p-1}p - 1 + 1$$

$$= p^2 \left[p^{p-2} - \binom{p}{1}p^{p-3} + \binom{p}{2}p^{p-4} - \cdots + 1 \right].$$

Therefore $p - 1 = 2$, $p = 3$, and we then obtain the pair $(3, 3)$.

The conclusion is that the required solutions are $(1, p)$, $(2, 2)$, and $(3, 3)$, where p is an arbitrary prime.

Remark. With a little bit more work, we can even erase the condition $n \leq 2p$.

5.1.2 The Pigeonhole Principle

Let S be a nonempty set and let $S_1, S_2, \ldots, S_n$ be a partition of S (that is, $S_1 \cup S_2 \cup \cdots \cup S_n = S$ and $S_i \cap S_j = \emptyset$ for $i \neq j$). If $a_1, a_2, \ldots, a_{n+1}$ are distinct elements in S, then there is a $k \in \{1, 2, \ldots, n\}$ such that at least two of these elements belong to S_k.

This simple observation is called the *pigeonhole principle* (or *Dirichlet's principle*).

Examples. (1) Let $m_1, m_2, \ldots, m_{n+1}$ be distinct integers. Then $m_i \equiv m_j \pmod{n}$ for some $i, j \in \{1, 2, \ldots, n+1\}$, $i \neq j$.

Indeed, let $S_t = \{x \in \mathbb{Z} : x \equiv t \pmod{n}\}$, $t = 1, 2, \ldots, n$. There is a $k \in \{1, 2, \ldots, n\}$ such that S_k contains at least two of the given integers, say m_i and m_j. Then $m_i \equiv m_j \pmod{n}$.

(2) (Erdős) Given $n + 1$ distinct positive integers $m_1, m_2, \ldots, m_{n+1}$ not exceeding $2n$, prove that there are two of them, m_i and m_j, such that $m_i \mid m_j$.

Indeed, for each $s \in \{1, 2, \ldots, n\}$ write $m_s = 2^{e_s} q_s$, where e_s is a nonnegative integer and q_s is an odd positive integer. Because $q_1, q_2, \ldots, q_{n+1} \in \{1, 2, \ldots, 2n\}$ and the set $\{1, 2, \ldots, 2n\}$ has exactly n odd elements, it follows that $q_i = q_j$ for some i and j. Without loss of generality, assume that $e_i < e_j$. Then $m_i \mid m_j$, as desired.

Problem 5.1.4. *Prove that among any integers $a_1, a_2, \ldots, a_n$, there are some whose sum is a multiple of n.*

Solution. Let $s_1 = a_1$, $s_2 = a_1 + a_2, \ldots, s_n = a_1 + a_2 + \cdots + a_n$. If at least one of the integers $s_1, s_2, \ldots, s_n$ is divisible by n, then we are done. If not, there are $n - 1$ possible remainders when $s_1, s_2, \ldots, s_n$ are divided by n. It follows that $s_i \equiv s_j \pmod{n}$ for some i and j, $i < j$. Then $s_j - s_i = a_{i+1} + \cdots + a_j$ is a multiple of n (see also Example (1) above).

Problem 5.1.5. *In a 10×10 table are written natural numbers not exceeding 10. Every pair of numbers that appear in adjacent or diagonally adjacent spaces of the table are relatively prime. Prove that some number appears in the table at least 17 times.*

(2001 St. Petersburg City Mathematical Olympiad)

Solution. In any 2×2 square, only one of the numbers can be divisible by 2 and only one can be divisible by 3, so if we tile the table with these 2×2 squares, at most 50 of the numbers in the table are divisible by 2 or 3. The remaining 50 numbers must be divided among the integers not divisible by 2 or 3, and thus the only ones available are 1, 5, and 7. By the pigeonhole principle, one of these numbers appears at least 17 times.

Problem 5.1.6. *Prove that from any set of 117 distinct three-digit numbers, it is possible to select 4 pairwise disjoint subsets such that the sums of the numbers in each subset are equal.*

(2001 Russian Mathematical Olympiad)

Solution. We examine subsets of exactly two numbers. Clearly, if two distinct subsets have the same sum, they must be disjoint. The number of two-element subsets is $\binom{117}{2} = 6786$. Furthermore, the lowest attainable sum is $100 + 101 = 201$, while the highest sum is $998 + 999 = 1997$, for a maximum of 1797 different sums. By the pigeonhole principle and the fact that $1797 \cdot 3 + 1 = 5392 < 6786$, we see that there are four two-element subsets with the required property.

Additional Problems

Problem 5.1.7. Let $n_1 < n_2 < \cdots < n_{2000} < 10^{100}$ be positive integers. Prove that one can find two nonempty disjoint subsets A and B of $\{n_1, n_2, \ldots, n_{2000}\}$

such that

$$|A| = |B|, \quad \sum_{x \in A} x = \sum_{x \in B} x, \quad \text{and} \quad \sum_{x \in A} x^2 = \sum_{x \in B} x^2.$$

(2001 Polish Mathematical Olympiad)

Problem 5.1.8. Find the greatest positive integer n for which there exist n nonnegative integers $x_1, x_2, \ldots, x_n$, not all zero, such that for any sequence $\varepsilon_1, \varepsilon_2, \ldots, \varepsilon_n$ of elements $\{-1, 0, 1\}$, not all zero, n^3 does not divide $\varepsilon_1 x_1 + \varepsilon_2 x_2 + \cdots + \varepsilon_n x_n$.

(1996 Romanian Mathematical Olympiad)

Problem 5.1.9. Given a positive integer n, prove that there exists $\varepsilon > 0$ such that for any n positive real numbers $a_1, a_2, \ldots, a_n$, there exists a real number $t > 0$ such that

$$\varepsilon < \{ta_1\}, \{ta_2\}, \ldots, \{ta_n\} < \tfrac{1}{2}.$$

(1998 St. Petersburg City Mathematical Olympiad)

Problem 5.1.10. We have 2^n prime numbers written on the blackboard in a line. We know that there are fewer than n different prime numbers on the blackboard. Prove that there is a subsequence of numbers in that line whose product is a perfect square.

Problem 5.1.11. Let $x_1 = x_2 = x_3 = 1$ and $x_{n+3} = x_n + x_{n+1} x_{n+2}$ for all positive integers n. Prove that for any positive integer m there is an integer $k > 0$ such that m divides x_k.

Problem 5.1.12. Prove that among seven arbitrary perfect squares there are two whose difference is divisible by 20.

(Mathematical Reflections)

5.2 Mathematical Induction

Mathematical induction is a powerful and elegant method for proving statements depending on nonnegative integers.

Let $(P(n))_{n \geq 0}$ be a sequence of propositions. The method of mathematical induction assists us in proving that $P(n)$ is true for all $n \geq n_0$, where n_0 is a given nonnegative integer.

Mathematical Induction (weak form): *Suppose that:*

- $P(n_0)$ *is true;*

- *For all $k \geq n_0$, $P(k)$ is true implies $P(k + 1)$ is true.*

Then $P(n)$ is true for all $n \geq n_0$.

Mathematical Induction (with step s): *Let s be a fixed positive integer. Suppose that:*

- *$P(n_0), P(n_0 + 1), \ldots, P(n_0 + s - 1)$ are true;*

- *For all $k \geq n_0$, $P(k)$ is true implies $P(k + s)$ is true.*

Then $P(n)$ is true for all $n \geq n_0$.

Mathematical Induction (strong form): *Suppose that*

- *$P(n_0)$ is true;*

- *For all $k \geq n_0$, $P(m)$ is true for all m with $n_0 \leq m \leq k$ implies $P(k + 1)$ is true.*

Then $P(n)$ is true for all $n \geq n_0$.

This method of proof is widely used in various areas of mathematics, including number theory.

Problem 5.2.1. *Prove that for any integer $n \geq 2$, there exist positive integers $a_1, a_2, \ldots, a_n$ such that $a_j - a_i$ divides $a_i + a_j$ for $1 \leq i < j \leq n$.*

(Kvant)

Solution. We will prove the statement by induction on the number of terms n. For $n = 2$, we can choose $a_1 = 1$ and $a_2 = 2$.

We assume that we can find integers $a_1, a_2, \ldots, a_n$ such that $a_j - a_i$ divides $a_i + a_j$ for $1 \leq i < j \leq n$, where n is a positive integer greater than 1. Let m be the least common multiple of numbers $a_1, a_2, \ldots, a_n, a_j - a_i$, for all $1 \leq i < j \leq n$. Then

$$(a_1', a_2', a_3', \ldots, a_{n+1}') = (m, m + a_1, m + a_2, \ldots, m + a_n)$$

is an $(n + 1)$-term sequence satisfying the conditions of the problem. Indeed, $a_i' - a_1' = a_{i-1}$ divides m and $a_i' + a_1' = 2m + a_{i-1}$ by the definition of m, and $a_j' - a_i' = a_{j-1} - a_{i-1}$ ($2 \leq i < j \leq n + 1$) divides m. Also, $a_j' + a_i' = 2m + (a_{j-1} + a_{i-1})$ by the definition of m and by the inductive hypothesis. Therefore our induction is complete.

Problem 5.2.2. *Prove that for each $n \geq 3$, the number $n!$ can be represented as the sum of n distinct divisors of itself.*

(Erdős)

Solution. The base case is $3! = 6 = 1 + 2 + 3$. Strengthening the statement by imposing the condition that one of the n divisors should be 1 puts us in a winning position. The question here is how we came to think of this. Indeed, there is just about one way to go in using the induction hypothesis $n! = d_1 + d_2 + \cdots + d_n$

(where $d_1, d_2, \ldots, d_n$ are the n divisors arranged in increasing order), namely, multiplying the above relation by $n + 1$. This yields

$$(n+1)! = (n+1)d_1 + (n+1)d_2 + \cdots + (n+1)d_n$$
$$= d_1 + nd_1 + (n+1)d_2 + \cdots + (n+1)d_n.$$

We split $(n+1)d_1$ into $d_1 + nd_1$, thus getting $n+1$ summands, as needed. Of them, only the second one might not be a divisor of $(n+1)!$. We would like to ensure that it is such a divisor, too. Hence the idea of insisting that $d_1 = 1$.

Problem 5.2.3. *Prove that there are infinitely many numbers not containing the digit 0 that are divisible by the sum of their digits.*

Solution. Let us prove by induction that $\underbrace{11 \ldots 1}_{3^n}$ is a good choice. The base case

is easily verified, and for the inductive step we have

$$\underbrace{11 \ldots 1}_{3^{n+1}} = \frac{10^3 - 1}{9} = \frac{(10^{3^n})^3 - 1}{9}$$

$$= \frac{10^{3^{n+1}} - 1}{9}(10^{2\cdot 3^n} + 10^{3^n} + 1)$$

$$= \underbrace{11 \ldots 1}_{3^n} \cdot N,$$

where N is a multiple of 3, and the conclusion follows.

Problem 5.2.4. *Let n be a positive integer. Let O_n be the number of $2n$-tuples $(x_1, \ldots, x_n, y_1, \ldots, y_n)$ with values in 0 or 1 for which the sum $x_1 y_1 + \cdots + x_n y_n$ is odd, and let E_n be the number of $2n$-tuples for which the sum is even. Prove that*

$$\frac{O_n}{E_n} = \frac{2^n - 1}{2^n + 1}.$$

(1997 Iberoamerican Mathematical Olympiad)

Solution. We prove by induction that $O_n = 2^{2n-1} - 2^{n-1}$ and $E_n = 2^{2n-1} + 2^{n-1}$, which will give the desired ratio.

The base case is $n = 1$. This case works because $O_1 = 1 = 2^1 - 2^0$ and $E_1 = 3 = 2^1 + 2^0$.

For the inductive step, we assume that this is true for $n = k$; then $x_1 y_1 + \cdots + x_k y_k$ is even for $(2^{2k-1} + 2^{k-1})$ $2k$-tuples and odd for $(2^{2k-1} - 2^{k-1})$ $2k$-tuples. Now, $x_1 y_1 + \cdots + x_{k+1} y_{k+1}$ is odd if and only if either $x_1 y_1 + \cdots + x_k y_k$ is odd and $x_{k+1}y_{k+1}$ is even or $x_1y_1 + \cdots + x_k y_k$ is even and $x_{k+1}y_{k+1}$ is odd. Clearly $x_{k+1}y_{k+1}$ can be odd in one way and even in three ways, so

$$O_{k+1} = 3(2^{2k-1} - 2^{k-1}) + 2^{2k-1} + 2^{k-1} = 2^{2(k+1)-1} - 2^{(k+1)-1}$$

and $E_{k+1} = 2^{2(k+1)} - O_{k+1}$, which completes the induction.

Problem 5.2.5. *Prove that for all integers $n \geq 3$, there exist odd positive integers x, y such that $7x^2 + y^2 = 2^n$.*

(1996 Bulgarian Mathematical Olympiad)

Solution. We will prove that there exist odd positive integers x_n, y_n such that $7x_n^2 + y_n^2 = 2^n$, $n \geq 3$.

For $n = 3$, we have $x_3 = y_3 = 1$. Now suppose that for a given integer $n \geq 3$ we have odd integers x_n, y_n satisfying $7x_n^2 + y_n^2 = 2^n$. We shall exhibit a pair (x_{n+1}, y_{n+1}) of odd positive integers such that $7x_{n+1}^2 + y_{n+1}^2 = 2^{n+1}$. In fact,

$$7\left(\frac{x_n \pm y_n}{2}\right)^2 + \left(\frac{7x_n \mp y_n}{2}\right)^2 = 2(7x_n^2 + y_n^2) = 2^{n+1}.$$

Precisely one of the numbers $\frac{x_n+y_n}{2}$ and $\frac{|x_n-y_n|}{2}$ is odd (since their sum is the larger of x_n and y_n, which is odd). If, for example, $\frac{x_n+y_n}{2}$ is odd, then

$$\frac{7x_n - y_n}{2} = 3x_n + \frac{x_n - y_n}{2}$$

is also odd (as the sum of an odd and an even number); hence in this case we may choose

$$x_{n+1} = \frac{x_n + y_n}{2} \quad \text{and} \quad y_{n+1} = \frac{7x_n - y_n}{2}.$$

If $\frac{x_n-y_n}{2}$ is odd, then

$$\frac{7x_n + y_n}{2} = 3x_n + \frac{x_n + y_n}{2},$$

so we can choose

$$x_{n+1} = \frac{|x_n - y_n|}{2} \quad \text{and} \quad y_{n+1} = \frac{7x_n + y_n}{2}.$$

Remark. This problem goes back to Euler.

Problem 5.2.6. *Let $f(x) = x^3 + 17$. Prove that for each natural number n, $n \geq 2$, there is a natural number x for which $f(x)$ is divisible by 3^n but not by 3^{n+1}.*

(1999 Japanese Mathematical Olympiad)

Solution. We prove the result by induction on n. If $n = 2$, then $x = 1$ suffices. Now suppose that the claim is true for some $n \geq 2$, that is, there is a natural number y such that $y^3 + 17$ is divisible by 3^n but not 3^{n+1}. We prove that the claim is true for $n + 1$.

Suppose we have integers a, m such that a is not divisible by 3 and $m \geq 2$. Then $a^2 \equiv 1 \pmod 3$ and thus $3^m a^2 \equiv 3^m \pmod{3^{m+1}}$. Also, because $m \geq 2$ we have $3m - 3 \geq 2m - 1 \geq m + 1$. Hence

$$(a + 3^{m-1})^3 \equiv a^3 + 3^m a^2 + 3^{2m-1} a + 3^{3m-3} \equiv a^3 + 3^m \pmod{3^{m+1}}.$$

Because $y^3 + 17$ is divisible by 3^n, it is congruent to either 0, 3^n, or $2 \cdot 3^n$ modulo 3^{n+1}. Because 3 does not divide 17, 3 cannot divide y either. Hence applying our result from the previous paragraph twice, once with $(a, m) = (y, n)$ and once with $(a, m) = (y + 3^{n-1}, n)$, we find that 3^{n+1} must divide either $(y+3^{n-1})^3+17$ or $(y + 2 \cdot 3^{n-1})^3 + 17$.

Hence there exists a natural number x' not divisible by 3 such that $3^{n+1} \mid x'^3 + 17$. If 3^{n+2} does not divide $x'^3 + 17$, we are done. Otherwise, we claim that the number $x = x' + 3^n$ suffices. Because $x = x' + 3^{n-1} + 3^{n-1} + 3^{n-1}$, the result from the previous paragraphs tells us that $x^3 \equiv x'^3 + 3^n + 3^n + 3^n \equiv x'^3 \pmod{3^{n+1}}$. Thus $3^{n+1} \mid x^3 + 17$ as well. On the other hand, because $x = x' + 3^n$, we have $x^3 \equiv x'^3 + 3^{n+1} \not\equiv x'^3 \pmod{3^{n+2}}$. It follows that 3^{n+2} does not divide $x^3 + 17$, as desired. This completes the inductive step.

Additional Problems

Problem 5.2.7. Let p be an odd prime. The sequence $(a_n)_{n \geq 0}$ is defined as follows: $a_0 = 0$, $a_1 = 1, \ldots, a_{p-2} = p - 2$, and for all $n \geq p - 1$, a_n is the least positive integer that does not form an arithmetic sequence of length p with any of the preceding terms. Prove that, for all n, a_n is the number obtained by writing n in base $p - 1$ and reading the result in base p.

(1995 USA Mathematical Olympiad)

Problem 5.2.8. Suppose that x, y, and z are natural numbers such that $xy = z^2 + 1$. Prove that there exist integers a, b, c, and d such that $x = a^2 + b^2$, $y = c^2 + d^2$, and $z = ac + bd$.

(Euler's problem)

Problem 5.2.9. Find all pairs of sets A, B that satisfy the following conditions:
  (i) $A \cup B = \mathbb{Z}$;
  (ii) if $x \in A$, then $x - 1 \in B$;
  (iii) if $x \in B$ and $y \in B$, then $x + y \in A$.

(2002 Romanian International Mathematical Olympiad Team Selection Test)

Problem 5.2.10. Find all positive integers n such that

$$n = \prod_{k=0}^{m}(a_k + 1),$$

where $\overline{a_m a_{m-1} \cdots a_0}$ is the decimal representation of n.

(2001 Japanese Mathematical Olympiad)

Problem 5.2.11. The sequence $(u_n)_{n\geq 0}$ is defined as follows: $u_0 = 2$, $u_1 = \frac{5}{2}$ and

$$u_{n+1} = u_n(u_{n-1}^2 - 2) - u_1 \text{ for } n = 1, 2, \ldots.$$

Prove that $[u_n] = 2^{\frac{2^n - (-1)^n}{3}}$, for all $n > 0$ ($\lfloor x \rfloor$ denotes the integer part of x).

(18th International Mathematical Olympiad)

5.3   Infinite Descent

Fermat[1] was the first mathematician to use a method of proof called *infinite descent*.

Let P be a property concerning the nonnegative integers and let $(P(n))_{n\geq 1}$ be the sequence of propositions

$$P(n): \text{``}n \text{ satisfies property } P.\text{''}$$

The following method is useful in proving that proposition $P(n)$ is false for all large enough n.

Let k be a nonnegative integer. Suppose that:

- *$P(k)$ is not true;*

- *if $P(m)$ is true for a positive integer $m > k$, then there is some smaller j, $m > j \geq k$, for which $P(j)$ is true.*

Then $P(n)$ is false for all $n \geq k$.

This is just the contrapositive of strong induction, applied to the negation of proposition $P(n)$. In the language of the ladder metaphor, if you know you cannot reach any rung without first reaching a lower rung, and you also know you cannot reach the bottom rung, then you cannot reach any of the rungs.

The above is often called the *finite descent method*.

Fermat's method of infinite descent (FMID) can be formulated as follows:

Let k be an integer. Suppose that:

---

[1] Pierre de Fermat (1601–1665), French lawyer and government official most remembered for his work in number theory, in particular for Fermat's last theorem. He is also important in the foundations of calculus, optics, and geometry.

- *if $P(m)$ is true for an integer $m > k$, then there must be some smaller integer j, $m > j > k$ for which $P(j)$ is true.*

Then $P(n)$ is false for all $n > k$.

That is, if there were an n for which $P(n)$ were true, one could construct a sequence $n > n_1 > n_2 > \cdots$ all of which would be greater than k. However, for the integers, no such sequence is possible.

Two special cases of FMID are particularly useful in solving number theory problems.

FMID Variant 1. *There is no sequence of nonnegative integers $n_1 > n_2 > \cdots$.*

In some situations it is convenient to replace FMID Variant 1 by the following equivalent form: If n_0 is the smallest integer n for which $P(n)$ is true, then $P(n)$ is false for all $n < n_0$. In fact, this is equivalent to an extremal argument.

FMID Variant 2. *If the sequence of integers $(n_i)_{i \geq 1}$ satisfies the inequalities $n_1 \geq n_2 \geq \cdots$, then there exists i_0 such that $n_{i_0} = n_{i_0+1} = \cdots$.*

Problem 5.3.1. *Find all triples (x, y, z) of nonnegative integers such that*

$$x^3 + 2y^3 = 4z^3.$$

Solution. Note that $(0, 0, 0)$ is such a triple. We will prove that there is no other. Assume that (x_1, y_1, z_1) is a nontrivial solution to the given equation. Because $\sqrt[3]{2}$, $\sqrt[3]{4}$ are both irrational, it is not difficult to see that $x_1 > 0$, $y_1 > 0$, $z_1 > 0$.

From $x_1^3 + 2y_1^3 = 4z_1^3$ it follows that $2 \mid x_1$, so $x_1 = 2x_2$, $x_2 \in \mathbb{Z}_+$. Then $4x_2^3 + y_1^3 = 2z_1^3$; hence $y_1 = 2y_2$, $y_2 \in \mathbb{Z}_+$. Similarly, $z_1 = 2z_2$, $z_2 \in \mathbb{Z}_+$. We obtain the "new" solution (x_2, y_2, z_2) with $x_1 > x_2$, $y_1 > y_2$, $z_1 > z_2$. Continuing this procedure, we construct a sequence of positive integral triples $(x_n, y_n, z_n)_{n \geq 1}$ such that $x_1 > x_2 > x_3 > \cdots$. But this contradicts FMID Variant 1.

Additional Problems

Problem 5.3.2. Find all primes p for which there exist positive integers x, y, and n such that $p^n = x^3 + y^3$.

(2000 Hungarian Mathematical Olympiad)

5.4 Inclusion–Exclusion

The main result in this section is contained in the following theorem.

Theorem 5.4.1. *Let $S_1, S_2, \ldots, S_n$ be finite sets. Then*

$$\left| \bigcup_{i=1}^{n} S_i \right| = \sum_{i=1}^{n} |S_i| - \sum_{1 \le i < j \le n} |S_i \cap S_j|$$

$$+ \sum_{1 \le i < j < k \le n} |S_i \cap S_j \cap S_k| - \cdots + (-1)^{n-1} \left| \bigcap_{i=1}^{n} S_i \right|,$$

where $|S|$ denotes the number of elements in S and $n \ge 2$.

Proof. We proceed by induction. For $n = 2$, we have to prove that $|S_1 \cup S_2| = |S_1| + |S_2| - |S_1 \cap S_2|$. This is clear because the number of elements in $S_1 \cup S_2$ is the number of elements in S_1 and S_2 less the ones in $S_1 \cap S_2$, since the latter elements were counted twice.

The inductive step uses the formula above for $S_1 \rightarrow \bigcup_{i=1}^{k} S_k$ and $S_2 \rightarrow S_{k+1}$.   □

The formula in the theorem is called the *inclusion–exclusion principle.*

Example. *How many positive integers not exceeding 1000 are divisible by 2, or 3, or 5?*

Solution. Consider the sets

$$S_1 = \{2m : 1 \le m \le 500\}, \quad S_2 = \{3n : 1 \le n \le 333\},$$
$$S_3 = \{5p : 1 \le p \le 200\}.$$

Then

$$S_1 \cap S_2 = \{6q : 1 \le q \le 166\}, \quad S_1 \cap S_3 = \{10r : 1 \le r \le 100\},$$
$$S_2 \cap S_3 = \{15s : 1 \le s \le 66\}, \quad \text{and} \quad S_1 \cap S_2 \cap S_3 = \{30u : 1 \le u \le 33\}.$$

Applying the inclusion–exclusion principle, we obtain

$$|S_1 \cup S_2 \cup S_3| = |S_1| + |S_2| + |S_3| - |S_1 \cap S_2| - |S_1 \cap S_3|$$

$$- |S_2 \cap S_3| + |S_1 \cap S_2 \cap S_3|$$

$$= 500 + 333 + 200 - 166 - 100 - 66 + 33 = 734.$$

The dual version of Theorem 5.4.1 is the following:

Theorem 5.4.2. *Let $S_1, S_2, \ldots, S_n$ be subsets of the finite set S and let $\overline{S}_i = S - S_i$ be the complementary set of S_i, $i = 1, 2, \ldots, n$. Then*

$$\left| \bigcap_{i=1}^{n} \overline{S}_i \right| = |S| - \sum_{i=1}^{n} |S_i| + \sum_{1 \le i < j \le n} |S_i \cap S_j|$$

$$- \sum_{1 \le i < j < k \le n} |S_i \cap S_j \cap S_k| + \cdots + (-1)^{n} \left| \bigcap_{i=1}^{n} S_i \right|.$$

Proof. Let

$$A = \bigcap_{i=1}^{n} \overline{S_i} \quad \text{and} \quad B = \bigcup_{i=1}^{n} S_i.$$

It is clear that $A \cup B = S$ and $A \cap B = \emptyset$. Hence $|S| = |A| + |B|$ and the conclusion follows from Theorem 5.4.1. $\qquad\square$

Example. *How many positive integers not exceeding* 120 *are not divisible by 2, 3, and 5?*

Solution. Consider the sets

$$S_1 = \{2m \mid 1 \le m \le 60\}, \quad S_2 = \{3n \mid 1 \le n \le 40\}, \quad S_3 = \{5p \mid 1 \le p \le 24\}.$$

We have

$$S_1 \cap S_2 = \{6q \mid 1 \le q \le 20\}, \quad S_1 \cap S_3 = \{10r \mid 1 \le r \le 12\},$$
$$S_2 \cap S_3 = \{15s \mid 1 \le s \le 8\}, \text{ and } S_1 \cap S_2 \cap S_3 = \{30u \mid 1 \le u \le 4\}.$$

Applying the formula in Theorem 5.3.2, we get

$$|\overline{S_1} \cap \overline{S_2} \cap \overline{S_3}| = 120 - (|S_1| + |S_2| + |S_3|) + |S_1 \cap S_2| + |S_1 \cap S_3|$$
$$+ |S_2 \cap S_3| - |S_1 \cap S_2 \cap S_3|$$
$$= 120 - (60 + 40 + 24) + 20 + 12 + 8 - 4 = 32.$$

Problem 5.4.1. *Let* $S = \{1, 2, 3, \ldots, 280\}$. *Find the smallest integer n such that each n-element subset of S contains five numbers that are pairwise relatively prime.*

(32nd International Mathematical Olympiad)

Solution. The solution is given in two steps.
First step. Consider the sets

$$M_2 = \{2, 4, 6, \ldots, 280\}, \quad M_3 = \{3, 6, 9, \ldots, 279\},$$
$$M_5 = \{5, 10, 15, \ldots, 280\}, \quad M_7 = \{7, 14, \ldots, 280\},$$

and let $M = M_2 \cup M_3 \cup M_5 \cup M_7$. The following cardinalities are obvious:

$$|M_2| = 140, \quad |M_3| = 93, \quad |M_5| = 56, \quad \text{and} \quad |M_7| = 40.$$

It is easy to prove that

$$|M_2 \cap M_3| = \left\lfloor \frac{280}{6} \right\rfloor = 46, \qquad |M_2 \cap M_5| = \left\lfloor \frac{280}{10} \right\rfloor = 28,$$

$$|M_2 \cap M_7| = \left\lfloor \frac{280}{14} \right\rfloor = 20, \qquad |M_3 \cap M_5| = \left\lfloor \frac{280}{15} \right\rfloor = 18,$$

$$|M_3 \cap M_7| = \left\lfloor \frac{280}{21} \right\rfloor = 13, \qquad |M_5 \cap M_7| = \left\lfloor \frac{280}{35} \right\rfloor = 8,$$

$$|M_2 \cap M_3 \cap M_5| = \left\lfloor \frac{280}{30} \right\rfloor = 9, \qquad |M_2 \cap M_3 \cap M_7| = \left\lfloor \frac{280}{42} \right\rfloor = 6,$$

$$|M_2 \cap M_5 \cap M_7| = \left\lfloor \frac{280}{70} \right\rfloor = 4, \qquad |M_4 \cap M_5 \cap M_7| = \left\lfloor \frac{28}{105} \right\rfloor = 2,$$

and

$$|M_2 \cap M_3 \cap M_5 \cap M_7| = \left\lfloor \frac{280}{210} \right\rfloor = 1.$$

By the principle of inclusion–exclusion, we obtain

$$\begin{aligned}
|M| &= |M_2 \cup M_3 \cup M_5 \cup M_7| \\
&= 140 + 93 + 56 + 40 - (46 + 28 + 20 + 18 + 13 + 8) \\
&\quad + (9 + 6 + 4 + 2) - 1 \\
&= 216.
\end{aligned}$$

By the pigeonhole principle, any five-element subset of M contains at least two elements from the same subset M_i, $i \in \{2, 3, 5, 7\}$. These elements are not relatively prime numbers. Thus, we have proved that $n > 216$.

Second step. We will prove that $n = 217$.

The set $S \setminus M$ contains $280 - 216 = 64$ elements. It contains prime numbers and composite numbers. Taking into account that $\lfloor \sqrt{280} \rfloor = 16$, we see that any composite numbers in $S \setminus M$ have one prime factor at most 16. Thus they are precisely

$$C = \{11^2; 11 \cdot 13; 11 \cdot 17; 11 \cdot 19; 11 \cdot 23; 13^2; 13 \cdot 17; 13 \cdot 19\}.$$

Observe that $|C| = 8$. The set $S \setminus M$ contains the number 1, 8 composite numbers, and 55 prime numbers. Also, taking into account the prime numbers 2, 3, 5, 7, we infer that the set S contains 59 prime numbers in all.

Let $p_1 = 2$, $p_2 = 3$, $p_3 = 5, \ldots, p_{59}$ be all these prime numbers and let $P = \{1, p_2, p_2, \ldots, p_{59}\}$. Thus, $|P| = 60$.

Let T be a subset containing 217 elements of S. If $|T \cap P| \geq 5$, it follows that T contains 5 elements that are relatively prime numbers. So, suppose $|T \cap P| \leq 4$. In this case, $|T \cap (S \setminus P)| \geq 217 - 4 = 213$. Since S contains 220 composite numbers, it follows that at most 7 composite numbers are not in T.

Consider the following five-element subsets of $S \setminus P$:

$$A_1 = \{2^2; 3^2; 5^2; 7^2; 13^2\},$$
$$A_2 = \{2 \cdot 23; 3 \cdot 19; 5 \cdot 17; 7 \cdot 13; 11 \cdot 11\},$$
$$A_3 = \{2 \cdot 29; 3 \cdot 23; 5 \cdot 19; 7 \cdot 17; 11 \cdot 13\},$$
$$A_4 = \{2 \cdot 31; 3 \cdot 29; 5 \cdot 23; 7 \cdot 19; 11 \cdot 17\},$$
$$A_5 = \{2 \cdot 37; 3 \cdot 31; 5 \cdot 29; 7 \cdot 23; 11 \cdot 19\},$$
$$A_6 = \{2 \cdot 41; 3 \cdot 37; 5 \cdot 31; 7 \cdot 29; 11 \cdot 23\},$$
$$A_7 = \{2 \cdot 43; 3 \cdot 41; 5 \cdot 37; 7 \cdot 23; 13 \cdot 17\},$$
$$A_8 = \{2 \cdot 47; 3 \cdot 43; 5 \cdot 41; 7 \cdot 37; 13 \cdot 19\}.$$

By the pigeonhole principle, there exists a set A_i, $1 \le i \le 8$, such that $A_i \subset T$; if not, the set $S \setminus T$ would contain eight composite numbers. Each A_i contains five relatively prime numbers and we are done.

Additional Problems

Problem 5.4.2. The numbers from 1 to $1{,}000{,}000$ can be colored black or white. A permissible move consists in selecting a number from 1 to $1{,}000{,}000$ and changing the color of that number and each number not relatively prime to it. Initially all of the numbers are black. Is it possible to make a sequence of moves after which all of the numbers are colored white?

(1999 Russian Mathematical Olympiad)

6

Arithmetic Functions

6.1 Multiplicative Functions

Arithmetic functions are defined on the positive integers and are complex valued. The arithmetic function $f \neq 0$ is called *multiplicative* if for every pair of relatively prime positive integers m and n,

$$f(mn) = f(m)f(n).$$

An arithmetic function $f \neq 0$ is called *completely multiplicative* if the relation above holds for all positive integers m and n.

Remarks. (1) If $f : \mathbb{Z}_+^* \to \mathbb{C}$ is multiplicative, then $f(1) = 1$. Indeed, if a is a positive integer for which $f(a) \neq 0$, then $f(a) = f(a \cdot 1) = f(a)f(1)$ and dividing by $f(a)$ yields $f(1) = 1$.

(2) If f is multiplicative and $n = p_1^{\alpha_1} \cdots p_k^{\alpha_k}$ is the prime factorization of the positive integer n, then $f(n) = f(p_1^{\alpha_1}) \cdots f(p_k^{\alpha_k})$; that is, in order to compute $f(n)$ it suffices to compute $f(p_i^{\alpha_i})$, $i = 1, \ldots, k$.

(3) If f is completely multiplicative and $n = p_1^{\alpha_1} \cdots p_k^{\alpha_k}$ is the prime factorization of n, then $f(n) = f(p_1)^{\alpha_1} \cdots f(p_k)^{\alpha_k}$; that is, in order to compute $f(n)$ it suffices to compute $f(p_i)$, $i = 1, \ldots, k$.

An important arithmetic function is the *Möbius*[1] *function* defined by

$$\mu(n) = \begin{cases} 1 & \text{if } n = 1, \\ 0 & \text{if } p^2 \mid n \text{ for some prime } p, \\ (-1)^k & \text{if } n = p_1 \cdots p_k, \text{ where } p_1, \ldots, p_k \text{ are distinct primes.} \end{cases}$$

For example, $\mu(2) = -1$, $\mu(6) = 1$, $\mu(12) = \mu(2^2 \cdot 3) = 0$.

---

[1] August Ferdinand Möbius (1790–1868), German mathematician best known for his work in topology, especially for his conception of the Möbius strip, a two-dimensional surface with only one side.

T. Andreescu and D. Andrica, *Number Theory*, DOI: 10.1007/b11856_6,
© Birkhäuser Boston, a part of Springer Science + Business Media, LLC 2009

Theorem 6.1.1. *The Möbius function μ is multiplicative.*

Proof. Let m, n be positive integers such that $\gcd(m, n) = 1$. If $p^2 \mid m$ for some $p > 1$, then $p^2 \mid mn$ and so $\mu(m) = \mu(mn) = 0$ and we are done. Consider now $m = p_1 \cdots p_k$, $n = q_1 \cdots q_h$, where $p_1, \ldots, p_k, q_1, \ldots, q_h$ are distinct primes. Then $\mu(m) = (-1)^k$, $\mu(n) = (-1)^h$, and $mn = p_1 \cdots p_k q_1 \cdots q_h$. It follows that $\mu(mn) = (-1)^{k+h} = (-1)^k(-1)^h = \mu(m)\mu(n)$. $\qquad \square$

For an arithmetic function f we define its *summation function F* by

$$F(n) = \sum_{d \mid n} f(d).$$

The connection between f and F is given by the following result.

Theorem 6.1.2. *If f is multiplicative, then so is its summation function F.*

Proof. Let m, n be positive integers such that $\gcd(m, n) = 1$ and let d be a divisor of mn. Then d can be uniquely represented as $d = kh$, where $k \mid m$ and $h \mid n$. Because $\gcd(m, n) = 1$, we have $\gcd(k, h) = 1$, so $f(kh) = f(k)f(h)$. Hence

$$F(mn) = \sum_{d \mid mn} f(d) = \sum_{\substack{k \mid m \\ h \mid n}} f(k)f(h)$$

$$= \left(\sum_{k \mid m} f(k) \right) \left(\sum_{h \mid n} f(h) \right) = F(m)F(n). \qquad \square$$

Remark. If f is a multiplicative function and $n = p_1^{\alpha_1} \cdots p_k^{\alpha_k}$, then

$$F(n) = \prod_{i=1}^{k} \left(1 + f(p_i) + \cdots + f(p_i^{\alpha_i}) \right). \tag{1}$$

Indeed, after multiplication on the right-hand side we get a sum having terms of the form $f(p_1^{\beta_1}) \cdots f(p_k^{\beta_k}) = f(p_1^{\beta_1} \cdots p_k^{\beta_k})$, where $0 \leq \beta_1 \leq \alpha_1, \ldots, 0 \leq \beta_k \leq \alpha_k$. This sum is obviously $F(n)$.

The function $g(n) = \mu(n)f(n)$ is multiplicative; hence applying (1), we get, for its summation function G,

$$G(n) = \prod_{i=1}^{k} \left(1 + \mu(p_i)f(p_i) \right) = \prod_{i=1}^{k} \left(1 - f(p_i) \right).$$

From (1) we also can derive the following formula:

$$\sum_{d \mid n} \mu(d)f(d) = (1 - f(p_1)) \cdots (1 - f(p_k)). \tag{2}$$

If we take $f = 1$ in formula (2), then we get the following basic property of the Möbius function: For any integer $n \geq 2$,

$$\sum_{d|n} \mu(d) = 0.$$

Theorem 6.1.3. (Möbius inversion formula) *Let f be an arithmetic function and let F be its summation function. Then*

$$f(n) = \sum_{d|n} \mu(d) F\left(\frac{n}{d}\right).$$ (3)

Proof. We have

$$\sum_{d|n} \mu(d) F\left(\frac{n}{d}\right) = \sum_{d|n} \mu(d) \left(\sum_{c|\frac{n}{d}} f(c)\right) = \sum_{d|n} \left(\sum_{c|\frac{n}{d}} \mu(d) f(c)\right)$$

$$= \sum_{c|n} \left(\sum_{d|\frac{n}{c}} \mu(d) f(c)\right) = \sum_{c|n} f(c) \left(\sum_{d|\frac{n}{c}} \mu(d)\right) = f(n),$$

since for $\frac{n}{c} > 1$ we have $\sum_{d|\frac{n}{c}} \mu(d) = 0$.

We have used the fact that the sets

$$\left\{(d, c) \mid d \mid n \text{ and } c \mid \frac{n}{d}\right\} \quad \text{and} \quad \left\{(d, c) \mid c \mid n \text{ and } d \mid \frac{n}{c}\right\}$$

are equal. □

They are both equal to $\{(c, d) \mid cd \mid n\}$.

Theorem 6.1.4. *Let f be an arithmetic function and let F be its summation function. If F is multiplicative, then so is f.*

Proof. Let m, n be positive integers such that $\gcd(m, n) = 1$ and let d be a divisor of mn. Then $d = kh$, where $k \mid m$, $h \mid n$, and $\gcd(k, h) = 1$. Applying the Möbius inversion formula, it follows that

$$f(mn) = \sum_{d|mn} \mu(d) F\left(\frac{mn}{d}\right) = \sum_{\substack{k|m \\ h|n}} \mu(kh) F\left(\frac{mn}{kh}\right)$$

$$= \sum_{\substack{k|m \\ h|n}} \mu(k) \mu(h) F\left(\frac{m}{k}\right) F\left(\frac{n}{h}\right)$$

$$= \left(\sum_{k|m} \mu(k) F\left(\frac{m}{k}\right)\right) \left(\sum_{h|n} \mu(h) F\left(\frac{n}{h}\right)\right)$$

$$= f(m) f(n).$$ □

Let f and g be two arithmetic functions. Define their *convolution product* or *Dirichlet[2] product* $f * g$ by

$$(f * g)(n) = \sum_{d|n} f(d)g\left(\frac{n}{d}\right).$$

Note that the convolution product can be written more symmetrically as

$$(f * g)(n) = \sum_{ab=n} f(a)g(b).$$

The following relation holds: $1 * f = F$, the summation function of f.

Problem 6.1.1. *(1) Prove that the convolution product is commutative and associative.*

(2) Prove that for any arithmetic function f,

$$f * \varepsilon = \varepsilon * f = f,$$

where $\varepsilon(n) = 1$ if $n = 1$ and 0 otherwise.

Solution. Let f and g be two arithmetic functions. Then

$$(f * g)(n) = \sum_{d|n} f(d)g\left(\frac{n}{d}\right) = \sum_{d_1|n} f\left(\frac{n}{d_1}\right)g(d_1) = (g * f)(n),$$

since if d runs through all divisors of, then so does $d_1 = \frac{n}{d}$. Therefore $f * g = g * f$.

Let f, g, h be arithmetic functions. To prove the associativity law, let $u = g * h$ and consider $f * u = f * (g * h)$. We have

$$(f * u)(n) = \sum_{a|n} f(a)u\left(\frac{n}{a}\right) = \sum_{ad=n} f(a) \sum_{bc=d} g(b)h(c)$$
$$= \sum_{abc=n} f(a)g(b)h(c).$$

Similarly, if we set $v = f * g$ and consider $v * h$, we have

$$(v * h)(n) = \sum_{dc=n} v(d)h(c) = \sum_{dc=n} \sum_{ab=d} f(a)g(b)h(c)$$
$$= \sum_{abc=n} f(a)g(b)h(c);$$

---

[2] Johann Peter Gustav Lejeune Dirichlet (1805–1859), German mathematician who proved in 1837 that there are infinitely many primes in any arithmetic progression of integers for which the common difference is relatively prime to the terms. Dirichlet made essential contributions in number theory, probability theory, functional analysis, and Fourier series.

hence $f * (g * h) = (f * g) * h$.

(2) We have

$$(\varepsilon * f)(n) = \sum_{d|n} \varepsilon(d) f\left(\frac{n}{d}\right) = f(n),$$

and we get $\varepsilon * f = f * \varepsilon = f$.

Problem 6.1.2. *Let f be an arithmetic function. If $f(1) \neq 0$, then there is a unique arithmetic function g such that*

$$f * g = \varepsilon.$$

Solution. We show by induction on n that $(f * g)(n) = \varepsilon(n)$ has a unique solution $g(1), \ldots, g(n)$.

For $n = 1$, we have $f(1)g(1) = 1$; hence $g(1) = 1/f(1)$.

Suppose $n > 1$ and assume that $g(1), \ldots, g(n-1)$ have been uniquely determined such that $(f * g)(k) = \varepsilon(k)$ holds for $k = 1, 2, \ldots, n-1$. Then

$$f(1)g(n) + \sum_{\substack{d|n \\ d>1}} f(d)g\left(\frac{n}{d}\right) = 0,$$

and we get

$$g(n) = -\frac{1}{f(1)} \sum_{\substack{d|n \\ d>1}} f(d)g\left(\frac{n}{d}\right)$$

i.e., the function g exists and is unique.

Remark. The unique function g satisfying $f * g = \varepsilon$, where $f(1) \neq 0$, is called the *convolution inverse* of f. It is not difficult to show that μ is the convolution inverse to the constant function 1.

Problem 6.1.3. *If f and g are multiplicative, so is their convolution product.*

Solution. Let $h = f * g$. We have

$$h(mn) = \sum_{c|mn} f(c)g\left(\frac{mn}{c}\right).$$

Set $c = ab$, where $a \mid m$ and $b \mid n$. Since $\gcd(m, n) = 1$, we can write c uniquely in this way. Hence we have

$$h(mn) = \sum_{a|m} \sum_{b|n} f(ab)g\left(\frac{m}{a}\frac{n}{b}\right)$$

$$= \left(\sum_{a|m} f(a)g\left(\frac{m}{a}\right)\right)\left(\sum_{b|n} f(b)g\left(\frac{n}{b}\right)\right) = h(m)h(n).$$

Problem 6.1.4. *(1) If both g and f * g are multiplicative, then f is also multiplicative.*

(2) If g is multiplicative, then so is its convolution inverse.

Solution. (1) Suppose f is not multiplicative. Let $h = f * g$. Since f is not multiplicative, there exist m and n, $\gcd(m, n) = 1$, such that $f(mn) \neq f(m)f(n)$. We choose mn as small as possible. If $mn = 1$, then we get $f(1) \neq f(1)f(1)$, so $f(1) \neq 1$. Since $h(1) = f(1)g(1) = f(1) \neq 1$, h is not multiplicative, a contradiction. If $mn > 1$, we have $f(ab) = f(a)f(b)$ for all $ab < mn$ with $\gcd(a, b) = 1$. Now

$$h(mn) = f(mn)g(1) + \sum_{\substack{a|m \\ b|n}} f(ab)g\left(\frac{mn}{ab}\right)$$

$$= f(mn) + \sum_{\substack{a|m \\ b|n \\ ab<mn}} f(a)f(b)g\left(\frac{m}{a}\right)g\left(\frac{n}{b}\right)$$

$$= f(mn) - f(m)f(n) + h(m)h(n).$$

Since $f(mn) \neq f(m)f(n)$, we have $h(mn) \neq h(m)h(n)$. Therefore, h is not multiplicative, a contradiction.

(2) Denote by g^{-1} the convolution inverse of g. Then $\varepsilon = g * g^{-1} = g^{-1} * g$ and g are both multiplicative. From the previous result it follows that g^{-1} is multiplicative.

Problem 6.1.5. *Let f be an arithmetic function that is not identically zero. Prove that it is completely multiplicative if and only if f * f = fτ, where τ(n) is the number of divisors of n.*

(American Mathematical Monthly)

Solution. If f is completely multiplicative, we have

$$(f * f)(n) = \sum_{d|n} f(d)f\left(\frac{n}{d}\right) = \sum_{d|n} f\left(d\frac{n}{d}\right) = \sum_{d|n} f(n)$$

$$= f(n)\sum_{d|n} 1 = f(n)\tau(n) = (f\tau)(n),$$

and the relation follows.

Conversely, take $n = 1$. We get $f^2(1) = f(1)\tau(1) = f(1)$. It follows that $f(1) = 0$ or $f(1) = 1$. Now suppose that $n \geq 2$ and let $n = p_1^{\alpha_1} \cdots p_k^{\alpha_k}$ be the prime factorization of n. Put $\alpha(n) = \alpha_1 + \cdots + \alpha_k$. It suffices to show that for any positive integer $n \geq 2$, the following relation holds:

$$f(n) = f(1)f(p_1)^{\alpha_1} \cdots f(p_k)^{\alpha_k}.$$

We proceed by induction on α. If $\alpha(n) = 1$, then n is a prime, say $n = p$, and the property follows from the fact that

$$2f(p) = \tau(p)f(p) = (f * f)(p) = f(1)f(p) + f(p)f(1) = 2f(1)f(p).$$

Suppose then that the property holds for all n with $\alpha(n) \leq k$. Take any n with $\alpha(n) = k + 1$. Then

$$\tau(n)f(n) = 2f(1)f(n) + \sum f(a)f(b),$$

where the sum runs over all a, b with $ab = n$ and $1 < a, b < n$. It follows that $\alpha(a) \leq k, \alpha(b) \leq k$, and from the inductive assumption we get

$$\tau(n)f(n) = 2f(1)f(n) + (\tau(n) - 2)\{f^2(1)f(p_1)^{\alpha_1} \cdots f(p_k)^{\alpha_k}\}.$$

Since n is not a prime, certainly $\tau(n) > 2$, and so for both $f(1) = 0$ and $f(1) = 1$, the desired result follows.

Additional Problems

Problem 6.1.6. Let f be a function from the positive integers to the integers satisfying $f(m + n) \equiv f(n) \pmod{m}$ for all $m, n \geq 1$ (e.g., a polynomial with integer coefficients). Let $g(n)$ be the number of values (including repetitions) of $f(1), f(2), \ldots, f(n)$ divisible by n, and let $h(n)$ be the number of these values relatively prime to n. Show that g and h are multiplicative functions related by

$$h(n) = n \sum_{d|n} \mu(d)\frac{g(d)}{d} = n \prod_{j=1}^{k}\left(1 - \frac{g(p_j)}{p_j}\right),$$

where $n = p_1^{\alpha_1} \cdots p_k^{\alpha_k}$ is the prime factorization of n.

(American Mathematical Monthly)

Problem 6.1.7. Define $\lambda(1) = 1$, and if $n = p_1^{\alpha_1} \cdots p_k^{\alpha_k}$, define

$$\lambda(n) = (-1)^{\alpha_1 + \cdots + \alpha_k}.$$

(1) Show that λ is completely multiplicative.
(2) Prove that

$$\sum_{d|n} \lambda(d) = \begin{cases} 1 & \text{if } n \text{ is a square,} \\ 0 & \text{otherwise.} \end{cases}$$

(3) Find the convolutive inverse of λ.

Problem 6.1.8. Let an integer $n > 1$ be factored into primes: $n = p_1^{\alpha_1} \cdots p_m^{\alpha_m}$ (p_i distinct) and let its own positive integral exponents be factored similarly. The process is to be repeated until it terminates with a unique "constellation" of prime numbers. For example, the constellation for 192 is $192 = 2^{2^2 \cdot 3} \cdot 3$ and for 10000 is $10000 = 2^{2^2} \cdot 5^2$. Call an arithmetic function g generally multiplicative if $g(ab) = g(a)g(b)$ whenever the constellations for a and b have no prime in common.

(1) Prove that every multiplicative function is generally multiplicative. Is the converse true?

(2) Let h be an additive function (i.e., $h(ab) = h(a) + h(b)$ whenever $\gcd(a, b) = 1$). Call a function k generally additive if $k(ab) = k(a) + k(b)$ whenever the constellations for a and b have no prime in common. Prove that every additive function is generally additive. Is the converse true?

(American Mathematical Monthly)

6.2   Number of Divisors

For a positive integer n denote by $\tau(n)$ the number of its divisors. It is clear that

$$\tau(n) = \sum_{d \mid n} 1,$$

that is, τ is the summation function of the multiplicative function $f(m) = 1$, $m \in \mathbb{Z}_+^*$. Applying Theorem 6.1.2, it follows that τ is multiplicative.

Theorem 6.2.1. If $n = p_1^{\alpha_1} \cdots p_k^{\alpha_k}$ is the prime factorization of n, then

$$\tau(n) = (\alpha_1 + 1) \cdots (\alpha_k + 1). \tag{4}$$

Proof. Using the fact that τ is multiplicative, we have

$$\tau(n) = \tau(p_1^{\alpha_1}) \cdots \tau(p_k^{\alpha_k}) = (\alpha_1 + 1) \cdots (\alpha_k + 1),$$

because $p_i^{\alpha_i}$ has exactly $\alpha_i + 1$ divisors, $i = 1, \ldots, k$.   □

Problem 6.2.1.

(1) For any $n \geq 1$,

$$\sum_{m=1}^{n} \tau(m) = \sum_{k=1}^{n} \left\lfloor \frac{n}{k} \right\rfloor.$$

(2) For any $n \geq 1$,

$$\tau(n) = \sum_{k=1}^{n} \left(\left\lfloor \frac{n}{k} \right\rfloor - \left\lfloor \frac{n-1}{k} \right\rfloor \right).$$

(3) Prove the formula

$$\frac{1}{n \log n} \sum_{m=1}^{n} \tau(m) = 1.$$

Solution. (1) Note that since k is a divisor of exactly $\lfloor n/k \rfloor$ of the numbers $\{1, 2, \ldots, n\}$, we have

$$\sum_{k=1}^{n} \left\lfloor \frac{n}{k} \right\rfloor = \sum_{m=1}^{n} \tau(m).$$

(2) Note that

$$\left\lfloor \frac{n}{k} \right\rfloor - \left\lfloor \frac{n-1}{k} \right\rfloor = \begin{cases} 1 \text{ if } k \mid n, \\ 0 \text{ otherwise.} \end{cases}$$

Hence

$$\sum_{k=1}^{n} \left(\left\lfloor \frac{n}{k} \right\rfloor - \left\lfloor \frac{n-1}{k} \right\rfloor \right) = \sum_{k \mid n} 1 = \tau(n).$$

Alternatively, we can derive this formula by taking a difference of the relation in (1).

(3) Using the inequalities $x - 1 < \lfloor x \rfloor \le x$, from the relation in (1) we get

$$\sum_{k=1}^{n} \frac{1}{k} - 1 < \frac{1}{n} \sum_{m=1}^{n} \tau(m) \le \sum_{k=1}^{n} \frac{1}{k},$$

i.e.,

$$\sum_{k=1}^{n} \frac{1}{k} - \log n + \log n - 1 < \frac{1}{n} \sum_{m=1}^{n} \tau(m) \le \sum_{k=1}^{n} \frac{1}{k} - \log n + \log n,$$

and the formula follows by dividing by $\log n$.

Remark. It is clear that n is a prime if and only if $\tau(n) = 2$. Hence

$$\sum_{k=1}^{n} \left(\left\lfloor \frac{n}{k} \right\rfloor - \left\lfloor \frac{n-1}{k} \right\rfloor \right) = 2$$

if and only if n is a prime.

Problem 6.2.2. *Find all positive integers d that have exactly* 16 *positive integral divisors $d_1, d_2, \ldots, d_{16}$ such that*

$$1 = d_1 < d_2 < \cdots < d_{16} = d,$$

$d_6 = 18$, *and* $d_9 - d_8 = 17$.

(1998 Irish Mathematical Olympiad)

Solution. Let $d = p_1^{\alpha_1} p_2^{\alpha_2} \cdots p_m^{\alpha_m}$ with $p_1, \ldots, p_m$ distinct primes. Then d has $(a_1 + 1)(a_2 + 1) \cdots (a_m + 1)$ divisors. Since $18 = 2 \cdot 3^2$, it has 6 divisors: 1, 2, 3, 6, 9, 18. Since d has 16 divisors, we know that $d = 2 \cdot 3^3 p$ or $d = 2 \cdot 3^7$. If $d = 2 \cdot 3^7$, then $d_8 = 54$, $d_9 = 81$, and $d_9 - d_8 \neq 17$. Thus $d = 2 \cdot 3^3 p$ for some prime $p > 18$. If $p < 27$, then $d_7 = p$, $d_8 = 27$, $d_9 = 2p = 27 + 17 + 44 \Rightarrow p = 22$, a contradiction. Thus $p > 27$. If $p < 54$, then $d_7 = 27$, $d_8 = p$, $d_9 = 54 = d_8 + 17 \Rightarrow p = 37$. If $p > 54$, then $d_7 = 27$, $d_8 = 54$, $d_9 = d_8 + 17 = 71$. We obtain two solutions to the problem: $2 \cdot 3^3 \cdot 37 = 1998$ and $2 \cdot 3^3 \cdot 71 = 3834$.

Problem 6.2.3. *For how many (a) even and (b) odd numbers n does n divide $3^{12} - 1$, yet n does not divide $3^k - 1$ for $k = 1, 2, \ldots, 11$?*

(1995 Austrian Mathematical Olympiad)

Solution. We note that

$$3^{12} - 1 = (3^6 - 1)(3^6 + 1)$$
$$= (3^2 - 1)(3^4 + 3^2 + 1)(3^2 + 1)(3^4 - 3^2 + 1)$$
$$= (2^3)(7 \cdot 13)(2 \cdot 5)(73).$$

Recall that the number of divisors of $p_1^{e_1} \cdots p_k^{e_k}$ is $(e_1 + 1) \cdots (e_k + 1)$. Therefore $3^{12} - 1$ has $2 \cdot 2 \cdot 2 \cdot 2 = 16$ odd divisors and $4 \cdot 16 = 64$ even divisors.

If $3^{12} \equiv 1 \pmod{m}$ for some integer m, then the smallest integer d such that $3^d \equiv 1 \pmod{m}$ divides 12. (Otherwise, we could write $12 = pq + r$ with $0 < r < d$ and obtain $3^r \equiv 1 \pmod{m}$.) Hence to ensure $n \nmid 3^k - 1$ for $k = 1, \ldots, 11$, we need only check $k = 1, 2, 3, 4, 6$. But

$$3^1 - 1 = 2,$$
$$3^2 - 1 = 2^3,$$
$$3^3 - 1 = 2 \cdot 13,$$
$$3^4 - 1 = 2^4 \cdot 5,$$
$$3^6 - 1 = 2^3 \cdot 7 \cdot 13.$$

The odd divisors we throw out are 1, 5, 7, 13, 91, while the even divisors are 2^i for $1 \leq i \leq 4$, $2^i \cdot 5$ for $1 \leq i \leq 4$, and each of $2^j \cdot 7$, $2^j \cdot 13$, and $2^j \cdot 7 \cdot 13$ for $1 \leq i \leq 3$. Since we are discarding 17 even divisors and 5 odd ones, we remain with 47 even divisors and 11 odd ones.

Problem 6.2.4. *Let $\tau(n)$ denote the number of divisors of the natural number n. Prove that the sequence $\tau(n^2 + 1)$ does not become strictly increasing from any given point onward.*

(1998 St. Petersburg City Mathematical Olympiad)

Solution. We first note that for n even, $\tau(n^2 + 1) \leq n$. Indeed, exactly half of the divisors of $n^2 + 1$ are less than n, and all are odd, so there are at most $2(n/2)$ in all.

Now if $\tau(n^2 + 1)$ becomes strictly increasing for $n \geq N$, then

$$\tau((n + 1)^2 + 1) \geq \tau(n^2 + 1) + 2$$

for $n \geq N$ (since $\tau(k)$ is even for k not a perfect square). Thus

$$\tau(n^2 + 1) \geq \tau(N^2 + 1) + 2(n - N)$$

, which exceeds n for large N, contradiction.

Additional Problems

Problem 6.2.5. Does there exist a positive integer such that the product of its proper divisors ends with exactly 2001 zeros?

(2001 Russian Mathematical Olympiad)

Problem 6.2.6. Prove that the number of divisors of the form $4k + 1$ of each positive integer is not less than the number of its divisors of the form $4k + 3$.

Problem 6.2.7. Let $d_1, d_2, \ldots, d_l$ be all positive divisors of a positive integer. For each $i = 1, 2, \ldots, l$ denote by a_i the number of divisors of d_i. Then

$$a_1^3 + a_2^3 + \cdots + a_l^3 = (a_1 + a_2 + \cdots + a_l)^2.$$

6.3   Sum of Divisors

For a positive integer n denote by $\sigma(n)$ the sum of its divisors. Clearly

$$\sigma(n) = \sum_{d|n} d,$$

that is, σ is the summation function of the multiplicative function $d(m) = m$, $m \in \mathbb{Z}_+^*$. Applying Theorem 6.1.2, it follows that σ is multiplicative.

Theorem 6.3.1. *If $n = p_1^{\alpha_1} \cdots p_k^{\alpha_k}$ is the prime factorization of n, then*

$$\sigma(n) = \frac{p_1^{\alpha_1+1} - 1}{p_1 - 1} \cdots \frac{p_k^{\alpha_k+1} - 1}{p_k - 1}.$$

Proof. Because σ is multiplicative, it suffices to compute $\sigma(p_i^{\alpha_i})$, $i = 1, \ldots, k$. The divisors of $p_i^{\alpha_i}$ are $1, p_i, \ldots, p_i^{\alpha_i}$; hence

$$\sigma(p_i^{\alpha_i}) = 1 + p_i + \cdots + p_i^{\alpha_i} = \frac{p_i^{\alpha_i+1} - 1}{p_i - 1},$$

and the conclusion follows. □

Problem 6.3.1. *(1) For any n $\geq$ 1,*

$$\sum_{k=1}^{n} k \left\lfloor \frac{n}{k} \right\rfloor = \sum_{m=1}^{n} \sigma(m).$$

(2) For any n $\geq$ 1,

$$\sigma(n) = \sum_{k=1}^{n} k \left(\left\lfloor \frac{n}{k} \right\rfloor - \left\lfloor \frac{n-1}{k} \right\rfloor \right).$$

(3) If

$$\lim_{n \to \infty} \frac{\sum_{m=1}^{n} \sigma(m)}{n^2} = a,$$

then prove that a $\in \left[\frac{1}{2}, 1 \right]$.

Solution. (1) Since k is a divisor of exactly $\lfloor n/k \rfloor$ of the numbers $\{1, 2, \ldots, n\}$, we have

$$\sum_{k=1}^{n} k \left\lfloor \frac{n}{k} \right\rfloor = \sum_{m=1}^{n} \sigma(m).$$

(2) We have

$$\left\lfloor \frac{n}{k} \right\rfloor - \left\lfloor \frac{n-1}{k} \right\rfloor = \begin{cases} 1 \text{ if } k \mid n, \\ 0 \text{ otherwise}; \end{cases}$$

hence

$$\sum_{k=1}^{n} k \left(\left\lfloor \frac{n}{k} \right\rfloor - \left\lfloor \frac{n-1}{k} \right\rfloor \right) = \sum_{k|n} k = \sigma(n).$$

Alternatively, we can apply the formula in (1) for n and $n-1$ and then take the difference.

(3) Since $x - 1 < \lfloor x \rfloor \leq x$, from the relation in (1) we get

$$n^2 - \frac{n(n+1)}{2} < \sum_{m=1}^{n} \sigma(m) \leq n^2,$$

i.e.,

$$\lim_{n \to \infty} \frac{\sum_{m=1}^{n} \sigma(m)}{n^2} \in \left[\frac{1}{2}, 1 \right].$$

Remarks. (1) The exact value of this interesting limit is $\pi^2/12$.

(2) It is clear that n is a prime if and only if $\sigma(n) = n + 1$. Hence

$$\sum_{k=1}^{n} k \left(\left\lfloor \frac{n}{k} \right\rfloor - \left\lfloor \frac{n-1}{k} \right\rfloor \right) = n + 1$$

if and only if n is a prime.

Problem 6.3.2. *If n is a composite positive integer, then*

$$\sigma(n) \geq n + \sqrt{n} + 1.$$

Solution. The integer n has a divisor d such that $d \neq 1$ and $d \leq \sqrt{n}$. Because $\frac{n}{d}$ is also a divisor of n, it follows that $\frac{n}{d} \geq \sqrt{n}$, and therefore

$$\sigma(n) = \sum_{k|n} k \geq 1 + n + \frac{n}{d} \geq n + \sqrt{n} + 1.$$

Problem 6.3.3. *For any $n \geq 7$,*

$$\sigma(n) < n \ln n.$$

Solution. Let $d_1, d_2, \ldots, d_k$ be all the divisors of n. They can be also written as

$$\frac{n}{d_1}, \quad \frac{n}{d_2}, \quad \cdots, \quad \frac{n}{d_k};$$

hence

$$\sigma(n) = n\left(\frac{1}{d_1} + \frac{1}{d_2} + \cdots + \frac{1}{d_k}\right) \leq n\left(1 + \frac{1}{2} + \cdots + \frac{1}{k}\right),$$

where $k = \tau(n)$. Inducting on k, we prove that for any $k \geq 2$,

$$1 + \frac{1}{2} + \cdots + \frac{1}{k} < 0.81 + \ln k.$$

Now we use the inequality $\tau(n) \leq 2\sqrt{n}$, $n \geq 1$. In order to prove it, let us consider $d_1 < d_2 < \cdots < d_k$, the divisors of n not exceeding $\sqrt{n}$. The other divisors are $n/d_1, n/d_2, \ldots, n/d_k$. We get $\tau(n) \leq 2k \leq 2\sqrt{n}$.

Using the inequality $\tau(n) \leq 2\sqrt{n}$, it follows that

$$1 + \frac{1}{2} + \cdots + \frac{1}{k} < 0.81 + \ln(2\sqrt{n}) < 1.51 + \frac{1}{2}\ln n.$$

For $n \geq 21$ we have $\ln n > 1.51 + \frac{1}{2}\ln n$, and checking directly the desired inequality for $n = 7, \ldots, 20$, the conclusion follows.

Problem 6.3.4. *For any $n \geq 1$,*

$$\frac{\sigma(n)}{\tau(n)} \geq \sqrt{n}.$$

Solution. Let $d_1, d_2, \ldots, d_{\tau(n)}$ be the divisors of n. They can be rewritten as

$$\frac{n}{d_1}, \quad \frac{n}{d_2}, \quad \cdots, \quad \frac{n}{d_{\tau(n)}}.$$

Hence

$$\sigma(n)^2 = n(d_1 + d_2 + \cdots + d_{\tau(n)})\left(\frac{1}{d_1} + \frac{1}{d_2} + \cdots + \frac{1}{d_{\tau(n)}}\right) \geq n\tau(n)^2,$$

and the conclusion follows because of the AM–HM inequality.

Remarks. (1) This means the average of divisors of n is at least $\sqrt{n}$.

(2) An alternative way to prove this inequality is the following: We start by noting that for any divisor d of n we have $d + \frac{n}{d} \geq 2\sqrt{n}$. This follows from the AM–GM inequality or is an easy calculus exercise. Then sum this result over all divisors d of n, noting that $\frac{n}{d}$ also varies over all divisors to get $2\sigma(1) \geq 2\sqrt{n}\tau(n)$.

Additional Problems

Problem 6.3.5. For any $n \geq 2$,

$$\sigma(n) < n\sqrt{2\tau(n)}.$$

(1999 Belarusian Mathematical Olympiad)

Problem 6.3.6. Find all the four-digit numbers whose prime factorization has the property that the sum of the prime factors is equal to the sum of the exponents.

Problem 6.3.7. Let m, n, k be positive integers with $n > 1$. Show that $\sigma(n)^k \neq n^m$.

(2001 St. Petersburg City Mathematical Olympiad)

6.4   Euler's Totient Function

For any positive integer n we denote by $\varphi(n)$ the number of integers m such that $m \leq n$ and $\gcd(m, n) = 1$. The arithmetic function φ is called *Euler's*[3] *totient function*. It is clear that $\varphi(1) = 1$ and for any prime p, $\varphi(p) = p - 1$. Moreover, if n is a positive integer such that $\varphi(n) = n - 1$ then n is a prime.

Theorem 6.4.1. (Gauss) *For any positive integer n,*

$$\sum_{d \mid n} \varphi(d) = n.$$

---

[3]Leonhard Euler (1707–1783), Swiss mathematician who worked at the Petersburg Academy and Berlin Academy of Science. Euler was one of the most prolific mathematicians of all time. Euler systematized mathematics by introducing the symbols e and i, and $f(x)$ for a function of x. He also made major contributions in optics, mechanics, electricity, and magnetism. Euler did important work in number theory, proving that the divergence of the harmonic series implies an infinite number of primes, factoring the fifth Fermat number, and introducing the totient function φ.

Proof. Let $d_1, d_2, \ldots, d_k$ be the divisors of n and let $S_i = \{m \mid m \leq n$ and $\gcd(m, n) = d_i\}$, $i = 1, \ldots, k$. If $m \in S_i$, then $m = d_i m'$, where $\gcd\left(m', \frac{n}{d_i}\right) = 1$. Because $m' \leq \frac{n}{d_i}$, from the definition of φ it follows that $|S_i| = \varphi\left(\frac{n}{d_i}\right)$. The sets $S_1, \ldots, S_k$ give a partition of $\{1, 2, \ldots, n\}$; hence

$$\sum_{i=1}^{k} \varphi\left(\frac{n}{d_i}\right) = \sum_{i=1}^{k} |S_i| = n.$$

But $\left\{\frac{n}{d_1}, \ldots, \frac{n}{d_k}\right\} = \{d_1, \ldots, d_k\}$, so $\sum_{d \mid n} \varphi(d) = n$. $\qquad\square$

Theorem 6.4.2. *The function φ is multiplicative.*

Proof. From Theorem 6.4.1 we obtain that the summation function of φ is $F(n) = n$, which is multiplicative.

The conclusion now follows from Theorem 6.1.4. $\qquad\square$

Theorem 6.4.3. *If $n = p_1^{\alpha_1} \cdots p_k^{\alpha_k}$ is the prime factorization of $n > 1$, then*

$$\varphi(n) = n\left(1 - \frac{1}{p_1}\right) \cdots \left(1 - \frac{1}{p_k}\right).$$

Proof. We first notice that for any prime p and for any positive integer α,

$$\varphi(p^\alpha) = p^\alpha - p^{\alpha-1} = p^\alpha\left(1 - \frac{1}{p}\right).$$

Indeed, the number of all positive integers not exceeding n that are divisible by p is $p^{\alpha-1}$; hence $\varphi(p^\alpha) = p^\alpha - p^{\alpha-1}$.

Using Theorem 6.4.3 we have

$$\varphi(n) = \varphi(p_1^{\alpha_1} \cdots p_k^{\alpha_k}) = \varphi(p_1^{\alpha_1}) \cdots \varphi(p_k^{\alpha_k})$$
$$= p_1^{\alpha_1}\left(1 - \frac{1}{p_1}\right) \cdots p_k^{\alpha_k}\left(1 - \frac{1}{p_k}\right) = p_1^{\alpha_1} \cdots p_k^{\alpha_k}\left(1 - \frac{1}{p_1}\right) \cdots \left(1 - \frac{1}{p_k}\right)$$
$$= n\left(1 - \frac{1}{p_1}\right) \cdots \left(1 - \frac{1}{p_k}\right).$$

Alternative proof. We employ the inclusion–exclusion principle. Let

$$T_i = \{d \mid d \leq n \text{ and } p_i \mid d\}, \quad i = 1, \ldots, k.$$

It follows that

$$T_1 \cup \cdots \cup T_k = \{m \mid m \leq n \text{ and } \gcd(m, n) > 1\}.$$

Hence

$$\varphi(n) = n - |T_1 \cup \cdots \cup T_k| = n - \sum_{i=1}^{k} |T_i| + \sum_{1 \le i < j \le k} |T_i \cap T_j|$$
$$- \cdots + (-1)^k |T_1 \cap \cdots \cap T_k|.$$

We have

$$|T_i| = \frac{n}{p_i}, \quad |T_i \cap T_j| = \frac{n}{p_i p_j}, \ldots, \quad |T_1 \cap \cdots \cap T_k| = \frac{n}{p_1 \cdots p_k}.$$

Finally,

$$\varphi(n) = n \left(1 - \sum_{i=1}^{n} \frac{1}{p_i} + \sum_{1 \le i < j \le k} \frac{1}{p_i p_j} - \cdots + (-1)^k \frac{1}{p_1 \cdots p_k} \right)$$
$$= n \left(1 - \frac{1}{p_1} \right) \cdots \left(1 - \frac{1}{p_k} \right). \qquad \square$$

Remarks. (1) A natural generalization of Euler's totient function is given in Problem 6.1.6; hence it is possible to derive the properties contained in Theorems 6.4.1-6.4.3 directly from the results of this problem.

(2) Writing the formula in Theorem 6.4.3 as

$$\varphi(n) = \frac{n}{p_1 \cdots p_k} (p_1 - 1) \cdots (p_k - 1),$$

it follows that $\varphi(n)$ is an even integer for any $n \ge 3$.

Problem 6.4.1. *Prove that there are infinitely many even positive integers k such that the equation $\varphi(n) = k$ has no solution.*

(Schinzel[4])

Solution. Take $k = 2 \cdot 7^m$, $m \ge 1$. If $n = p_1^{\alpha_1} \cdots p_h^{\alpha_h}$, then

$$\varphi(n) = p_1^{\alpha_1} \left(1 - \frac{1}{p_1} \right) \cdots p_h^{\alpha_h} \left(1 - \frac{1}{p_h} \right)$$
$$= p_1^{\alpha_1 - 1} \cdots p_h^{\alpha_h - 1} (p_1 - 1) \cdots (p_h - 1).$$

If at least two of the primes $p_1, \ldots, p_h$ are odd, then $4 \mid \varphi(n)$ and $\varphi(n) \ne k$.

---

[4] Andrzej Schinzel (1935–), Polish mathematician with important work on exponential congruences, Euler's φ-function, Diophantine equations, and applications of transcendental number theory to arithmetic problems.

If $p_i = 7$, for some i, then $3 \mid \phi(n)$ and $\phi(n) \neq k$. If any odd prime $p_i \neq 7$ has $\alpha_i > 1$, then $p_i \mid \phi(n)$ and again $\phi(n) \neq k$. Thus the only remaining possibilities are $n = 2^\alpha$ or $2^\alpha p$ for some $p \geq 3$. In the first case $\phi(n) = 2^{\alpha-1} \neq k$. In the second case, if $\alpha > 1$, then again $4 \mid \phi(n)$ and $\phi(n) \neq k$. If $\alpha \leq 1$, then $\phi(n) = p - 1$. For this to be k we need $p - 1 = 2 \cdot 7^m$ or $p = 2 \cdot 7^m + 1$. However, one easily checks that $3 \mid 2 \cdot 7^m + 1$, so this forces $p = 3$ and $m = 0$, contrary to our assumption.

Problem 6.4.2. *Prove that there are infinitely many positive integers n such that*

$$\varphi(n) = \frac{n}{3}.$$

Solution. Let $n = 2 \cdot 3^m$, where m is a positive integer. Then

$$\varphi(n) = \varphi(2 \cdot 3^m) = \varphi(2)\varphi(3^m) = 3^m - 3^{m-1} = 2 \cdot 3^{m-1} = \frac{n}{3}$$

for infinitely many values of n, as desired.

Problem 6.4.3. *If n is a composite positive integer, then*

$$\varphi(n) \leq n - \sqrt{n}.$$

Solution. because n is composite, it has a prime factor $p_j \leq \sqrt{n}$. We have

$$\varphi(n) = n\left(1 - \frac{1}{p_1}\right) \cdots \left(1 - \frac{1}{p_k}\right) \leq n\left(1 - \frac{1}{p_j}\right) \leq n\left(1 - \frac{1}{\sqrt{n}}\right) = n - \sqrt{n}.$$

Problem 6.4.4. *For any positive integer n, $n \neq 2$, $n \neq 6$,*

$$\varphi(n) \geq \sqrt{n}.$$

Solution. If $n = 2^m$, where $m \geq 2$, then

$$\varphi(n) = 2^m - 2^{m-1} = 2^{m-1} \geq \sqrt{2^m} = \sqrt{n}.$$

If $n = p^m$, where p is an odd prime and $m \geq 2$, then

$$\varphi(n) = p^m - p^{m-1} = p^{m-1}(p - 1) \geq \sqrt{2p^m} = \sqrt{2n}.$$

If $n = p^m$, where p is a prime greater than or equal to 5, then $\varphi(n) \geq \sqrt{2n}$. If n is odd or $4 \mid n$, then

$$\varphi(n) = \varphi(p_1^{\alpha_1}) \cdots \varphi(p_k^{\alpha_k}) \geq \sqrt{p_1^{\alpha_1}} \cdots \sqrt{p_k^{\alpha_k}} = \sqrt{n}.$$

If $n = 2t$, where t is odd, then since $n \neq 6$, we see that t has at least one prime power factor that is at least 5. Hence $\varphi(n) = \varphi(t) \geq \sqrt{2t}$.

Additional Problems

Problem 6.4.5. For a positive integer n, let $\psi(n)$ be the number of prime factors of n. Show that if $\varphi(n)$ divides $n - 1$ and $\psi(n) \leq 3$, then n is prime.

(1998 Korean Mathematical Olympiad)

Problem 6.4.6. Show that the equation $\varphi(n) = \tau(n)$ has only the solutions $n = 1, 3, 8, 10, 18, 24, 30$.

Problem 6.4.7. Let $n > 6$ be an integer and let $a_1, a_2, \ldots, a_k$ be all positive integers less than n and relatively prime to n. If

$$a_2 - a_1 = a_3 - a_2 = \cdots = a_k - a_{k-1} > 0,$$

prove that n must be either a prime number or a power of 2.

(32nd International Mathematical Olympiad)

6.5 Exponent of a Prime and Legendre's Formula

Let p be a prime and let us denote by $v_p(a)$ the exponent of p in the decomposition of a. Of course, if p doesn't divide a, then $v_p(a) = 0$.

It is easy to prove the following properties of v_p:

(1) $\min\{v_p(a), v_p(b)\} \leq v_p(a + b)$. If $v_p(a) \neq v_p(b)$, then $v_p(a + b) = \min\{v_p(a), v_p(b)\}$.

(2) $v_p(ab) = v_p(a) + v_p(b)$.

(3) $v_p(\gcd(a_1, a_2, \ldots, a_n)) = \min\{v_p(a_1), v_p(a_2), \ldots, v_p(a_n)\}$.

(4) $v_p(\mathrm{lcm}(a_1, a_2, \ldots, a_n)) = \max\{v_p(a_1), v_p(a_2), \ldots, v_p(a_n)\}$.

If we have to prove that $a \mid b$, then it is enough to prove that the exponent of any prime number in the decomposition of a is at least the exponent of that prime in the decomposition of b. Now let us repeat the above idea in terms of the function v_p. We have $a \mid b$ if and only if for every prime p we have $v_p(a) \leq v_p(b)$. Also, we have $a = b$ if and only if for every prime p, $v_p(a) = v_p(b)$.

For each positive integer n, let $e_p(n)$ be the exponent of the prime p in the prime factorization of $n!$.

The arithmetic function e_p is called the *Legendre*[5] *function* associated with the prime p, and it is connected to the function v_p by the relation $e_p(n) = v_p(n!)$.

The following result gives a formula for the computation of $e_p(n)$.

---

[5] Adrien-Marie Legendre (1752–1833), French mathematician who was a disciple of Euler and Lagrange. In number theory, he studied the function e_p, and he proved the unsolvability of Fermat's last theorem for $n = 5$.

Theorem 6.5.1. (Legendre's formula) *For any prime p and any positive integer n,*

$$e_p(n) = \sum_{i \geq 1} \left\lfloor \frac{n}{p^i} \right\rfloor = \frac{n - S_p(n)}{p - 1},$$

where $S_p(n)$ is the sum of the digits of n when written in base p.

Proof. For $n < p$ it is clear that $e_p(n) = 0$. If $n \geq p$, then in order to determine $e_p(n)$ we need to consider only the multiples of p in the product $1 \cdot 2 \cdots n$, that is, $(1 \cdot p)(2 \cdot p) \cdots (kp) = p^k k!$, where $k = \left\lfloor \frac{n}{p} \right\rfloor$. Hence

$$e_p(n) = \left\lfloor \frac{n}{p} \right\rfloor + e_p \left(\left\lfloor \frac{n}{p} \right\rfloor \right).$$

Replacing n by $\left\lfloor \frac{n}{p} \right\rfloor$ and taking into account that

$$\left\lfloor \frac{\left\lfloor \frac{n}{p} \right\rfloor}{p} \right\rfloor = \left\lfloor \frac{n}{p^2} \right\rfloor,$$

we obtain

$$e_p \left(\left\lfloor \frac{n}{p} \right\rfloor \right) = \left\lfloor \frac{n}{p^2} \right\rfloor + e_p \left(\left\lfloor \frac{n}{p^2} \right\rfloor \right).$$

Continuing this procedure, we get

$$e_p \left(\left\lfloor \frac{n}{p^2} \right\rfloor \right) = \left\lfloor \frac{n}{p^3} \right\rfloor + e_p \left(\left\lfloor \frac{n}{p^3} \right\rfloor \right),$$

$$\cdots$$

$$e_p \left(\left\lfloor \frac{n}{p^{m-1}} \right\rfloor \right) = \left\lfloor \frac{n}{p^m} \right\rfloor + e_p \left(\left\lfloor \frac{n}{p^m} \right\rfloor \right),$$

where m is the least positive integer such that $n < p^{m+1}$, that is, $m = \left\lfloor \frac{\ln n}{\ln p} \right\rfloor$. Summing up the relations above yields

$$e_p(n) = \left\lfloor \frac{n}{p} \right\rfloor + \left\lfloor \frac{n}{p^2} \right\rfloor + \cdots + \left\lfloor \frac{n}{p^m} \right\rfloor.$$

The other relation is not difficult. Indeed, let us write

$$n = a_0 + a_1 p + \cdots + a_k p^k,$$

where $a_0, a_1, \ldots, a_k \in \{0, 1, \ldots, p - 1\}$ and $a_k \neq 0$. Then

$$\left\lfloor \frac{n}{p} \right\rfloor + \left\lfloor \frac{n}{p^2} \right\rfloor + \cdots = a_1 + a_2 p + \cdots + a_k p^{k-1} + a_2 + a_3 p + \cdots + a_k p^{k-2} + \cdots + a_k.$$

The coefficient of a_i on the right-hand side is $1 + p + \cdots + p^{i-1}$, so using the formula

$$1 + p + \cdots + p^{i-1} = \frac{p^i - 1}{p - 1},$$

we obtain exactly the second part in the expression of $e_p(n)$.   □

Remark. An alternative proof of the first half of Legendre's formula is to note that

$$e_p(n) = \sum_{k=1}^{n} v_p(k) = \sum_{k=1}^{n} \sum_{m=1}^{v_p(k)} 1.$$

Now look at the total contribution of a particular value of m to this double sum. You will get a contribution of 1 for every multiple of p^m in the set $\{1, 2, \ldots, n\}$. The number of such multiples is $\lfloor \frac{n}{p^m} \rfloor$; hence

$$e_p(n) = \sum_{m \geq 1} \left\lfloor \frac{n}{p^m} \right\rfloor.$$

Examples. (1) Let us find the exponent of 7 in 400!. Applying Legendre's formula, we have

$$e_7(400) = \left\lfloor \frac{400}{7} \right\rfloor + \left\lfloor \frac{400}{7^2} \right\rfloor + \left\lfloor \frac{400}{7^3} \right\rfloor = 57 + 8 + 1 = 66.$$

(2) Let us determine the exponent of 3 in $((3!)!)!$. We have $((3!)!)! = (6!)! = 720!$. Applying Legendre's formula yields

$$e_3(720) = \left\lfloor \frac{720}{3} \right\rfloor + \left\lfloor \frac{720}{3^2} \right\rfloor + \left\lfloor \frac{720}{3^3} \right\rfloor + \left\lfloor \frac{720}{3^4} \right\rfloor + \left\lfloor \frac{720}{3^5} \right\rfloor$$
$$= 240 + 80 + 26 + 8 + 2 = 356.$$

Problem 6.5.1. *Let p be a prime. Find the exponent of p in the prime factorization of $(p^m)!$.*

Solution. Using the first half of Legendre's formula, we have

$$e_p(p^m) = \sum_{i \geq 1} \left\lfloor \frac{p^m}{p^i} \right\rfloor = p^{m-1} + p^{m-2} + \cdots + p + 1 = \frac{p^m - 1}{p - 1}.$$

An easier argument for the above formula follows directly from the second version of Legendre's formula. Then it is just $S_p(p^m) = 1$, so $e_p(p^m) = (p^m - 1)/(p - 1)$.

Problem 6.5.2. *Find all positive integers n such that $n!$ ends in exactly 1000 zeros.*

First solution. There are clearly more 2's than 5's in the prime factorization of $n!$; hence it suffices to solve the equation

$$\left\lfloor \frac{n}{5} \right\rfloor + \left\lfloor \frac{n}{5^2} \right\rfloor + \cdots = 1000.$$

But

$$\left\lfloor \frac{n}{5} \right\rfloor + \left\lfloor \frac{n}{5^2} \right\rfloor + \cdots < \frac{n}{5} + \frac{n}{5^2} + \cdots = \frac{n}{5}\left(1 + \frac{1}{5} + \cdots\right)$$
$$= \frac{n}{5} \cdot \frac{1}{1 - \frac{1}{5}} = \frac{n}{4};$$

hence $n > 4000$.

On the other hand, using the inequality $\lfloor a \rfloor > a - 1$, we have

$$1000 > \left(\frac{n}{5} - 1\right) + \left(\frac{n}{5^2} - 1\right) + \left(\frac{n}{5^3} - 1\right) + \left(\frac{n}{5^4} - 1\right) + \left(\frac{n}{5^5} - 1\right)$$
$$= \frac{n}{5}\left(1 + \frac{1}{5} + \frac{1}{5^2} + \frac{1}{5^3} + \frac{1}{5^4}\right) - 5 = \frac{n}{5} \cdot \frac{1 - \left(\frac{1}{5}\right)^5}{1 - \frac{1}{5}} - 5,$$

so

$$n < \frac{1005 \cdot 4 \cdot 3125}{3124} < 4022.$$

We have narrowed n down to $\{4001, 4002, \ldots, 4021\}$. Using Legendre's formula we find that 4005 is the first positive integer with the desired property and that 4009 is the last. Hence $n = 4005, 4006, 4007, 4008, 4009$.

Second solution. It suffices to solve the equation $e_5(n) = 1000$. Using the second form of Legendre's formula, this becomes $n - S_5(n) = 4000$. Hence $n > 4000$. We work our way upward from 4000 looking for a solution. Since $e_5(n)$ can change only at multiples of 5, we step up 5 each time:

$$e_5(4000) = \frac{4000 - 4}{5 - 1} = 999,$$
$$e_5(4005) = \frac{4005 - 5}{5 - 1} = 1000,$$
$$e_5(4010) = \frac{4010 - 6}{5 - 1} = 1001.$$

Any $n > 4010$ will clearly have $e_5(n) \geq e_5(4010) = 1001$. Hence the only solutions are $n = 4005, 4006, 4007, 4008, 4009$.

Problem 6.5.3. *Prove that for any positive integer n, 2^n does not divide n!.*

First solution. The exponent of 2 in the prime factorization of $n!$ is

$$k = e_2(n) = \left\lfloor \frac{n}{2} \right\rfloor + \left\lfloor \frac{n}{2^2} \right\rfloor + \cdots .$$

We have

$$k < \frac{n}{2} + \frac{n}{2^2} + \cdots = \frac{n}{2}\left(1 + \frac{1}{2} + \cdots\right) = \frac{n}{2} \cdot \frac{1}{1 - \frac{1}{2}} = n,$$

and we are done.

Second solution. Using the second version of Legendre's formula, we have $e_2(n) = n - S_2(n) < n$, and we are done.

Remark. Similarly, for any prime p, p^n does not divide $((p-1)n)!$.

Problem 6.5.4. *Find all positive integers n such that 2^{n-1} divides $n!$.*

First solution. If $n = 2^s$, $s = 0, 1, 2, \ldots$, then

$$e_2(n) = 2^{s-1} + \cdots + 2 + 1 = 2^s - 1;$$

hence 2^{n-1} divides $n!$.

Assume that n is odd, $n = 2n_1 + 1$. Then from $2^{n-1} = 2^{2n_1} \mid (2n_1 + 1)! = (2n_1)!(2n_1 + 1)$ it follows that $2^{2n_1} \mid (2n_1)!$, which is not possible, by Problem 6.5.3. We get $n = 2m_1$. If m_1 is odd, $m_1 = 2n_2 + 1$, we have

$$2^{n-1} = 2^{4n_2+1} \mid (4n_2 + 2)! = (4n_2)!(4n_2 + 1) \cdot 2 \cdot (2n_2 + 1),$$

and we obtain $2^{4n_2} \mid (4n_2)!$, a contradiction. Continuing this procedure, we get $n = 2^s$.

Second solution. We use the second version of Legendre's formula. It is just $e_2(n) = n - S_2(n) = n - 1$ if and only if $S_2(n) = 1$, that is, if and only if n is a power of 2.

Problem 6.5.5. *Let p be an odd prime. Prove that the exponent of p in the prime factorization of $1 \cdot 3 \cdot 5 \cdots (2m + 1)$ is*

$$\sum_{k \geq 1} \left(\left\lfloor \frac{2m + 1}{p^k} \right\rfloor - \left\lfloor \frac{m}{p^k} \right\rfloor \right).$$

Solution. We have

$$1 \cdot 3 \cdot 5 \cdots (2m + 1) = \frac{(2m + 1)!}{m! \cdot 2^m}.$$

Because p is odd, the desired exponent is

$$e_p(2m + 1) - e_p(m) = \sum_{k \geq 1} \left\lfloor \frac{2m + 1}{p^k} \right\rfloor - \sum_{k \geq 1} \left\lfloor \frac{m}{p^k} \right\rfloor,$$

and the conclusion follows.

Problem 6.5.6. *If p is a prime and $p^\alpha \mid \binom{n}{m}$, then $p^\alpha \le n$.*

Solution. Because

$$\binom{n}{m} = \frac{n!}{m!(n-m)!},$$

the exponent of p in the prime factorization of $\binom{n}{m}$ is

$$\beta = e_p(n) - e_p(m) - e_p(n-m) = \sum_{k \ge 1} \left(\left\lfloor \frac{n}{p^k} \right\rfloor - \left\lfloor \frac{m}{p^k} \right\rfloor - \left\lfloor \frac{n-m}{p^k} \right\rfloor \right).$$

This sum has at most s nonzero terms, where $p^s \le n < p^{s+1}$. Using the inequality $\lfloor x + y \rfloor - \lfloor x \rfloor - \lfloor y \rfloor \le 1$ for $x = \frac{m}{p^k}$ and $y = \frac{n-m}{p^k}$, it follows that $\beta \le s$. Because $p^\alpha \mid \binom{n}{m}$, we obtain $\alpha \le \beta \le s$; hence $p^\alpha \le p^s \le n$.

Additional Problems

Problem 6.5.7. (a) If p is a prime, prove that for any positive integer n,

$$-\left\lfloor \frac{\ln n}{\ln p} \right\rfloor + n \sum_{k=1}^{\left\lfloor \frac{\ln n}{\ln p} \right\rfloor} \frac{1}{p^k} < e_p(n) < \frac{n}{p-1}.$$

(b) Prove that

$$\lim_{n \to \infty} \frac{e_p(n)}{n} = \frac{1}{p-1}.$$

Problem 6.5.8. Show that for all nonnegative integers m, n, the number

$$\frac{(2m)!(2n)!}{m!n!(m+n)!}$$

is also an integer.

(14th International Mathematical Olympiad)

Problem 6.5.9. Prove that $\frac{(3a+3b)!(2a)!(3b)!(2b)!}{(2a+3b)!(a+2b)!(a+b)!a!(b!)^2}$ is an integer for any positive integers a, b.

(American Mathematical Monthly)

Problem 6.5.10. Prove that there exists a constant c such that for any positive integers a, b, n for which $a! \cdot b! \mid n!$, we have $a + b < n + c \ln n$.

(Paul Erdős)

Problem 6.5.11. Prove that for any integer $k \geq 2$, the equation

$$\frac{1}{10^n} = \frac{1}{n_1!} + \frac{1}{n_2!} + \cdots + \frac{1}{n_k!}$$

does not have integer solutions such that $1 \leq n_1 < n_2 < \cdots < n_k$.

(Tuymaada Olympiad)

7
More on Divisibility

7.1 Congruences Modulo a Prime: Fermat's Little Theorem

In this section, p will denote a prime number. We begin by noticing that it makes sense to consider a polynomial with integer coefficients

$$f(x) = a_0 + a_1 x + \cdots + a_d x^d,$$

but reduced modulo p. If for each j, $a_j \equiv b_j \pmod{p}$, we write

$$a_0 + a_1 x + \cdots + a_d x^d \equiv b_0 + b_1 x + \cdots + b_d x^d \pmod{p},$$

and talk about the residue class of a polynomial modulo p. We will denote the residue class of $f(x)$ by $f(x)_p$. We say that $f(x)$ has *degree d modulo p* if $a_d \not\equiv 0 \pmod{p}$.

For an integer c, we can evaluate $f(c)$ and reduce the answer modulo p to obtain $f(c)_p$. If $f(c)_p = 0$ modulo p, then c is said to be a *root of $f(x)$ modulo p*.

Theorem 7.1.1. (Lagrange) *If $f(x)$ has degree d modulo p, then the number of distinct roots of $f(x)$ modulo p is at most p.*

Proof. Begin by noticing that if c is root of $f(x)$ modulo p, then

$$f(x) \equiv f(x) - f(c) \pmod{p}.$$

Hence

$$f(x) \equiv [a_1 + a_2(x + c) + \cdots$$
$$+ a_d(x^{d-1} + x^{d-2}c + \cdots + xc^{d-2} + c^{d-1})](x - c) \pmod{p},$$

T. Andreescu and D. Andrica, *Number Theory*, DOI: 10.1007/b11856_7,
© Birkhäuser Boston, a part of Springer Science + Business Media, LLC 2009

and we get $f(x) \equiv f_1(x)(x - c)$, where $f_1(x)$ is a polynomial of degree $d - 1$ modulo p with integer coefficients. If c' is another root of $f(x)$ modulo p such that $c' \not\equiv c \pmod{p}$, then since $f_1(c')(c' - c) \equiv 0 \pmod{p}$, we have $p \mid f_1(c')(c' - c)$. Hence, by Euclid's lemma (Proposition 1.3.2), $p \mid f_1(c')$. Thus c' is a root of $f_1(x)$ modulo p.

If now the integers $c_1, c_2, \ldots, c_k$ are the distinct roots of $f(x)$ modulo p, then

$$f(x) \equiv (x - c_1)(x - c_2) \cdots (x - c_k)g(x) \pmod{p}.$$

In fact, the degree modulo p of $g(x)$ is $d - k$. This implies $0 \le k \le d$.   □

Theorem 7.1.2. (Fermat's little theorem) *Let a be a positive integer and let p be a prime. Then*

$$a^p \equiv a \pmod{p}.$$

Proof. We induct on a. For $a = 1$ everything is clear. Assume that $p \mid a^p - a$. Then

$$(a + 1)^p - (a + 1) = (a^p - a) + \sum_{k=1}^{p-1} \binom{p}{k}a^k.$$

Using the fact that $p \mid \binom{p}{k}$ for $1 \le k \le p - 1$ and the inductive hypothesis, it follows that $p \mid (a + 1)^p - (a + 1)$, that is, $(a + 1)^p \equiv (a + 1) \pmod{p}$.

Alternative proof. Suppose that $\gcd(a, p) = 1$ and let us show that $a^{p-1} \equiv 1 \pmod{p}$. Consider the integers $a, 2a, \ldots, (p - 1)a$, whose remainders when divided by p are distinct (otherwise, if $ia \equiv ja \pmod{p}$, then $p \mid (i - j)a$, that is, $p \mid i - j$, which holds only if $i = j$). Hence the remainders are $1, 2, \ldots, p - 1$ in some order and

$$a \cdot (2a) \cdots (p - 1)a \equiv 1 \cdot 2 \cdots (p - 1) \pmod{p},$$

i.e.,

$$a^{p-1}(p - 1)! \equiv (p - 1)! \pmod{p}.$$

Because p and $(p - 1)!$ are relatively prime, the conclusion follows.   □

Remark. The converse is not true. For example, $3 \cdot 11 \cdot 17$ divides $a^{3 \cdot 11 \cdot 17} - a$, since 3, 11, 17 each divide $a^{3 \cdot 11 \cdot 17} - a$ (for instance, if 11 did not divide a, then from Fermat's little theorem, we would have $11 \mid a^{10} - 1$; hence $11 \mid a^{10 \cdot 56} - 1$, i.e., $11 \mid a^{561} - a$ and $561 = 3 \cdot 11 \cdot 17$).

The composite integers n satisfying $a^n \equiv a \pmod{n}$ for any integer a are called *Carmichael numbers*. There exist such integers, for example $n = 2 \cdot 73 \cdot 1103$. For other comments see the remark after Problem 7.3.11.

For problems involving x^n it might be good to work modulo a prime p with $p \equiv 1 \pmod{n}$.

Problem 7.1.1. *(1) Let a be a positive integer. Prove that any prime factor > 2 of $a^2 + 1$ is of the form $4m + 1$.*

(2) Prove that there are infinitely many primes of the form $4m + 1$.

Solution. (1) Assume that $p \mid a^2 + 1$ and $p = 4m + 3$ for some integer m. Then $a^2 \equiv -1 \pmod{p}$ and $a^{p-1} = (a^2)^{2m+1} \equiv (-1)^{2m+1} \equiv -1 \pmod{p}$, contradicting Fermat's little theorem.

(2) The integer $(n!)^2 + 1$ is of the form $4m + 1$; hence all its prime factors are of this form. It follows that for any prime p of the form $4m + 1$, $(p!)^2 + 1$ is a prime or has a prime factor $p_1 > p$ and we are done.

Problem 7.1.2. *For any prime p, $p^{p+1} + (p + 1)^p$ is not a perfect square.*

Solution. For $p = 2$ the property holds. Assume by way of contradiction that $p \geq 3$ and $p^{p+1} + (p + 1)^p = t^2$ for some positive integer t. It follows that $(t + p^{\frac{p+1}{2}})(t - p^{\frac{p+1}{2}}) = (p+1)^p$; hence $t \pm p^{\frac{p+1}{2}} = 2^{p-1}u^p$ and $t \mp p^{\frac{p+1}{2}} = 2v^p$, for some positive integers u, v such that $2uv = p + 1$ and $\gcd(u, v) = 1$. We obtain $p^{\frac{p+1}{2}} = |2^{p-2}u^p - v^p|$. Using Fermat's little theorem we have $u^p \equiv u$ (mod p), $v^p \equiv v$ (mod p), and $2^{p-1} \equiv 1$ (mod p), so $u \equiv 2v$ (mod p). From $2uv = p+1$ we get $u = 2v$ and since $\gcd(u, v) = 1$, this gives $v = 1$ and $p = 3$. This leads to $t^2 = 145$, a contradiction.

Problem 7.1.3. *Let $n \geq 2$, $a > 0$ be integers and p a prime such that $a^p \equiv 1$ (mod p^n). Show that if $p > 2$, then $a \equiv 1$ (mod p^{n-1}), and if $p = 2$, then $a \equiv \pm 1$ (mod 2^{n-1}).*

(1995 UNESCO Mathematical Contest)

Solution. We have $a^p \equiv 1$ (mod p^n) with $n \geq 2$, so $a^p \equiv 1$ (mod p). But from Fermat's little theorem, $a^p \equiv a$ (mod p); hence $a \equiv 1$ (mod p). For $a = 1$, the result is obvious; otherwise, put $a = 1 + kp^d$, where $d \geq 1$ and $p \nmid k$. Expanding $a^p = (1 + kp^d)^p$ using the binomial theorem and using the fact that $\binom{p}{m}$ is a multiple of p for $1 \leq m \leq p-1$, we see that for $p > 2$, $a^p = 1 + kp^{d+1} + Mp^{2d+1}$ for M an integer. Therefore $d + 1 \geq n$, and so $a \equiv 1$ (mod p^{n-1}). In case $p = 2$, we have $2^n \mid a^2 - 1 = (a - 1)(a + 1)$. Since these differ by 2, both cannot be multiples of 4. Hence either $a + 1$ or $a - 1$ is divisible by 2^{n-1}, i.e., $a \equiv \pm 1$ (mod 2^{n-1}), as desired.

Problem 7.1.4. *Find the smallest integer n such that among any n integers, with repetition allowed, there exist 18 integers whose sum is divisible by 18.*

(1997 Ukrainian Mathematical Olympiad)

Solution. The minimum is $n = 35$; the 34-element collection of 17 zeros and 17 ones shows that $n \geq 35$, so it remains to show that among 35 integers, there are 18 whose sum is divisible by 18. In fact, one can show that for any n, among $2n - 1$ integers there are n whose sum is divisible by n.

We prove this claim by induction on n; it is clear for $n = 1$. If n is composite, say $n = pq$, we can assemble collections of p integers whose sum is divisible by p as long as at least $2p - 1$ numbers remain; this gives $2q - 1$ collections, and again by the induction hypothesis, some q of these have sum divisible by q.

Now suppose $n = p$ is prime. The number x is divisible by p if and only if $x^{p-1} \not\equiv 1 \pmod{p}$. Thus if the claim is false, then the sum of $(a_1 + \cdots + a_p)^{p-1}$ over all subsets $\{a_1, \ldots, a_p\}$ of the given numbers is congruent to $\binom{2p-1}{p-1} \equiv 1$ (mod p). On the other hand, the sum of $a_1^{e_1} \cdots a_p^{e_p}$ for $e_1 + \cdots + e_p \le p - 1$ is always divisible by p: if $k \le p - 1$ of the e_i are nonzero, then each product is repeated $\binom{2p-1-k}{p-k}$ times, and the latter is a multiple of p. This contradiction shows that the claim holds in this case. (Note: to solve the original problem, of course, it suffices to prove the cases $p = 2, 3$ directly.)

Remark. The fact that for any n, among $2n - 1$ integers there are n whose sum is divisible by n is a famous theorem of Erdős and Ginzburg.

Problem 7.1.5. *Several integers are given (some of them may be equal) whose sum is equal to 1492. Decide whether the sum of their seventh powers can equal*
    (a) 1996;
    (b) 1998.

(1997 Czech–Slovak Match)

Solution. (a) Consider a set of 1492 1's, 4 2's, and 8 -1's. Their sum is 1492, and the sum of their seventh powers is $1492(1) + 4(128) + 8(-1) = 1996$.

(b) By Fermat's little theorem, $x^7 \equiv x \pmod{7}$. Thus, the sum of the numbers' seventh powers must be congruent to the sum of the numbers modulo 7. But $1998 \not\equiv 1492 \pmod{7}$, so the numbers' seventh powers cannot add up to 1998.

Problem 7.1.6. *Find the number of integers $n > 1$ for which the number $a^{25} - a$ is divisible by n for each integer a.*

(1995 Bulgarian Mathematical Olympiad)

Solution. Let n have the required property. Then p^2 (p prime) cannot divide n, since p^2 does not divide $p^{25} - p$. Hence n is the product of distinct prime numbers. On the other hand, $2^{25} - 2 = 2 \cdot 3^2 \cdot 5 \cdot 7 \cdot 13 \cdot 17 \cdot 241$. But n is not divisible by 17 or 241, since $3^{25} \equiv -3 \pmod{17}$ and $3^{25} \equiv 32 \pmod{241}$. Fermat's little theorem implies that $a^{25} \equiv a \pmod{p}$ when $p = 2, 3, 5, 7, 13$. Thus n should be equal to the divisors of $2 \cdot 3 \cdot 5 \cdot 7 \cdot 13$ that are different from 1, and there are $2^5 - 1 = 31$ of them.

Problem 7.1.7. *(a) Find all positive integers n such that 7 divides $2^n - 1$.*
    (b) Prove that for any positive integer n the number $2^n + 1$ cannot be divisible by 7.

(6th International Mathematical Olympiad)

Solution. Fermat's little theorem gives

$$2^{6k} \equiv 1 \pmod 7.$$

It follows from the divisibility $7 \mid (2^{3k} - 1)(2^{3k} + 1)$ that $2^{3k} \equiv 1 \pmod 7$, since $2^{3k} + 1 = 8^k + 1 \equiv 2 \pmod 7$. Hence all numbers n that are divisible by 3 answer the question.

Let $n = 3k + r$, where $r = 1$ or $r = 2$. Then

$$2^n \equiv 2^{3k+r} \equiv (2^3)^k \cdot 2^r = 2 \text{ or } 4 \pmod 7.$$

Hence, we cannot obtain $2^n \equiv -1 \pmod 7$.

Problem 7.1.8. *Consider the sequence $a_1, a_2, \ldots$ defined by*

$$a_n = 2^n + 3^n + 6^n - 1 \quad (n = 1, 2, \ldots).$$

Determine all positive integers that are relatively prime to every term of the sequence.

(46th International Mathematical Olympiad)

Solution. If $p > 3$, then $2^{p-2} + 3^{p-2} + 6^{p-2} \equiv 1 \pmod p$. To see this, multiply both sides by 6 to get

$$3 \cdot 2^{p-1} + 2 \cdot 3^{p-1} + 6^{p-1} \equiv 6 \pmod p,$$

which is a consequence of Fermat's little theorem. Therefore p divides a_{p-2}. Also, 2 divides a_1 and 3 divides a_2. So, there is no number other than 1 that is relatively prime to all the terms in the sequence.

Problem 7.1.9. *Prove that the sequence $\{2^n - 3 \mid n = 2, 3, \ldots\}$ contains a subsequence whose members are all relatively prime.*

(13th International Mathematical Olympiad)

Solution. We use induction. The numbers $2^2 - 3$, $2^3 - 3$, $2^4 - 3$ are pairwise relatively prime numbers. We shall prove that if $n_1, n_2, \ldots, n_k$ are positive integers such that the members of the sequence

$$2^{n_1} - 3, \quad 2^{n_2} - 3, \quad \ldots, \quad 2^{n_k} - 3 \tag{1}$$

are relatively prime to each other, then there exists n_{k+1} such that $2^{n_{k+1}} - 3$ is relatively prime to each number of the sequence (1).

Let $\{p_1, p_2, \ldots, p_r\}$ be the set of all prime divisors of numbers from the sequence (1). Then $p_1, p_2, \ldots, p_r$ are odd prime numbers, and by Fermat's little theorem,

$$2^{p_i - 1} \equiv 1 \pmod{p_i}.$$

It follows that

$$2^{(p_1-1)(p_2-1)\cdots(p_r-1)} \equiv 1 \pmod{p_i}, \ \forall \, i = 1, \ldots, r.$$

Let $n_{k+1} = \prod_{i=1}^{r}(p_i - 1)$. We shall prove that $2^{n_i} - 3$ and $2^{n_{k+1}} - 3$ are relatively prime for all $i = 1, \ldots, r$. Let p be a common prime divisor of $2^{n_i} - 3$ and $2^{n_{k+1}} - 3$. Then $2^{n_{k+1}} - 3 \equiv 1 - 3 \pmod{p} \equiv 0 \pmod{p}$; this is a contradiction.

Problem 7.1.10. *Let $p \geq 2$ be a prime number such that $3 \mid (p - 2)$. Let*

$$S = \{y^2 - x^3 - 1 \mid x \text{ and } y \text{ are integers, } 0 \leq x, y \leq p - 1\}.$$

Prove that at most p elements of S are divisible by p.

(1999 Balkan Mathematical Olympiad)

First solution. We need the following lemma.

Lemma. *Given a prime p and a positive integer $k > 1$, if k and $p-1$ are relatively prime, then $x^k \equiv y^k \pmod{p} \Rightarrow x \equiv y \pmod{p}$ for all x, y.*

Proof. If $y \equiv 0 \pmod{p}$ the claim is obvious. Otherwise, note that $x^k \equiv y^k \Rightarrow (xy^{-1})^k \equiv 1 \pmod{p}$, so it suffices to prove that $a^k \equiv 1 \pmod{p} \Rightarrow a \equiv 1 \pmod{p}$.

Because $\gcd(p - 1, k) = 1$, there exist integers b and c such that $b(p - 1) + ck = 1$. Thus, $a^k \equiv 1 \pmod{p} \Rightarrow a^c \equiv 1 \pmod{p} \Rightarrow a^{1-b(p-1)} \equiv 1 \pmod{p}$. If $a = 0$ this is impossible. Otherwise, by Fermat's little theorem, $(a^{-b})^{p-1} \equiv 1 \pmod{p}$, so that $a \equiv 1 \pmod{p}$, as desired.

Alternatively, again note that clearly $a \not\equiv 0 \pmod{p}$. Then let d be the order of a, the smallest positive integer such that $a^d \equiv 1 \pmod{p}$; we have $d \mid k$. Take the set $\{1, a, a^2, \ldots, a^{d-1}\}$. If it does not contain all of $1, 2, \ldots, p - 1$ then pick some other element b and consider the set $\{b, ba, ba^2, \ldots, ba^{d-1}\}$. These two sets are disjoint, because otherwise $ba^i \equiv a^j \Rightarrow b \equiv a^{j-1} \pmod{p}$, a contradiction. Continuing similarly, we can partition $\{1, 2, \ldots, p - 1\}$ into d-element subsets, and hence $d \mid p - 1$. However, $d \mid k$ and $\gcd(k, p - 1) = 1$, implying that $d = 1$. Therefore $a \equiv a^d \equiv 1 \pmod{p}$, as desired. $\square$

Because $3 \mid p - 2$, $\gcd(3, p - 1) = 1$. Then from the claim, it follows that the set of elements $\{1^3, 2^3, \ldots, p^3\}$ equals $\{1, 2, \ldots, p\}$ modulo p. Hence, for each y with $0 \leq y \leq p - 1$, there is exactly one x between 0 and $p - 1$ such that $x^3 \equiv y^2 - 1 \pmod{p}$, that is, such that $p \mid y^2 - x^3 - 1$. Therefore S contains at most p elements divisible by p, as desired.

Second solution. Note that applying Fermat's little theorem repeatedly, we get that for p prime, $a^{m(p-1)+1} \equiv a \pmod{p}$. Since $\gcd(k, p-1) = 1$ by the lemma from Problem 1.3.2, there are positive integers a and b with $ak - b(p-1) = 1$ or $ak = b(p-1) + 1$. Since $x^k \equiv y^k \pmod{p}$, we have $x^{ak} \equiv y^{ak} \pmod{p}$. Since $x^{ak} = x^{b(p-1)+1} \equiv x \pmod{p}$, and similarly for y, we conclude that $x \equiv y \pmod{p}$.

Additional Problems

Problem 7.1.11. Let $3^n - 2^n$ be a power of a prime for some positive integer n. Prove that n is a prime.

Problem 7.1.12. Let $f(x_1, \ldots, x_n)$ be a polynomial with integer coefficients of total degree less than n. Show that the number of ordered n-tuples $(x_1, \ldots, x_n)$ with $0 \le x_i \le 12$ such that $f(x_1, \ldots, x_n) \equiv 0 \pmod{13}$ is divisible by 13.

(1998 Turkish Mathematical Olympiad)

Problem 7.1.13. Find all pairs (m, n) of positive integers, with $m, n \ge 2$, such that $a^n - 1$ is divisible by m for each $a \in \{1, 2, \ldots, n\}$.

(2001 Romanian International Mathematical Olympiad Team Selection Test)

Problem 7.1.14. Let p be a prime and b_0 an integer, $0 < b_0 < p$. Prove that there exists a unique sequence of base-p digits $b_0, b_1, b_2, \ldots, b_n, \ldots$ with the following property: If the base-p representation of a number x ends in the group of digits $b_n b_{n-1} \ldots b_1 b_0$, then so does the representation of x^p.

Problem 7.1.15. Determine all integers $n > 1$ such that $\frac{2^n + 1}{n^2}$ is an integer.

(31st International Mathematical Olympiad)

Problem 7.1.16. Prove that for any $n > 1$ we cannot have $n \mid 2^{n-1} + 1$.

(Sierpiński)

Problem 7.1.17. Prove that for any natural number n, $n!$ is a divisor of

$$\prod_{k=0}^{n-1} (2^n - 2^k).$$

7.2 Euler's Theorem

Theorem 7.2.1. (Euler's theorem) *Let a and n be relatively prime positive integers. Then $a^{\varphi(n)} \equiv 1 \pmod{n}$.*

Proof. Consider the set $S = \{a_1, a_2, \ldots, a_{\varphi(n)}\}$ consisting of all positive integers less than n that are relatively prime to n. Because $\gcd(a, n) = 1$, it follows that $aa_1, aa_2, \ldots, aa_{\varphi(n)}$ is a permutation of $a_1, a_2, \ldots, a_{\varphi(n)}$. Then

$$(aa_1)(aa_2) \cdots (aa_{\varphi(n)}) \equiv a_1 a_2 \cdots a_{\varphi(n)} \pmod{n}.$$

Using that $\gcd(a_k, n) = 1$, $k = 1, 2, \ldots, \varphi(n)$, the conclusion now follows. $\square$

Remark. Euler's theorem also follows from Fermat's little theorem. Indeed, let $n = p_1^{\alpha_1} \cdots p_k^{\alpha_k}$ be the prime factorization of n. We have $a^{p_i-1} \equiv 1 \pmod{p_i}$; hence $a^{p_i(p_i-1)} \equiv 1 \pmod{p_i^2}$, $a^{p_i^2(p_i-1)} \equiv 1 \pmod{p_i^3}, \ldots, a^{p_i^{\alpha_i-1}(p_i-1)} \equiv 1 \pmod{p_i^{\alpha_i}}$. That is, $a^{\varphi(p_i^{\alpha_i})} \equiv 1 \pmod{p_i^{\alpha_i}}$, $i = 1, \ldots, k$. Applying this property to each prime factor, the conclusion follows.

Problem 7.2.1. *Prove that for any positive integer s, there is a positive integer n whose sum of digits is s and $s \mid n$.*

(Sierpiński)

Solution. If $\gcd(s, 10) = 1$ then let $n = 10^{s\varphi(s)} + 10^{(s-1)\varphi(s)} + \cdots + 10^{\varphi(s)}$. It is clear that the sum of digits of n is s and that

$$n = (10^{s\varphi(s)} - 1) + (10^{(s-1)\varphi(s)} - 1) + \cdots + (10^{\varphi(s)} - 1) + s$$

is divisible by s, by Euler's theorem.

If $\gcd(s, 10) > 1$, then let $s = 2^a 5^b t$ with $\gcd(t, 10) = 1$ and take $n = 10^{a+b}(10^{s\varphi(t)} + 10^{(s-1)\varphi(t)} + \cdots + 10^{\varphi(t)})$.

Remark. The integers divisible by the sum of its digits are called Niven numbers. For some information about these numbers see the remark after Problem 4.2.12.

Problem 7.2.2. *Let $n > 3$ be an odd integer with prime factorization $n = p_1^{\alpha_1} \cdots p_k^{\alpha_k}$ (each p_i is prime). If*

$$m = n \left(1 - \frac{1}{p_1}\right) \left(1 - \frac{1}{p_2}\right) \cdots \left(1 - \frac{1}{p_k}\right),$$

prove that there is a prime p such that p divides $2^m - 1$, but does not divide m.

(1995 Iranian Mathematical Olympiad)

Solution. Because $m = \varphi(n)$ is Euler's phi function and n is odd, we know by Euler's theorem that n divides $2^m - 1$. We consider two cases.

First let $n = p^r > 3$ for some odd prime p. Then $m = p^r - p^{r-1}$ is even and $m \geq 4$. Since p divides

$$2^m - 1 = (2^{m/2} - 1)(2^{m/2} + 1),$$

is must also divide one of the factors on the right. Any prime divisor of the other factor (note that this factor exceeds 1) will also divide $2^m - 1$ but will not divide $n = p^r$.

If n has at least two distinct prime factors, then $m \equiv 0 \pmod 4$ and $p - 1$ divides $m/2$ for each prime factor of n. Hence, by Fermat's theorem, p also divides $2^{m/2} - 1$. It follows that no prime factor of n divides $2^{m/2} + 1$. Hence any prime factor of $2^{m/2} + 1$ is a factor of $2^m - 1$ but not a factor of n.

Problem 7.2.3. *Let $a > 1$ be an integer. Show that the set*

$$\{a^2 + a - 1, a^3 + a^2 - 1, \ldots\}$$

contains an infinite subset any two members of which are relatively prime.

(1997 Romanian International Mathematical Olympiad Team Selection Test)

Solution. We show that any set of n elements of the set that are pairwise coprime can be extended to a set of $n+1$ elements. For $n = 1$, note that any two consecutive terms in the sequence are relatively prime. For $n > 1$, let N be the product of the numbers in the set so far; then $a^{\varphi(N)+1} + a^{\varphi(N)} - 1 \equiv a \pmod{N}$, and so this number can be added (since every element of the sequence is coprime to a, N is as well).

Problem 7.2.4. *Let m and n be integers greater than 1 such that $\gcd(m, n - 1) = \gcd(m, n) = 1$. Prove that the first $m - 1$ terms of the sequence $n_1, n_2, \ldots$, where $n_1 = mn + 1$ and $n_{k+1} = n \cdot n_k + 1$, $k \geq 1$, cannot all be primes.*

Solution. It is straightforward to show that

$$n_k = n^k m + n^{k-1} + \cdots + n + 1 = n^k m + \frac{n^k - 1}{n - 1}$$

for every positive integer k. Hence

$$n_{\varphi(m)} = n^{\varphi(m)} \cdot m + \frac{n^{\varphi(m)} - 1}{n - 1}.$$

From Euler's theorem, $m \mid (n^{\varphi(m)} - 1)$, and since $\gcd(m, n-1) = 1$, it follows that

$$m \mid \frac{n^{\varphi(m)} - 1}{n - 1}.$$

Consequently, m divides $n_{\varphi(m)}$. Because $\varphi(m) \leq m - 1$, $n_{\varphi(m)}$ is not a prime, and we are done.

Additional Problems

Problem 7.2.5. Prove that for every positive integer n, there exists a polynomial with integer coefficients whose values at $1, 2, \ldots, n$ are distinct powers of 2.

(1999 Hungarian Mathematical Olympiad)

Problem 7.2.6. Let $a > 1$ be an odd positive integer. Find the least positive integer n such that 2^{2000} is a divisor of $a^n - 1$.

(2000 Romanian International Mathematical Olympiad Team Selection Test)

Problem 7.2.7. Let $n = p_1^{r_1} \cdots p_k^{r_k}$ be the prime factorization of the positive integer n and let $r \geq 2$ be an integer. Prove that the following are equivalent:

(a) The equation $x^r \equiv a \pmod{n}$ has a solution for every a.

(b) $r_1 = r_2 = \cdots = r_k = 1$ and $\gcd(p_i - 1, r) = 1$ for every $i \in \{1, 2, \ldots, k\}$.

(1995 UNESCO Mathematical Contest)

7.3   The Order of an Element

Given are the positive integer $n > 1$ and an integer a such that $\gcd(a, n) = 1$. The smallest positive integer d for which $n \mid a^d - 1$ is called the *order of a modulo n*. Observe first of all that the definition is well defined, since from Euler's theorem we have $n \mid a^{\varphi(n)} - 1$, so such numbers d indeed exist. In what follows we will denote by $o_n(a)$ the order of a modulo n. The following properties hold:

(1) If $a^m \equiv 1 \pmod{n}$, then $o_n(a) \mid m$;

(2) $o_n(a) \mid \varphi(n)$; if p is a prime, then $o_p(n) \mid p - 1$ for any n.

(3) If $a^l \equiv a^m \pmod{n}$, then $l \equiv m \pmod{o_n(a)}$.

In order to prove property (1) let us consider $d = o_n(a)$. Indeed, because $n \mid a^m - 1$ and $n \mid a^d - 1$, we find that $n \mid a^{\gcd(m,d)} - 1$ (see also Proposition 1.3.4). But from the definition of d it follows that $d \leq \gcd(m, d)$, which cannot hold unless $d \mid m$.

The positive integer a is called a *primitive root modulo n* if we have $\gcd(a, n) = 1$ and $o_n(a) = \varphi(n)$. One can show that there are primitive roots modulo n if and only if $n \in \{2, 4, p^\alpha, 2p^\alpha\}$, where $p \geq 3$ is any prime and α is any positive integer.

Problem 7.3.1. *Prove that $n \mid \varphi(a^n - 1)$ for all positive integers a, n.*

(Saint Petersburg Mathematical Olympiad)

Solution. What is $o_{a^n-1}(a)$? It may seem a silly question, since of course $o_{a^n-1}(a) = n$. Using the observation in the introduction, we obtain exactly $n \mid \varphi(a^n - 1)$.

Problem 7.3.2. *Prove that any prime factor of the nth Fermat number $2^{2^n} + 1$ is congruent to 1 modulo 2^{n+1}. Show that there are infinitely many prime numbers of the form $2^n k + 1$ for any fixed n.*

Solution. Let us consider a prime p such that $p \mid 2^{2^n} + 1$. Then $p \mid 2^{2^{n+1}} - 1$ and consequently $o_p(2) \mid 2^{n+1}$. This ensures the existence of a positive integer $k \leq n + 1$ such that $o_p(2) = 2^k$. We will prove that in fact $k = n + 1$. The proof is easy. Indeed, if this is not the case, then $o_p(2) \mid 2^n$ and so $p \mid 2^{o_p(2)} - 1 \mid 2^{2^n} - 1$.

But this is impossible, since $p \mid 2^{2^n} + 1$ and p is odd. Therefore, we have found that $o_p(2) = 2^{n+1}$ and we have to prove that $o_p(2) \mid p - 1$ to finish the first part of the question. But this follows from the introduction.

The second part is a direct consequence of the first. Indeed, it is enough to prove that there exists an infinite set of Fermat numbers $(2^{2^{n_k}} + 1)_{n_k > a}$ any two relatively prime. Then we could take a prime factor of each such Fermat number and apply the first part to obtain that each such prime is of the form $2^n k + 1$. Not only is it easy to find such a sequence of coprime Fermat numbers, but in fact any two distinct Fermat numbers are relatively prime. Indeed, suppose that $d \mid \gcd(2^{2^n} + 1, 2^{2^{n+k}} + 1)$. Then $d \mid 2^{2^{n+1}} - 1$ and so $d \mid 2^{2^{n+k}} - 1$. Combining this with $d \mid 2^{2^{n+k}} + 1$, we obtain a contradiction. Hence both parts of the problem are solved.

Problem 7.3.3. *For a prime p, let $f_p(x) = x^{p-1} + x^{p-2} + \cdots + x + 1$.*

(a) If $p \mid m$, prove that there exists a prime factor of $f_p(m)$ that is relatively prime with $m(m - 1)$.

(b) Prove that there are infinitely many numbers n such that $pn + 1$ is prime, for any fixed n.

(2003 Korean International Mathematical Olympiad Team Selection Test)

Solution. Part (a) is straightforward. In fact, we will prove that any prime factor of $f_p(m)$ is relatively prime to $m(m-1)$. Take such a prime divisor q. Because $q \mid 1 + m + \cdots + m^{p-1}$, it is clear that $\gcd(q, m) = 1$. Moreover, if $\gcd(q, m-1) \neq 1$, then $q \mid m - 1$, and because $q \mid 1 + m + \cdots + m^{p-1}$, it follows that $q \mid p$. But $p \mid m$ and we find that $q \mid m$, which is clearly impossible.

More difficult is (b). But we are tempted to use (a) and to explore the properties of $f_p(m)$, just as in the previous problem. So, let us take a prime $q \mid f_p(m)$ for a certain positive integer m divisible by p. By part (a), $\gcd(q, m - 1) = 1$. Since $q \mid m^p - 1$, we must have $o_q(m) = p$ and hence $q \equiv 1 \pmod{p}$. Now we need to find a sequence $(m_k)_{k \geq 1}$ of multiples of p such that $f_p(m_k)$ are pairwise relatively prime. This is not as easy as in the first example. Anyway, just by trial and error, it is not difficult to find such a sequence. There are many other approaches, but we like the following one: take $m_1 = p$ and $m_k = pf(m_1)f_p(m_2) \cdots f_p(m_{k-1})$. Let us prove that $f_p(m_k)$ is relatively prime to $f_p(m_1), f_p(m_2), \ldots, f_p(m_{k-1})$. Fortunately, this is easy, since $f_p(m_1)f_p(m_2) \cdots f_p(m_{k-1}) \mid f_p(m_k) - f_p(0) = f_p(m_k) - 1$. The solution ends here.

Problem 7.3.4. *Find the smallest number n with the property that*

$$2^{2005} \mid 17^n - 1.$$

Solution. The problem actually asks for $o_{2^{2005}}(17)$. We know that $o_{2^{2005}}(17) \mid \varphi(2^{2005}) = 2^{2004}$, so $o_{2^{2005}}(17) = 2^k$, where $k \in \{1, 2, \ldots, 2004\}$. The order of an element has done its job. Now it is time to work with exponents. We have $2^{2005} \mid 17^{2^k} - 1$. Using the factorization

$$17^{2^k} - 1 = (17 - 1)(17 + 1)(17^2 + 1) \cdots (17^{2^{k-1}} + 1),$$

we proceed by finding the exponent of 2 in each factor of this product. But this is not difficult, because for all $i \geq 0$ the number $17^{2^i} + 1$ is a multiple of 2, but not a multiple of 4. Thus, $v_2(17^{2^k} - 1) = 4 + k$, and the order is found by solving the equation $k + 4 = 2005$. Thus, $o_{2^{2005}}(17) = 2^{2001}$ is the answer to the problem.

Problem 7.3.5. *Find all prime numbers p, q such that $p^2 + 1 \mid 2003^q + 1$ and $q^2 + 1 \mid 2003^p + 1$.*

Solution. Let us suppose that $p \leq q$. We discuss first the trivial case $p = 2$. In this case, $5 \mid 2003^q + 1$, and it is easy to deduce that q is even; hence $q = 2$, which is a solution of the problem. Now, suppose that $p > 2$ and let r be a prime factor of $p^2 + 1$. Because $r \mid 2003^{2q} - 1$, it follows that $o_r(2003) \mid 2q$. Suppose that $(q, o_r(2003)) = 1$. Then $o_r(2003) \mid 2$ and $r \mid 2003^2 - 1$, but we cannot have $r \mid 2003 - 1$, since this would give $2003^q \equiv 1 \pmod r$, contrary to the hypotheses. Thus $r \mid 2003 + 1 = 2^2 \cdot 3 \cdot 167$. It seems that this is a dead end, since there are too many possible values for r. Another simple observation narrows the number of possible cases: because $r \mid p^2 + 1$, r must be of the form $4k + 1$ or equal to 2, and now we do not have many possibilities: $r \in \{2, 13\}$. The case $r = 13$ is also impossible, because $2003^q + 1 \equiv 2 \pmod{13}$ and $r \mid 2003^q + 1$. So, we have found that for any prime factor r of $p^2 + 1$, we have either $r = 2$ or $q \mid o_r(2003)$, which in turn implies $q \mid r - 1$. Because $p^2 + 1$ is even but not divisible by 4 and because any odd prime factor of it is congruent to 1 modulo q, we must have $p^2 + 1 \equiv 2 \pmod q$. This implies that $p^2 + 1 \equiv 2 \pmod q$, that is, $q \mid (p - 1)(p + 1)$. Combining this with the assumption that $p \leq q$ yields $q \mid p+1$ and in fact $q = p+1$. It follows that $p = 2$, contradicting the assumption $p > 2$. Therefore the only pair is $(2, 2)$.

Additional Problems

Problem 7.3.6. Find all ordered triples of primes (p, q, r) such that

$$p \mid q^r + 1, \quad q \mid r^p + 1, \quad r \mid p^q + 1.$$

(2003 USA International Mathematical Olympiad Team Selection Test)

Problem 7.3.7. Find all primes p, q such that $pq \mid 2^p + 2^q$.

Problem 7.3.8. Prove that for any positive integer $n \geq 2$, $3^n - 2^n$ is not divisible by n.

Problem 7.3.9. Find all positive integers m, n such that $n \mid 1 + m^{3^n} + m^{2 \cdot 3^n}$.

(Bulgarian International Mathematical Olympiad Team Selection Test)

Problem 7.3.10. Let $a, n > 2$ be positive integers such that $n \mid a^{n-1} - 1$ and n does not divide any of the numbers $a^x - 1$, where $x < n - 1$ and $x \mid n - 1$. Prove that n is a prime number.

Problem 7.3.11. Find all prime numbers p, q for which the congruence

$$\alpha^{3pq} \equiv \alpha \pmod{3pq}$$

holds for all integers α.

(1996 Romanian Mathematical Olympiad)

Problem 7.3.12. Let p be a prime number. Prove that there exists a prime number q such that for every integer n, the number $n^p - p$ is not divisible by q.

(44th International Mathematical Olympiad)

7.4 Wilson's Theorem

Theorem 7.4.1. (Wilson's[1] theorem) *For any prime p, $p \mid (p - 1)! + 1$.*

Proof. The property holds for $p = 2$ and $p = 3$, so we may assume that $p \geq 5$. Let $S = \{2, 3, \ldots, p - 2\}$. For any h in S, the integers $h, 2h, \ldots, (p - 1)h$ yield distinct remainders when divided by p. Hence there is a unique $h' \in \{1, 2, \ldots, p - 1\}$ such that $hh' \equiv 1 \pmod{p}$. Moreover, $h' \neq 1$ and $h' \neq p - 1$; hence $h' \in S$. In addition, $h' \neq h$; otherwise, $h^2 \equiv 1 \pmod{p}$, implying $p \mid h - 1$ or $p \mid h + 1$, which is not possible, since $h + 1 < p$. It follows that we can group the elements of S in $\frac{p-3}{2}$ distinct pairs (h, h') such that $hh' \equiv 1 \pmod{p}$. Multiplying these congruences gives $(p - 2)! \equiv 1 \pmod{p}$, and the conclusion follows. $\qquad\square$

Alternative proof. The property is trivially true when $p = 2$, so assume that p is odd. By Fermat's little theorem, the polynomial $x^{p-1} - 1$ has for its $p - 1$ distinct roots modulo p the numbers $1, 2, \ldots, p-1$. According to Theorem 7.1.1 we have

$$(x - 1)(x - 2) \cdots (x - p + 1) \equiv x^{p-1} - 1 \pmod{p}.$$

By setting $x = 0$ we obtain

$$(-1)^{p-1}(p - 1)! \equiv -1 \pmod{p}.$$

Since $p - 1$ is even, the result follows. $\qquad\square$

Remark. The converse is true, that is, if $n \mid (n - 1)! + 1$ for an integer $n \geq 2$, then n is a prime. Indeed, if n were equal to $n_1 n_2$ for some integers $n_1, n_2 \geq 2$, we would have $n_1 \mid 1 \cdot 2 \cdots n_1 \cdots (n - 1) + 1$, which is not possible.

---

[1] John Wilson (1741–1793), English mathematician who published this result without proof. It was first proved in 1773 by Lagrange, who showed that the converse is also true.

Problem 7.4.1. *If p is an odd prime, then the remainder when $(p-1)!$ is divided by $p(p-1)$ is $p-1$.*

Solution. We need to show that $(p-1)! \equiv p-1 \pmod{p(p-1)}$.

From Wilson's theorem we obtain $(p-1)! - (p-1) \equiv 0 \pmod{p}$. Because $(p-1)! - (p-1) \equiv 0 \pmod{p-1}$ and $\gcd(p, p-1) = 1$, we get

$$(p-1)! - (p-1) \equiv 0 \pmod{p(p-1)}.$$

Remark. Generally, if $a \equiv b \pmod{u}$, $a \equiv b \pmod{v}$, and $\gcd(u, v) = 1$, then $a \equiv b \pmod{uv}$.

Problem 7.4.2. *Let p be an odd prime and $a_1, a_2, \ldots, a_p$ an arithmetic sequence whose common difference is not divisible by p. Prove that there is an $i \in \{1, 2, \ldots, p\}$ such that $a_i + a_1 a_2 \cdots a_p \equiv 0 \pmod{p^2}$.*

Solution. Note that $a_1, a_2, \ldots, a_p$ give distinct remainders when divided by p. Take i such that $a_i \equiv 0 \pmod{p}$. It follows that

$$\frac{a_1 a_2 \cdots a_p}{a_i} \equiv (p-1)! \pmod{p}.$$

From Wilson's theorem, we have $(p-1)! \equiv -1 \pmod{p}$, and the conclusion follows.

Problem 7.4.3. *Let a and n be positive integers such that $n \geq 2$ and $\gcd(a, n) = 1$. Prove that*

$$a^{n-1} + (n-1)! \equiv 0 \pmod{n}$$

if and only if n is a prime.

Solution. If n is a prime, the conclusion follows from Fermat's little theorem and Wilson's theorem.

For the converse, assume by way of contradiction that $n = n_1 n_2$, where $n_1 \geq n_2 \geq 2$.

Because $n \mid a^{n-1} + (n-1)!$, it follows that $n_1 \mid a^{n-1} + (n-1)!$, that is, $n_1 \mid a^{n-1}$, contradicting the hypothesis $\gcd(a, n) = 1$.

Problem 7.4.4. *(1) If p is a prime, then for any positive integer $n < p$,*

$$(n-1)!(p-n)! \equiv (-1)^n \pmod{p}.$$

(2) If p is a prime, then $\binom{p-1}{k} \equiv (-1)^k \pmod{p}$ for all $k = 0, 1, \ldots, p-1$.

Solution. (1) The property is obvious for $p = 2$, so assume that p is odd. From Wilson's theorem, $(p-1)! \equiv -1 \pmod{p}$; hence

$$(n-1)!n(n+1) \cdots (p-1) \equiv -1 \pmod{p}.$$

This is equivalent to

$$(n - 1)!(p - (p - n))(p - (p - n - 1)) \cdots (p - 1) \equiv -1 \quad (\mathrm{mod} \ p).$$

But $p - k \equiv -k \pmod{p}$, $k = 1, 2, \ldots, p - n$; hence

$$(n - 1)!(-1)^{p-n}(p - n)! \equiv -1 \quad (\mathrm{mod} \ p),$$

and taking into account that p is odd, the conclusion follows.

(2) We have

$$\binom{p - 1}{k} = \frac{(p - 1)!}{k!(p - k - 1)!},$$

hence $k!(p - k - 1)!\binom{p-1}{k} = (p - 1)! \equiv -1 \pmod{p}$. Applying the result in (1) for $n = k + 1$, it follows that $k!(p - k - 1)! \equiv (-1)^{k+1} \pmod{p}$ and we are done.

Additional Problems

Problem 7.4.5. Let p be an odd prime. Prove that

$$1^2 \cdot 3^2 \cdots (p - 2)^2 \equiv (-1)^{\frac{p+1}{2}} \quad (\mathrm{mod} \ p)$$

and

$$2^2 \cdot 4^2 \cdots (p - 1)^2 \equiv (-1)^{\frac{p+1}{2}} \quad (\mathrm{mod} \ p).$$

Problem 7.4.6. Show that there do not exist nonnegative integers k and m such that $k! + 48 = 48(k + 1)^m$.

(1996 Austrian–Polish Mathematics Competition)

Problem 7.4.7. For each positive integer n, find the greatest common divisor of $n! + 1$ and $(n + 1)!$.

(1996 Irish Mathematical Olympiad)

Problem 7.4.8. Let $p \geq 3$ be a prime and let σ be a permutation of $\{1, 2, \ldots, p - 1\}$. Prove that there are $i \neq j$ such that $p \mid i\sigma(i) - j\sigma(j)$.

(1986 Romanian International Mathematical Olympiad Team Selection Test)

8

Diophantine Equations

8.1 Linear Diophantine Equations

An equation of the form

$$a_1 x_1 + \cdots + a_n x_n = b, \tag{1}$$

where $a_1, a_2, \ldots, a_n, b$ are fixed integers, is called a *linear Diophantine*[1] *equation*. We assume that $n \geq 1$.

The main result concerning linear Diophantine equations is the following:

Theorem 8.1.1. *Equation (1) is solvable if and only if*

$$\gcd(a_1, \ldots, a_n) \mid b.$$

In case of solvability, one can choose $n - 1$ solutions such that any solution is a integer linear combination of those $n - 1$.

Proof. Let $d = \gcd(a_1, \ldots, a_n)$. If b is not divisible by d, then (1) is not solvable, since for any integers $x_1, \ldots, x_n$ the left-hand side of (1) is divisible by d and the right-hand side is not.

Actually, we need to prove that $\gcd(x_1, x_2, \ldots, x_n)$ is a linear combination with integer coefficients of $x_1, x_2, \ldots, x_n$. For $n = 2$ this follows from Proposition 1.3.1. Since

$$\gcd(x_1, \ldots, x_n) = \gcd(\gcd(x_1, \ldots, x_{n-1}), x_n),$$

we obtain that $\gcd(x_1, \ldots, x_n)$ is a linear combination of x_n and $\gcd(x_1, \ldots, x_{n-1})$; thus by the induction hypothesis, it is a linear combination of $x_1, \ldots, x_{n-1}, x_n$. $\square$

---

[1] Diophantus of Alexandria (ca. 200–284), Greek mathematician sometimes known as "the father of algebra" who is best known for his book *Arithmetica*. This book had an enormous influence on the development of number theory.

T. Andreescu and D. Andrica, *Number Theory*, DOI: 10.1007/b11856_8,
© Birkhäuser Boston, a part of Springer Science + Business Media, LLC 2009

Corollary 8.1.2. *Let a_1, a_2 be relatively prime integers. If (u, v) is a solution to the equation*

$$a_1 x_1 + a_2 x_2 = b, \tag{2}$$

then all of its solutions are given by

$$\begin{cases} x_1 = u + a_2 t, \\ x_2 = v - a_1 t, \end{cases} \tag{3}$$

where $t \in \mathbb{Z}$.

Example. *Solve the equation*

$$3x + 4y + 5z = 6.$$

Solution. Working modulo 5, we have $3x + 4y \equiv 1 \pmod 5$; hence

$$3x + 4y = 1 + 5s, \quad s \in \mathbb{Z}.$$

A solution to this equation is $x = -1 + 3s$, $y = 1 - s$. Applying (3) we obtain $x = -1 + 3s + 4t$, $y = 1 - s - 3t$, $t \in \mathbb{Z}$, and substituting back into the original equation yields $z = 1 - s$. Hence all solutions are

$$(x, y, z) = (-1 + 3s + 4t, 1 - s - 3t, 1 - s), \quad s, t \in \mathbb{Z}.$$

Problem 8.1.1. *Solve in nonnegative integers the equation*

$$x + y + z + xyz = xy + yz + zx + 2.$$

Solution. We have

$$xyz - (xy + yz + zx) + x + y + z - 1 = 1,$$

and consequently,

$$(x - 1)(y - 1)(z - 1) = 1.$$

Because x, y, z are nonnegative integers, we obtain

$$x - 1 = y - 1 = z - 1 = 1,$$

so $x = y = z = 2$.

Problem 8.1.2. *Find all triples (x, y, z) of integers such that*

$$x^2(y - z) + y^2(z - x) + z^2(x - y) = 2.$$

Solution. The equation is equivalent to

$$(x - y)(x - z)(y - z) = 2.$$

Observe that $(x - y) + (y - z) = x - z$. On the other hand, 2 can be written as a product of three integers in the following ways:

(i) $2 = (-1) \cdot (-1) \cdot 2$,

(ii) $2 = 1 \cdot 1 \cdot 2$,

(iii) $2 = (-1) \cdot 1 \cdot (-2)$.

Since in the first case no two factors add up to the third, we have only three possibilities up to cyclic rotation:

(a) $\begin{cases} x - y = 1 \\ x - z = 2, \\ y - z = 1 \end{cases}$ so $(x, y, z) = (k + 1, k, k - 1)$ for some integer k;

(b) $\begin{cases} x - y = -2 \\ x - z = -1, \\ y - z = 1 \end{cases}$ so $(x, y, z) = (k - 1, k + 1, k)$ for some integer k;

(c) $\begin{cases} x - y = 1 \\ x - z = -1, \\ y - z = -2 \end{cases}$ so $(x, y, z) = (k, k - 1, k + 1)$ for some integer k.

Problem 8.1.3. *Let p and q be prime numbers. Find all positive integers x and y such that*

$$\frac{1}{x} + \frac{1}{y} = \frac{1}{pq}.$$

Solution. The equation is equivalent to

$$(x - pq)(y - pq) = p^2 q^2.$$

We have the cases:

(1) $x - pq = 1$, $y - pq = p^2 q^2$, so $x = 1 + pq$, $y = pq(1 + pq)$.

(2) $x - pq = p$, $y - pq = pq^2$, so $x = p(1 + q)$, $y = pq(1 + q)$.

(3) $x - pq = q$, $y - pq = p^2 q$, so $x = q(1 + p)$, $y = pq(1 + p)$.

(4) $x - pq = p^2$, $y - pq = q^2$, so $x = p(p + q)$, $y = q(p + q)$.

(5) $x - pq = pq$, $y - pq = pq$, so $x = 2pq$, $y = 2pq$.

The equation is symmetric, so we have also:

(6) $x = pq(1 + pq)$, $y = 1 + pq$.

(7) $x = pq(1 + q)$, $y = p(1 + q)$.

(8) $x = pq(1 + p)$, $y = q(1 + p)$.

(9) $x = q(1 + q)$, $y = p(p + q)$.

Additional Problems

Problem 8.1.4. Solve in integers the equation

$$(x^2 + 1)(y^2 + 1) + 2(x - y)(1 - xy) = 4(1 + xy).$$

Problem 8.1.5. Determine the side lengths of a right triangle if they are integers and the product of the legs' lengths equals three times the perimeter.

(1999 Romanian Mathematical Olympiad)

Problem 8.1.6. Let a, b, and c be positive integers each two of them being relatively prime. Show that $2abc - ab - bc - ca$ is the largest integer that cannot be expressed in the form $xbc + yca + zab$, where x, y, and z are nonnegative integers.

(24th International Mathematical Olympiad)

8.2   Quadratic Diophantine Equations

8.2.1   The Pythagorean Equation

One of the most celebrated Diophantine equations is the so-called *Pythagorean equation*

$$x^2 + y^2 = z^2.$$ (1)

Studied in detail by Pythagoras[2] in connection with right-angled triangles whose side lengths are all integers, this equation was known even to the ancient Babylonians.

Note first that if the triple of integers (x_0, y_0, z_0) satisfies equation (1), then all triples of the form (kx_0, ky_0, kz_0), $k \in \mathbb{Z}$, also satisfy (1). That is why it is sufficient to find solutions (x, y, z) to (1) with $\gcd(x, y, z) = 1$. This is equivalent to the condition that x, y, z be pairwise relatively prime, since any prime that divides two of x, y, z also divides the third.

A solution (x_0, y_0, z_0) to (1) where x_0, y_0, z_0 are pairwise relatively prime is called a *primitive solution*.

Theorem 8.2.1. *Any primitive solution (x, y, z) in positive integers to equation (1) up to the symmetry of x and y is of the form*

$$x = m^2 - n^2, \quad y = 2mn, \quad z = m^2 + n^2,$$ (2)

where m and n are relatively prime positive integers such that $m > n$ and $m \not\equiv n$ (mod 2) (i.e., exactly one is odd).

Proof. The identity

$$(m^2 - n^2)^2 + (2mn)^2 = (m^2 + n^2)^2$$

---

[2]Pythagoras of Samos (ca. 569–475 B.C.E.), Greek philosopher who made fundamental developments in mathematics, astronomy, and the theory of music. The theorem now known as the Pythagorean theorem was known to the Babylonians 1000 years earlier, but Pythagoras may have been the first to prove it.

shows that the triple given by (2) is indeed a solution to equation (1) and y is even.
The integers x and y cannot be both odd, for otherwise

$$z^2 = x^2 + y^2 \equiv 2 \pmod 4,$$

a contradiction. Hence exactly one of the integers x and y is even.

Assume that m is odd and n is even. If $\gcd(x, y, z) = d \geq 2$, then d divides

$$2m^2 = (m^2 + n^2) + (m^2 - n^2)$$

and d divides

$$2n^2 = (m^2 + n^2) - (m^2 - n^2).$$

Since m and n are relatively prime, it follows that $d = 2$. Hence $m^2 + n^2$ is even, in contradiction to the fact that exactly one of m and n is odd. It follows that $d = 1$, so the solution (2) is primitive.

Conversely, let (x, y, z) be a primitive solution to (1) with $y = 2a$. Then x and z are odd, and consequently the integers $z + x$ and $z - x$ are even. Let $z + x = 2b$ and $z - x = 2c$. We may assume that b and c are relatively prime, for otherwise z and x would have a nontrivial common divisor. On the other hand, $4a^2 = y^2 = z^2 - z^2 = (z + x)(z - x) = 4bc$, i.e., $a^2 = bc$. Since b and c are relatively prime, it follows that $b = m^2$ and $c = n^2$ for some positive integers m and n with $m + n$ odd. We obtain

$$x = b - c = m^2 - n^2, \quad y = 2mn, \quad z = b + c = m^2 + n^2. \qquad \square$$

A triple (x, y, z) of the form (2) is called a *Pythagorean triple*.

In order to list systematically all the primitive solutions to equation (1) with (2) giving a primitive Pythagorean triple, we assign values $2, 3, 4, \ldots$ for the number m successively, and then for each of these values we take those integers n that are relatively prime to m, less than m, and even whenever m is odd.

Here is the table of the first twenty primitive solutions listed according to the above-mentioned rule.

m	n	x	y	z	area	m	n	x	y	z	area
2	1	3	4	5	6	7	6	13	84	85	546
3	2	5	12	13	30	8	1	63	16	65	504
4	1	15	8	17	60	8	3	55	48	73	1320
4	3	7	24	25	84	8	5	39	80	89	1560
5	2	21	20	29	210	8	7	15	112	113	840
5	4	9	40	41	180	9	2	77	36	85	1386
6	1	35	12	37	210	9	4	65	72	97	2340
6	5	11	60	61	330	9	8	17	144	145	1224
7	2	45	28	53	630	10	1	99	20	101	990
7	4	33	56	65	924	10	3	91	60	109	2730

Corollary 8.2.2. *The general integral solution to (1) is given by*

$$x = k(m^2 - n^2), \quad y = 2kmn, \quad z = k(m^2 + n^2), \tag{3}$$

where $k, m, n \in \mathbb{Z}$.

Problem 8.2.1. *Solve the following equation in positive integers:*

$$x^2 + y^2 = 1997(x - y).$$

(1998 Bulgarian Mathematical Olympiad)

Solution. The solutions are

$$(x, y) = (170, 145) \text{ or } (1827, 145).$$

We have

$$x^2 + y^2 = 1997(x - y),$$
$$2(x^2 + y^2) = 2 \cdot 1997(x - y),$$
$$x^2 + y^2 + (x^2 + y^2 - 2 \cdot 1997(x - y)) = 0,$$
$$(x + y)^2 + ((x - y)^2 - 2 \cdot 1997(x - y)) = 0,$$
$$(x + y)^2 + (1997 - x + y)^2 = 1997^2.$$

Since x and y are positive integers, $0 < x + y < 1997$ and $0 < 1997 - x + y < 1997$. Thus the problem reduces to solving $a^2 + b^2 = 1997^2$ in positive integers. Since 1997 is a prime, $\gcd(a, b) = 1$. By Pythagorean substitution, there are positive integers $m > n$ such that $\gcd(m, n) = 1$ and

$$1997 = m^2 + n^2, \quad a = 2mn, \quad b = m^2 - n^2.$$

Since $m^2, n^2 \equiv 0, 1, -1 \pmod 5$ and $1997 \equiv 2 \pmod 5$, we have $m, n \equiv \pm 1 \pmod 5$. Since $m^2, n^2 \equiv 0, 1 \pmod 3$ and $1997 \equiv 2 \pmod 3$, we have $m, n \equiv \pm 1 \pmod 3$. Therefore $m, n \equiv 1, 4, 11, 14 \pmod{15}$. Since $m > n$, $1997/2 \leq m^2 \leq 1997$. Thus we need to consider only $m = 34, 41, 44$. The only solution is $(m, n) = (34, 29)$. Thus

$$(a, b) = (1972, 315),$$

which leads to our two solutions.

Problem 8.2.2. *Let p, q, r be primes and let n be a positive integer such that*

$$p^n + q^n = r^2.$$

Prove that $n = 1$.

(2004 Romanian Mathematical Olympiad)

Solution. Assume that $n \geq 2$ satisfies the relation in the problem. Clearly one of the primes p, q, and r is equal to 2. If $r = 2$ then $p^n + q^n = 4$, false, so assume that $p > q = 2$.

Consider the case that $n > 1$ is odd; we have

$$(p + 2)(p^{n-1} - 2p^{n-2} + 2^2 p^{n-3} - \cdots + 2^{n-1}) = r^2.$$

Notice that

$$p^{n-1} - 2p^{n-2} + 2^2 p^{n-3} - \cdots + 2^{n-1}$$
$$= 2^{n-1} + (p - 2)(p^{n-2} + 2^2 p^{n-4} + \cdots + 1) > 1$$

and $p + 2 > 1$ hence both factors are equal to r. This can be written as $p^n + 2^n = (p + 2)^2 = p^2 + 4p + 4$, which is false for $n \geq 3$.

Consider the case that $n > 1$ is even and let $n = 2m$. From Theorem 8.2.1 it follows that $p^m = a^2 - b^2$, $2^m = 2ab$ and $r = a^2 + b^2$, for some integers a, b with $(a, b) = 1$. Therefore, a and b are powers of 2, so $b = 1$ and $a = 2^{m-1}$. This implies $p^m = 4^{m-1} - 1 < 4^m$, so p must be equal to 3. The equality $3^m = 4^{m-1} - 1$ fails for $m = 1$ and also for $m \geq 2$, since $4^{m-1} > 3^m + 1$, by induction.

Consequently $n = 1$. For example, in this case we can take $p = 23, q = 2$, and $r = 5$.

Additional Problems

Problem 8.2.3. Find all Pythagorean triangles whose areas are numerically equal to their perimeters.

Problem 8.2.4. Prove that for every positive integer n there is a positive integer k such that k appears in exactly n nontrivial Pythagorean triples.

(American Mathematical Monthly)

Problem 8.2.5. Find the least perimeter of a right-angled triangle whose sides and altitude are integers.

(Mathematical Reflections)

8.2.2 Pell's Equation

A special quadratic equation is

$$u^2 - Dv^2 = 1, \tag{1}$$

where D is a positive integer that is not a perfect square. Equation (1) is called *Pell's*[3] *equation*, and it has numerous applications in various fields of mathemat-

---

[3] John Pell (1611–1685), English mathematician best known for *Pell's equation*, which in fact he had little to do with.

ics. We will present an elementary approach to solving this equation, due to Lagrange.

Theorem 8.2.3. *If D is a positive integer that is not a perfect square, then equation (1) has infinitely many solutions in positive integers, and the general solution is given by $(u_n, v_n)_{n \geq 1}$,*

$$u_{n+1} = u_1 u_n + D v_1 v_n, \quad v_{n+1} = v_1 u_n + u_1 v_n, \tag{2}$$

where (u_1, v_1) is its fundamental solution, i.e., the minimal solution different from $(1, 0)$.

Proof. First, we will prove that equation (1) has a fundamental solution.

Let c_1 be an integer greater than 1. We will show that there exist integers $t_1, w_1 \geq 1$ such that

$$\left| t_1 - w_1 \sqrt{D} \right| < \frac{1}{c_1}, \quad w_1 \leq c_1.$$

Indeed, considering $l_k = [k\sqrt{D}+1], k = 0, 1, \ldots, c_1$, yields $0 < l_k - k\sqrt{D} \leq 1, k = 0, 1, \ldots, c_1$, and since $\sqrt{D}$ is an irrational number, it follows that $l_{k'} \neq l_{k''}$ whenever $k' \neq k''$.

There exist $i, j, p \in \{0, 1, 2, \ldots, c_1\}, i \neq j, p \neq 0$, such that

$$\frac{p-1}{c_1} < l_i - i\sqrt{D} \leq \frac{p}{c_1} \quad \text{and} \quad \frac{p-1}{c_1} < l_j - j\sqrt{D} \leq \frac{p}{c_1}$$

because there are c_1 intervals of the form $\left(\frac{p-1}{c_1}, \frac{p}{c_1} \right), p = 0, 1, \ldots, c_1$, and $c_1 + 1$ numbers of the form $l_k - k\sqrt{D}, k = 0, 1, \ldots, c_1$.

Assume $j > i$ and note that $l_j > l_i$.

From the inequalities above it follows that $|(l_k - l_i) - (j - i)\sqrt{D}| < \frac{1}{c_1}$, and setting $|l_j - l_i| = t_1$ and $|j - i| = w_1$ yields $|t_1 - w_1\sqrt{D}| < \frac{1}{c_1}$ and $w_1 \leq c_1$.

Multiplying this inequality by $t_1 + w_1\sqrt{D} < 2w_1\sqrt{D} + 1$ gives

$$|t_1^2 - Dw_1^2| < 2\frac{w_1}{c_1}\sqrt{D} + \frac{1}{c_1} < 2\sqrt{D} + 1.$$

Choosing a positive integer $c_2 > c_1$ such that $|t_1 - w_1\sqrt{D}| > \frac{1}{c_2}$, we obtain positive integers t_2, w_2 with the properties

$$|t_2^2 - Dw_2^2| < 2\sqrt{D} + 1 \quad \text{and} \quad |t_1 - t_2| + |w_1 - w_2| \neq 0.$$

By continuing this procedure, we obtain a sequence of distinct pairs $(t_n, w_n)_{n \geq 1}$ satisfying the inequalities $|t_n^2 - Dw_n^2| < 2\sqrt{D} + 1$ for all positive integers n. It follows that the interval $(-2\sqrt{D} - 1, 2\sqrt{D} + 1)$ contains a nonzero integer k

such that there exists a subsequence of $(t_n, w_n)_{n \geq 1}$ satisfying the equation $t^2 - Dw^2 = k$. This subsequence contains at least two pairs (t_s, w_s), (t_r, w_r) for which $t_s \equiv t_r \pmod{k}$, $w_s \equiv w_r \pmod{k}$, and $t_s w_r - t_r w_s \neq 0$, otherwise $t_s = t_r$ and $w_s = w_r$, in contradiction to $|t_s - t_r| + |w_s - w_r| \neq 0$.

Let $t_0 = t_s t_r - D w_s w_r$ and let $w_0 = t_s w_r - t_r w_s$. Then

$$t_0^2 - D w_0^2 = k^2. \tag{3}$$

On the other hand, $t_0 = t_s t_r - D w_s w_r \equiv t_s^2 - D w_0^2 \equiv 0 \pmod{k}$, and it follows immediately that $w_0 \equiv 0 \pmod{k}$. The pair (t, w), where $t_0 = t|k|$ and $w_0 = wk$, is a nontrivial solution to equation (1).

We show now that the pair (u_n, v_n) defined by (2) satisfies Pell's equation (1). We proceed by induction with respect to n. By definition, (u_1, v_1) is a solution to equation (1). If (u_n, v_n) is a solution to this equation, then

$$
\begin{aligned}
u_{n+1}^2 - D v_{n+1}^2 &= (u_1 u_n + D v_1 v_n)^2 - D(v_1 u_n + u_1 v_n)^2 \\
&= (u_1^2 - D v_1^2)(u_n^2 - D v_n^2) = 1,
\end{aligned}
$$

i.e., the pair (u_{n+1}, v_{n+1}) is also a solution to equation (1).

It is not difficult to see that for all positive integers n,

$$u_n + v_n \sqrt{D} = (u_1 + v_1 \sqrt{D})^n. \tag{4}$$

Let $z_n = u_n + v_n \sqrt{D} = (u_1 + v_1 \sqrt{D})^n$ and note that $z_1 < z_2 < \cdots < z_n < \cdots$. We will prove now that all solutions to equation (1) are of the form (4). Indeed, if equation (1) had a solution (u, v) such that $z = u + v\sqrt{D}$ is not of the form (4), then $z_m < z < z_{m+1}$ for some integer m. Also, we have $1/z_m = u_m - v_m \sqrt{D}$. Then $1 < (u + v\sqrt{D})(u_m - v_m \sqrt{D}) < u_1 + v_1 \sqrt{D}$, and therefore $1 < (u u_m - D v v_m) + (u_m v - u v_m)\sqrt{D} < u_1 + v_1 \sqrt{D}$. On the other hand, $(u u_m - D v v_m)^2 - D(u_m v - u v_m)^2 = (u^2 - D v^2)(u_m^2 - D v_m^2) = 1$, i.e., $(u u_m - D v v_m, u_m v - u v_m)$ is a solution of (1) smaller than (u_1, v_1), in contradiction to the assumption that (u_1, v_1) was the minimal one.

In order to complete the proof we have only to show that $a = u u_m - D v v_m$ and $b = u_m v - u v_m$ are positive. We have $1 < a + b\sqrt{D}$ and $a^2 - D b^2 = 1$; hence $0 < 1/(a + b\sqrt{D}) = a - b\sqrt{D} < 1$. Now we obtain

$$a = \frac{1}{2}(a + b\sqrt{D}) + \frac{1}{2}(a - b\sqrt{D}) > \frac{1}{2} + 0 > 0,$$

$$b\sqrt{D} = \frac{1}{2}(a + b\sqrt{D}) - \frac{1}{2}(a - b\sqrt{D}) > \frac{1}{2} - \frac{1}{2} = 0,$$

so a, b are positive integers. $\qquad\square$

Remarks. (1) The identity $u_m + v_m\sqrt{D} = (u_1 + v_1\sqrt{D})^m$, $m = 0, 1, 2, \ldots$, shows how the fundamental solution generates all solutions.

(2) The relations (1) could be written in the following useful matrix form:

$$\begin{pmatrix} u_{n+1} \\ v_{n+1} \end{pmatrix} = \begin{pmatrix} u_1 & Dv_1 \\ v_1 & u_1 \end{pmatrix} \begin{pmatrix} u_n \\ v_n \end{pmatrix},$$

whence

$$\begin{pmatrix} u_n \\ v_n \end{pmatrix} = \begin{pmatrix} u_1 & Dv_1 \\ v_1 & u_1 \end{pmatrix}^n \begin{pmatrix} u_0 \\ v_0 \end{pmatrix} = \begin{pmatrix} u_1 & Dv_1 \\ v_1 & u_1 \end{pmatrix} \begin{pmatrix} 1 \\ 0 \end{pmatrix}. \tag{5}$$

If

$$\begin{pmatrix} u_1 & Dv_1 \\ v_1 & u_1 \end{pmatrix}^n = \begin{pmatrix} a_n & b_n \\ c_n & d_n \end{pmatrix}$$

then it is well known that each of a_n, b_n, c_n, d_n is a linear combination of λ_1^n, λ_2^n, where λ_1, λ_2 are the eigenvalues of the matrix $\begin{pmatrix} u_1 & Dv_1 \\ v_1 & u_1 \end{pmatrix}$. Using (5), after an easy computation it follows that

$$u_n = \tfrac{1}{2}[(u_1 + v_1\sqrt{D})^n + (u_1 - v_1\sqrt{D})^n],$$

$$v_n = \tfrac{1}{2\sqrt{D}}[(u_1 + v_1\sqrt{D})^n - (u_1 - v_1\sqrt{D})^n]. \tag{6}$$

(3) The solutions to Pell's equation given in the form (4) or (6) may be used in the approximation of the square roots of positive integers that are not perfect squares. Indeed, if (u_n, v_n) are the solutions of equation (1), then

$$u_n - v_n\sqrt{D} = \frac{1}{u_n + v_n\sqrt{D}},$$

and so

$$\frac{u_n}{v_n} - \sqrt{D} = \frac{1}{v_n(u_n + v_n\sqrt{D})} < \frac{1}{\sqrt{D}v_n^2} < \frac{1}{v_n^2},$$

i.e., the fractions u_n/v_n approximate $\sqrt{D}$ with an error less than $1/v_n^2$.

It follows that

$$\lim_{n\to\infty} \frac{u_n}{v_n} = \sqrt{D}. \tag{7}$$

Problem 8.2.4. *Consider the sequences $(u_n)_{n\geq 0}$, $(v_n)_{n\geq 0}$ defined by $u_0 = 1$, $v_0 = 0$, and $u_{n+1} = 3u_n + 4v_n$, $v_{n+1} = 2u_n + 3v_n$, $n \geq 1$. Define $x_n = u_n + v_n$, $y_n = u_n + 2v_n$, $n \geq 0$. Prove that $y_n = \lfloor x_n\sqrt{2}\rfloor$ for all $n \geq 0$.*

Solution. We prove by induction that

$$u_n^2 - 2v_n^2 = 1, \quad n \geq 1. \tag{1}$$

For $n = 1$ the claim is true. Assuming that the equality is true for some n, we have

$$u_{n+1}^2 - 2v_{n+1}^2 = (3u_n + 4v_n)^2 - 2(2u_n + 3v_n)^2 = u_n^2 - 2v_n^2 = 1;$$

hence (1) is true for all $n \geq 1$.

We prove now that

$$2x_n^2 - y_n^2 = 1, \quad n \geq 1. \tag{2}$$

Indeed,

$$2x_n^2 - y_n^2 = 2(u_n + v_n)^2 - (u_n + 2v_n)^2 = u_n^2 - 2v_n^2 = 1,$$

as claimed. It follows that

$$\left(x_n\sqrt{2} - y_n\right)\left(x_n\sqrt{2} + y_n\right) = 1, \quad n \geq 1.$$

Notice that $x_n\sqrt{2} + y_n > 1$, so

$$0 < x_n\sqrt{2} - y_n < 1, \quad n \geq 1.$$

Hence $y_n = \left\lfloor x_n\sqrt{2} \right\rfloor$, as claimed.

Problem 8.2.5. *Show that there exist infinitely many systems of positive integers* (x, y, z, t) *that have no common divisor greater than 1 and such that*

$$x^3 + y^3 + z^2 = t^4.$$

(2000 Romanian International Mathematical Olympiad Team Selection Test)

First solution. Let us consider the identity

$$[1^3 + 2^3 + \cdots + (n-2)^3] + (n-1)^3 + n^3 = \left(\frac{n(n+1)}{2}\right)^2.$$

We may write it in the form

$$(n-1)^3 + n^3 + \left(\frac{(n-1)(n-2)}{2}\right)^2 = \left(\frac{n(n+1)}{2}\right)^2.$$

It is sufficient to find positive integers n for which $n(n+1)/2$ is a perfect square. Such a goal can be attained.

Let us remark that the equality

$$(2n + 1)^2 - 2(2x)^2 = 1$$

can be realized by taking the solutions (u_k, v_k) of Pell's equation $u^2 - 2v^2 = 1$, where $u_0 = 3$, $v_0 = 2$, and u_k, v_k are obtained from the identity

$$(u_0 + \sqrt{2}v_0)^k (u_0 - \sqrt{2}v_0)^k = (u_k + \sqrt{2}v_k)(u_k - \sqrt{2}v_k) = 1.$$

Second solution. Consider the following identity:

$$(a + 1)^4 - (a - 1)^4 = 8a^3 + 8a,$$

where a is a positive integer. Take $a = b^3$, where b is an even integer. From the above identity one obtains

$$(b^3 + 1)^4 = (2b^3)^3 + (2b)^3 + [(b^3 - 1)^2]^2.$$

Since b is an even number, $b^3 + 1$ and $b^3 - 1$ are odd numbers. It follows that the numbers $x = 2b^3$, $y = 2b$, $z = (b^3 - 1)^2$, and $t = b^3 + 1$ have no common divisor greater than 1.

Additional Problems

Problem 8.2.6. Let p be a prime number congruent to 3 modulo 4. Consider the equation

$$(p + 2)x^2 - (p + 1)y^2 + px + (p + 2)y = 1.$$

Prove that this equation has infinitely many solutions in positive integers, and show that if $(x, y) = (x_0, y_0)$ is a solution of the equation in positive integers, then $p \mid x_0$.

(2001 Bulgarian Mathematical Olympiad)

Problem 8.2.7. Determine all integers a for which the equation

$$x^2 + axy + y^2 = 1$$

has infinitely many distinct integer solutions (x, y).

(1995 Irish Mathematical Olympiad)

Problem 8.2.8. Prove that the equation

$$x^3 + y^3 + z^3 + t^3 = 1999$$

has infinitely many integral solutions.

(1999 Bulgarian Mathematical Olympiad)

8.2.3 Other Quadratic Equations

There are many other general quadratic equations that appear in concrete situations. Here is an example.

Consider the equation

$$axy + bx + cy + d = 0, \qquad (1)$$

where a is a nonzero integer and b, c, d are integers such that $ad - bc \neq 0$.

Theorem 8.2.4. *If* $\gcd(a, b) = \gcd(a, c) = 1$, *then equation (1) is solvable if and only if there is a divisor m of $ad - bc$ such that $a \mid m - b$ or $a \mid m - c$.*

Proof. We can write (1) in the following equivalent form:

$$(ax + c)(ay + b) = bc - ad. \qquad (2)$$

If such a divisor m exists and $a \mid m - c$, then we take $ax + c = m$ and $ay + b = m'$, where $mm' = bc - ad$. In order to have solutions, it suffices to show that $a \mid m' - b$. Indeed, the relation $mm' = bc - ad$ implies $(ax + c)m' = bc - ad$, which is equivalent to $a(m'x + d) = -c(m' - b)$. Taking into account that $\gcd(a, c) = 1$, we get $a \mid m' - b$.

The converse is clearly true. $\qquad \square$

Remarks. In case of solvability, equation (1) has only finitely many solutions. These solutions depend upon the divisors m of $ad - bc$.

Example. Solve the equation

$$3xy + 4x + 7y + 6 = 0.$$

Solution. We have $ad - bc = -10$, whose integer divisors are $-10, -5, -2, -1, 1, 2, 5, 10$. The conditions in Theorem 8.2.4 are satisfied only for $m = -5, -2, 1, 10$. We obtain the solutions $(x, y) = (-4, -2), (-3, -3), (-2, 2), (1, -1)$, respectively.

In what follows you can find several nonstandard quadratic equations.

Problem 8.2.9. *For any given positive integer n, determine (as a function of n) the number of ordered pairs (x, y) of positive integers such that*

$$x^2 - y^2 = 10^2 \cdot 30^{2n}.$$

Prove further that the number of such pairs is never a perfect square.

(1999 Hungarian Mathematical Olympiad)

Solution. Because $10^2 \cdot 30^{2n}$ is even, x and y must have the same parity. Then (x, y) is a valid solution if and only if $(u, v) = \left(\frac{x+y}{2}, \frac{x-y}{2}\right)$ is a pair of positive integers that satisfies $u > v$ and $uv = 5^2 \cdot 30^{2n}$. Now $5^2 \cdot 30^{2n} = 2^{2n} \cdot 3^{2n} \cdot 5^{2n+2}$

has exactly $(2n + 1)^2(2n + 3)$ factors. Thus without the condition $u > v$ there are exactly $(2n + 1)^2(2n + 3)$ such pairs (u, v). Exactly one pair has $u = v$, and by symmetry half of the remaining pairs have $u > v$. It follows that there are $\frac{1}{2}((2n + 1)^2(2n + 3) - 1) = (n + 1)(4n^2 + 6n + 1)$ valid pairs.

Now suppose that $(n + 1)(4n^2 + 6n + 1)$ were a square. Because $n + 1$ and $4n^2 + 6n + 1 = (4n + 2)(n + 1) - 1$ are coprime, $4n^2 + 6n + 1$ must be a square as well. However, $(2n + 1)^2 < 4n^2 + 6n + 1 < (2n + 2)^2$, a contradiction.

Problem 8.2.10. *Prove that the equation $a^2 + b^2 = c^2 + 3$ has infinitely many integer solutions $\{a, b, c\}$.*

(1996 Italian Mathematical Olympiad)

Solution. Let a be any odd number, let $b = (a^2 - 5)/2$ and $c = (a^2 - 1)/2$. Then

$$c^2 - b^2 = (c + b)(c - b) = a^2 - 3.$$

Remark. Actually one can prove that any integer n can be represented in infinitely many ways in the form $a^2 + b^2 - c^2$ with $a, b, c \in \mathbb{Z}$.

Additional Problems

Problem 8.2.11. Prove that the equation

$$x^2 + y^2 + z^2 + 3(x + y + z) + 5 = 0$$

has no solutions in rational numbers.

(1997 Bulgarian Mathematical Olympiad)

Problem 8.2.12. Find all integers x, y, z such that $5x^2 - 14y^2 = 11z^2$.

(2001 Hungarian Mathematical Olympiad)

Problem 8.2.13. Let n be a nonnegative integer. Find the nonnegative integers a, b, c, d such that
$$a^2 + b^2 + c^2 + d^2 = 7 \cdot 4^n.$$

(2001 Romanian JBMO Team Selection Test)

Problem 8.2.14. Prove that the equation

$$x^2 + y^2 + z^2 + t^2 = 2^{2004},$$

where $0 \le x \le y \le x \le t$, has exactly two solutions in the set of integers.

(2004 Romanian Mathematical Olympiad)

Problem 8.2.15. Let n be a positive integer. Prove that the equation

$$x + y + \frac{1}{x} + \frac{1}{y} = 3n$$

does not have solutions in positive rational numbers.

8.3 Nonstandard Diophantine Equations

8.3.1 Cubic Equations

Problem 8.3.1. *Find all pairs (x, y) of nonnegative integers such that $x^3 + 8x^2 - 6x + 8 = y^3$.*

(1995 German Mathematical Olympiad)

Solution. Note that for all real x,

$$0 < 5x^2 - 9x + 7 = (x^3 + 8x^2 - 6x + 8) - (x + 1)^3.$$

Therefore if (x, y) is a solution, we must have $y \geq x + 2$. In the same vein, we note that for $x \geq 1$,

$$0 > -x^2 - 33x + 15 = (x^3 + 8x^2 - 6x + 8) - (x^3 + 9x^2 + 27x + 27).$$

Hence we have either $x = 0$, in which case $y = 2$ is a solution, or $x \geq 1$, in which case we must have $y = x + 2$. But this means that

$$0 = (x^3 + 8x^2 - 6x + 8) - (x^3 + 6x^2 + 12x + 8) = 2x^2 - 18x.$$

Hence the only solutions are $(0, 2)$, $(9, 11)$.

Problem 8.3.2. *Find all pairs (x, y) of integers such that*

$$x^3 = y^3 + 2y^2 + 1.$$

(1999 Bulgarian Mathematical Olympiad)

Solution. When $y^2 + 3y > 0$, $(y+1)^3 > x^3 > y^3$. Thus we must have $y^2 + 3y \leq 0$, and $y = -3, -2, -1$, or 0, yielding the solutions $(x, y) = (1, 0)$, $(1, -2)$, and $(-2, -3)$.

Problem 8.3.3. *Find all the triples (x, y, z) of positive integers such that*

$$xy + yz + zx - xyz = 2.$$

First solution. Let $x \leq y \leq z$. We consider the following cases:

(1) For $x = 1$, we obtain $y + z = 2$, and then

$$(x, y, z) = (1, 1, 1).$$

(2) If $x = 2$, then $2y + 2z - yz = 2$, which gives $(z - 2)(y - 2) = 2$. The solutions are $z = 4$, $y = 3$ and $z = 3$, $y = 4$. Due to the symmetry of the relations, the solutions (x, y, z) are

$$(2, 3, 4), \ (2, 4, 3), \ (3, 2, 4), \ (4, 2, 3), \ (3, 4, 2), \ (4, 3, 2).$$

(3) If $x \geq 3$, $y \geq 3$, $z \geq 3$, then $xyz \geq 3yz$, $xyz \geq 3xz$, $xyz \geq 3xy$. Thus $xy + xz + yz - xyz \leq 0$, so there are no solutions.

Second solution. Let $x' = x - 1$, $y' = y - 1$, $z' = z - 1$. The equation is equivalent to $x'y'z' = x' + y' + z'$. If $x' = 0$, then $y' = z' = 0$, and we get the solution $(x, y, z) = (1, 1, 1)$ for the initial equation. If $x' \neq 0$, $y' \neq 0$, and $x' \neq 0$, then

$$\frac{1}{x'y'} + \frac{1}{y'z'} + \frac{1}{z'x'} = 1$$

forces one of $x'y'$ or the other two to be at most 3.

It follows that $(x', y', z') = (1, 2, 3)$ and all corresponding permutations and we get all solutions in case (2).

Problem 8.3.4. *Determine a positive constant c such that the equation*

$$xy^2 - y^2 - x + y = c$$

has exactly three solutions (x, y) in positive integers.

(1999 United Kingdom Mathematical Olympiad)

Solution. When $y = 1$ the left-hand side is 0. Thus we can rewrite our equation as

$$x = \frac{y(y - 1) + c}{(y + 1)(y - 1)}.$$

Note that from the offset equation $y - 1 \mid c$ and writing $c = (y - 1)d$ we get $x = \frac{y + d}{y + 1}$. Hence $d \equiv 1 \pmod{y + 1}$, and thus $c \equiv y - 1 \pmod{y^2 - 1}$. Conversely, any such c makes x an integer.

Thus we want c to satisfy exactly three congruences $c \equiv y - 1 \pmod{y^2 - 1}$. Every c always satisfies this congruence for $y = c + 1$, so we need two others. The first two nontrivial congruences for $y = 2, 3$ give $c \equiv 1 \pmod 3$ and $c \equiv 2 \pmod 8$. Hence $c = 10$ is the least solution to both these congruences and also works for $y = 11$. It does not satisfy any others, since we would have $y - 1 \mid 10$; hence $y = 2, 3, 6, 11$. We have already seen that 2, 3, 11 all work, but trying $y = 6$ gives $x = 2/7$. Thus there are exactly three solutions with $c = 10$, namely $(x, y) = (4, 2)$, $(2, 3)$, and $(1, 11)$.

Additional Problems

Problem 8.3.5. Find all triples (x, y, z) of natural numbers such that y is a prime number, y and 3 do not divide z, and $x^3 - y^3 = z^2$.

(1999 Bulgarian Mathematical Olympiad)

Problem 8.3.6. Find all positive integers a, b, c such that

$$a^3 + b^3 + c^3 = 2001.$$

(2001 Junior Balkan Mathematical Olympiad)

Problem 8.3.7. Determine all ordered pairs (m, n) of positive integers such that

$$\frac{n^3 + 1}{mn - 1}$$

is an integer.

(35th International Mathematical Olympiad)

8.3.2 High-Order Polynomial Equations

Problem 8.3.8. *Prove that there are no integers x, y, z such that*

$$x^4 + y^4 + z^4 - 2x^2y^2 - 2y^2z^2 - 2z^2x^2 = 2000.$$

Solution. Suppose by way of contradiction that such numbers exist. Assume without loss of generality that x, y, z are nonnegative integers.

First we prove that the numbers are distinct. For this, consider that $y = z$. Then $x^4 - 4x^2y^2 = 2000$; hence x is even.

Setting $x = 2t$ yields $t^2(t^2 - y^2) = 125$. It follows that $t^2 = 25$ and $y^2 = 20$, a contradiction.

Let now $x > y > z$. Since $x^4 + y^4 + z^4$ is even, at least one of the numbers x, y, z is even and the other two have the same parity. Observe that

$$
\begin{aligned}
x^4 + y^4 + z^4 &- 2x^2y^2 - 2y^2z^2 - 2z^2x^2 \\
&= (x^2 - y^2)^2 - 2(x^2 - y^2)z^2 + z^4 - 4y^2z^2 \\
&= (x^2 - y^2 - z^2 - 2yz)(x^2 - y^2 - z^2 + 2yz) \\
&= (x + y + z)(x - y - z)(x - y + z)(x + y - z),
\end{aligned}
$$

each of the four factors being even. Since $2000 = 16 \cdot 125 = 2^4 \cdot 125$, we deduce that each factor is divisible by 2, but not by 4. Moreover, the factors are distinct:

$$x + y + z > x + y - z > x - y + z > x - y - z.$$

The smallest even divisors of 2000 that are not divisible by 4 are 2, 10, 50, 250. But $2 \cdot 10 \cdot 50 \cdot 250 > 2000$, a contradiction.

Problem 8.3.9. *Find the smallest value for n for which there exist positive integers* $x_1, \ldots, x_n$ *with*

$$x_1^4 + x_2^4 + \cdots + x_n^4 = 1998.$$

Solution. Observe that for any integer x we have $x^4 = 16k$ or $x^4 = 16k + 1$ for some k.

Since $1998 = 16 \cdot 124 + 14$, it follows that $n \geq 14$.

If $n = 14$, all the numbers $x_1, x_2, \ldots, x_{14}$ must be odd, so let $x_k^4 = 16a_k + 1$. Then $a_k = \frac{x_k^4 - 1}{16}$, $k = 1, 2, \ldots, 14$; hence $a_k \in \{0, 5, 39, 150, \ldots\}$ and $a_1 + a_2 + \cdots + a_{14} = 124$. It follows that $a_k \in \{0, 5, 39\}$ for all $k = 1, 2, \ldots, 14$, and since $124 = 5 \cdot 24 + 4$, the number of terms a_k equal to 39 is 1 or at least 6. A simple analysis shows that the claim fails in both cases; hence $n \geq 15$. Any of the equalities

$$1998 = 5^4 + 5^4 + 3^4 + 3^4 + 3^4 + 3^4 + 3^4 + 3^4 + 3^4 + 3^4 + 3^4 + 2^4$$
$$+ 1^4 + 1^4 + 1^4$$
$$= 5^4 + 5^4 + 4^4 + 3^4 + 3^4 + 3^4 + 3^4 + 3^4 + 3^4$$
$$+ 1^4 + 1^4 + 1^4 + 1^4 + 1^4 + 1^4$$

proves that $n = 15$.

Problem 8.3.10. *Find all positive integer solutions* (x, y, z, t) *of the equation*

$$(x + y)(y + z)(z + x) = txyz$$

such that $\gcd(x, y) = \gcd(y, z) = \gcd(z, x) = 1$.

(1995 Romanian International Mathematical Olympiad Team Selection Test)

Solution. It is obvious that $(x, x + y) = (x, x + z) = 1$, and x divides $y + z$, y divides $z + x$, and z divides $x + y$. Let a, b, and c be integers such that

$$x + y = cz,$$
$$y + z = ax,$$
$$z + x = by.$$

We may assume that $x \geq y \geq z$. If $y = z$, then $y = z = 1$ and then $x \in \{1, 2\}$. If $x = y$, then $x = y = 1$ and $z = 1$. So assume that $x > y > z$. Since $a = \frac{y+z}{x} < 2$, we have $a = 1$ and $x = y + z$. Thus, $y \mid y + 2z$ and $y \mid 2z$. Since $y > z$, $y = 2z$ and since $\gcd(y, z) = 1$, one has $z = 1$, $y = 2$, $x = 3$.

Finally, the solutions are $(1, 1, 1, 8)$, $(2, 1, 1, 9)$, $(3, 2, 1, 10)$ and those obtained by permutations of x, y, z.

Problem 8.3.11. *Determine all triples of positive integers a, b, c such that $a^2 + 1$, $b^2 + 1$ are prime and $(a^2 + 1)(b^2 + 1) = c^2 + 1$.*

(2002 Polish Mathematical Olympiad)

Solution. Of course, we may assume that $a \le b$. Since $a^2(b^2+1) = (c-b)(c+b)$ and $b^2 + 1$ is a prime, we have $b^2 + 1 \mid c - b$ or $b^2 + 1 \mid c + b$. If $b^2 + 1 \mid c - b$, then $a^2 \ge c + b \ge b^2 + 2b + 1$; impossible, since $a \le b$. So there is k such that $c+b = k(b^2+1)$ and $a^2 = k(b^2+1) - 2b$. Thus, $b^2 \ge k(b^2+1) - 2b > kb^2 - 2b$, whence $k \le 2$. If $k = 2$, then $b^2 \ge 2b^2 - 2b + 2$; thus $(b-1)^2 + 1 \le 0$, false. Thus $k = 1$ and $a = b - 1$. But then $b^2 + 1$ and $(b-1)^2 + 1$ are primes and at least one of them is even, forcing $b - 1 = 1$ and $b = 2$, $a = 1$, $c = 3$. By symmetry, we obtain $(a, b, c) = (1, 2, 3)$ or $(2, 1, 3)$.

Additional Problems

Problem 8.3.12. Prove that there are no positive integers x and y such that

$$x^5 + y^5 + 1 = (x + 2)^5 + (y - 3)^5.$$

Problem 8.3.13. Prove that the equation $y^2 = x^5 - 4$ has no integer solutions.

(1998 Balkan Mathematical Olympiad)

Problem 8.3.14. Let $m, n > 1$ be integers. Solve in positive integers the equation

$$x^n + y^n = 2^m.$$

(2003 Romanian Mathematical Olympiad)

Problem 8.3.15. For a given positive integer m, find all triples (n, x, y) of positive integers such that m, n are relatively prime and $(x^2+y^2)^m = (xy)^n$, where n, x, y can be represented in terms of m.

(1995 Korean Mathematical Olympiad)

8.3.3 Exponential Diophantine Equations

Problem 8.3.16. *Find the integer solutions to the equation*

$$9^x - 3^x = y^4 + 2y^3 + y^2 + 2y.$$

Solution. We have successively

$$4((3^x)^2 - 3^x) + 1 = 4y^4 + 8y^3 + 4y^2 + 8y + 1,$$

then

$$(2t - 1)^2 = 4y^4 + 8y^3 + 4y^2 + 8y + 1,$$

where $3^x = t \geq 1$, since it is clear that there are no solutions with $x < 0$.
   Observe that

$$(2y^2 + 2y)^2 < E \leq (2y^2 + 2y + 1)^2.$$

Since $E = (2t - 1)^2$ is a square, then

$$E = (2y^2 + 2y + 1)^2$$

if and only if

$$4y(y - 1) = 0,$$

so $y = 0$ or $y = 1$.
   If $y = 0$, then $t = 1$ and $x = 0$.
   If $y = 1$, then $t = 3$ and $x = 1$.
   Hence the solutions (x, y) are $(0, 0)$ and $(1, 1)$.

Problem 8.3.17. *The positive integers x, y, z satisfy the equation $2x^x = y^y + z^z$. Prove that $x = y = z$.*

(1997 St. Petersburg City Mathematical Olympiad)

Solution. We note that $(x + 1)^{x+1} \geq x^{x+1} + (x + 1)x^x > 2x^x$. Thus we cannot have $y > x$ or $z > x$, since otherwise, the right side of the equation will exceed the left. But then $2x^x \geq y^y + z^z$, with equality if and only if $x = y = z$.

Problem 8.3.18. *Find all solutions in nonnegative integers x, y, z of the equation*

$$2^x + 3^y = z^2.$$

(1996 United Kingdom Mathematical Olympiad)

Solution. If $y = 0$, then $2^x = z^2 - 1 = (z+1)(z-1)$, so $z+1$ and $z-1$ are powers of 2. The only powers of 2 that differ by 2 are 4 and 2, so $(x, y, z) = (3, 0, 3)$.
   If $y > 0$, then taking the equation mod 3, it follows that x is even. Now we have $3^y = z^2 - 2^x = (z + 2^{x/2})(z - 2^{x/2})$. The factors are powers of 3, say $z + 2^{x/2} = 3^m$ and $z - 2^{x/2} = 3^n$, but then $3^m - 3^n = 2^{x/2+1}$. Since the right side is not divisible by 3, we must have $n = 0$ and

$$3^m - 1 = 2^{x/2+1}.$$

   If $x = 0$, we have $m = 1$, yielding $(x, y, z) = (0, 1, 2)$. Otherwise, $3^m - 1$ is divisible by 4, so m is even and $2^{x/2+1} = (3^{m/2} + 1)(3^{m/2} - 1)$. The two factors on the right are powers of 2 differing by 2, so they are 2 and 4, giving $x = 4$ and $(x, y, z) = (4, 2, 5)$.

Additional Problems

Problem 8.3.19. Determine all triples (x, k, n) of positive integers such that

$$3^k - 1 = x^n.$$

(1999 Italian Mathematical Olympiad)

Problem 8.3.20. Find all pairs of nonnegative integers x and y that satisfy the equation

$$p^x - y^p = 1,$$

where p is a given odd prime.

(1995 Czech–Slovak Match)

Problem 8.3.21. Let x, y, z be integers with $z > 1$. Show that

$$(x + 1)^2 + (x + 2)^2 + \cdots + (x + 99)^2 \neq y^z.$$

(1998 Hungarian Mathematical Olympiad)

Problem 8.3.22. Determine all solutions (x, y, z) of positive integers such that

$$(x + 1)^{y+1} + 1 = (x + 2)^{z+1}.$$

(1999 Taiwanese Mathematical Olympiad)

9

Some Special Problems in Number Theory

9.1 Quadratic Residues; the Legendre Symbol

Let a and m be positive integers such that $m \neq 0$ and $\gcd(a, m) = 1$. We say that a is a *quadratic residue* mod m if the congruence $x^2 \equiv a \pmod{m}$ has a solution. Otherwise, we say that a is a quadratic nonresidue.

Let p be a prime and let a be a positive integer not divisible by p. The *Legendre symbol* of a with respect to p is defined by

$$\left(\frac{a}{p}\right) = \begin{cases} 1 & \text{if } a \text{ is a quadratic residue (mod } p), \\ -1 & \text{otherwise.} \end{cases}$$

It is clear that the perfect squares are quadratic residues mod p. It is natural to ask how many integers among $1, 2, \ldots, p - 1$ are quadratic residues. The answer is given in the following theorem.

Theorem 9.1.1. *Let p be an odd prime. There are $\frac{p-1}{2}$ quadratic residues in the set $\{1, 2, \ldots, p - 1\}$.*

Proof. Consider the numbers k^2, $k = 1, 2, \ldots, \frac{p-1}{2}$. These are quadratic residues, and moreover, they are distinct. Indeed, if $i^2 \equiv j^2 \pmod{p}$, then it follows that $p \mid (i - j)(i + j)$, and since $i + j < p$, this implies $p \mid i - j$; hence $i = j$.

Conversely, if $\gcd(a, p) = 1$ and the congruence $x^2 \equiv a \pmod{p}$ has a solution x, then $x = qp + i$, where $-\frac{p-1}{2} \leq i \leq \frac{p-1}{2}$ and so $i^2 \equiv q \pmod{p}$. $\square$

The basic properties of the Legendre symbol are as follows:

(1) (Euler's criterion) If p is an odd prime and a an integer not divisible by p, then

$$a^{\frac{p-1}{2}} \equiv \left(\frac{a}{p}\right) \pmod{p}.$$

T. Andreescu and D. Andrica, *Number Theory*, DOI: 10.1007/b11856_9,
© Birkhäuser Boston, a part of Springer Science + Business Media, LLC 2009

(2) If $a \equiv b \pmod{p}$, then $\left(\frac{a}{p}\right) = \left(\frac{b}{p}\right)$.

(3) (multiplicity) $\left(\frac{a_1 \cdots a_n}{p}\right) = \left(\frac{a_1}{p}\right) \cdots \left(\frac{a_n}{p}\right)$.

(4) $\left(\frac{-1}{p}\right) = (-1)^{\frac{p-1}{2}}$.

For Euler's criterion, suppose that $\left(\frac{a}{p}\right) = 1$. Then $i^2 \equiv a \pmod{p}$ for some integer i. We have $\gcd(i, p) = 1$, and from Fermat's little theorem, $i^{p-1} \equiv 1 \pmod{p}$. Hence $a^{\frac{p-1}{2}} \equiv 1 \pmod{p}$ and we are done.

If $\left(\frac{a}{p}\right) = -1$, then each of the congruences

$$x^{\frac{p-1}{2}} - 1 \equiv 0 \pmod{p} \quad \text{and} \quad x^{\frac{p-1}{2}} + 1 \equiv 0 \pmod{p}$$

has $\frac{p-1}{2}$ distinct solutions in the set $\{1, 2, \ldots, p-1\}$. The $\frac{p-1}{2}$ quadratic residues correspond to the first congruence, and the $\frac{p-1}{2}$ quadratic nonresidues correspond to the second. Hence if a is a quadratic nonresidue, we have $a^{\frac{p-1}{2}} \equiv -1 \pmod{p}$, and we are done.

Remark. From Fermat's little theorem, $a^{p-1} \equiv 1 \pmod{p}$, and hence $p \mid (a^{\frac{p-1}{2}} - 1)(a^{\frac{p-1}{2}} + 1)$. From Euler's criterion, $p \mid a^{\frac{p-1}{2}} - 1$ if and only if a is a quadratic residue mod p.

Property (2) is clear. For (3) we apply Euler's criterion:

$$\left(\frac{a_i}{p}\right) \equiv a_i^{\frac{p-1}{2}} \pmod{p}, \quad i = 1, \ldots, n.$$

Therefore

$$\left(\frac{a_1}{p}\right) \cdots \left(\frac{a_n}{p}\right) \equiv a_1^{\frac{p-1}{2}} \cdots a_n^{\frac{p-1}{2}} = (a_1 \cdots a_n)^{\frac{p-1}{2}}$$

$$\equiv \left(\frac{a_1 \cdots a_n}{p}\right) \pmod{p}.$$

In order to prove (4), note that $(-1)^{\frac{p-1}{2}}, \left(\frac{-1}{p}\right) \in \{-1, 1\}$. Hence $p \mid (-1)^{\frac{p-1}{2}} - \left(\frac{-1}{p}\right)$ is just Euler's criterion for $a = -1$.

The following theorem gives necessary and sufficient conditions under which 2 is a quadratic residue.

Theorem 9.1.2. *For any odd prime p,*

$$\left(\frac{2}{p}\right) = (-1)^{\frac{p^2-1}{8}}.$$

Proof. We need the following lemma.

Lemma. (Gauss[1]) *If a is a positive integer that is not divisible by p, then from the division algorithm,*

$$ka = pq_k + r_k, \quad k = 1, \ldots, \frac{p-1}{2}.$$

Let $b_1, \ldots, b_m$ be the distinct remainders $r_1, \ldots, r_{(p-1)/2}$ that are less than $\frac{p}{2}$ and let $c_1, \ldots, c_n$ be the other distinct remaining remainders. Then

$$\left(\frac{a}{p}\right) = (-1)^n.$$

Proof of lemma. We have

$$\prod_{i=1}^{m} b_i \prod_{j=1}^{n} c_j = \prod_{k=1}^{\frac{p-1}{2}} r_k = \prod_{k=1}^{\frac{p-1}{2}} (ka - pq_k) \equiv \prod_{k=1}^{\frac{p-1}{2}} ka = a^{\frac{p-1}{2}} \left(\frac{p-1}{2}\right)! \quad (\mathrm{mod}\ p).$$

Because $\frac{p}{2} < c_j \le p - 1$, $j = 1, \ldots, n$, we have $1 \le p - c_j \le (p-1)/2$. It is not possible to have $p - c_j = b_i$ for some i and j. Indeed, if $b_i + c_j = p$, then $p = as - pq_s + at - pq_t$, so $p \mid s + t$, which is impossible, since $1 \le s, t \le (p-1)/2$.

Therefore the integers $b_1, \ldots, b_m, p - c_1, \ldots, p - c_n$ are distinct and

$$\{b_1, \ldots, b_m, p - c_1, \ldots, p - c_n\} = \left\{1, 2, \ldots, \frac{p-1}{2}\right\}.$$

We obtain

$$\prod_{i=1}^{m} b_i \prod_{j=1}^{n} (p - c_j) = \left(\frac{p-1}{2}\right)!.$$

Finally,

$$(-1)^n \prod_{i=1}^{m} b_i \prod_{j=1}^{n} c_j \equiv \left(\frac{p-1}{2}\right)! \quad (\mathrm{mod}\ p);$$

hence $a^{(p-1)/2} \equiv (-1)^n \ (\mathrm{mod}\ p)$. The conclusion now follows from Euler's criterion. $\qquad \square$

In order to prove the theorem we use Gauss's lemma for $a = 2$. We have $\{r_1, r_2, \ldots, r_{(p-1)/2}\} = \{2, 4, \ldots, p - 1\}$. The number of integers k such that $p/2 < 2k < p$ is $n = \lfloor \frac{p}{2} \rfloor - \lfloor \frac{p}{4} \rfloor$.

---

[1] Karl Friedrich Gauss (1777–1855), German mathematician who is sometimes called the "prince of mathematics". Gauss proved in 1801 the fundamental theorem of arithmetic, and he published one of the most brilliant achievements in mathematics, *Disquisitiones Arithmeticae*. In this book he systematized the study of number theory and developed the algebra of congruences.

If $p = 4u + 1$, then $n = 2u - u = u$ and $\frac{p^2-1}{8} = 2u^2 + u$. We have $n \equiv \frac{p^2-1}{8}$ (mod 2) and we are done.

If $p = 4v + 3$, then $n = 2v + 1 - v = v + 1$ and $\frac{p^2-1}{8} = 2v^2 + 3v + 1$, and again $n \equiv \frac{p^2-1}{8}$ (mod 2). $\qquad \square$

The central result concerning the Legendre symbol is the so-called *quadratic reciprocity law* of Gauss:

Theorem 9.1.3. *If p and q are distinct odd primes, then*

$$\left(\frac{q}{p}\right)\left(\frac{p}{q}\right) = (-1)^{\frac{p-1}{2} \cdot \frac{q-1}{2}}.$$

Proof. In Gauss's lemma we take $a = q$ and we get $\left(\frac{q}{p}\right) = (-1)^n$. Let $\sum_{i=1}^{m} b_i = b$ and $\sum_{j=1}^{n} c_j = c$. Then using the equality

$$\{b_1, \ldots, b_m, p - c_1, \ldots, p - c_n\} = \left\{1, 2, \ldots, \frac{p-1}{2}\right\},$$

it follows that

$$b + np - c = \sum_{k=1}^{\frac{p-1}{2}} k = \frac{p^2 - 1}{8}.$$

But from Gauss's lemma we have $q_k = \lfloor \frac{kq}{p} \rfloor$, $k = 1, 2, \ldots, p - 1$; hence

$$q \frac{p^2 - 1}{8} = p \sum_{k=1}^{\frac{p-1}{2}} \left\lfloor \frac{kq}{p} \right\rfloor + b + c.$$

Summing up the last two relations gives

$$2c + p \sum_{k=1}^{\frac{p-1}{2}} \left\lfloor \frac{kq}{p} \right\rfloor + \frac{p^2 - 1}{8}(1 - q) - np = 0.$$

Because $2c$ and $1 - q$ are even, it follows that

$$n \equiv \sum_{k=1}^{\frac{p-1}{2}} \left\lfloor \frac{kq}{p} \right\rfloor \quad \text{(mod 2)},$$

and applying Gauss's lemma again we obtain

$$\left(\frac{q}{p}\right) = (-1)^{\sum_{k=1}^{\frac{p-1}{2}} \lfloor \frac{kq}{p} \rfloor}.$$

Similarly, we derive the relation

$$\left(\frac{p}{q}\right) = (-1)^{\sum_{j=1}^{\frac{p-1}{2}} \left\lfloor \frac{jp}{q} \right\rfloor}.$$

Multiplying the last two equalities and taking into account Landau's identity in Problem 3.2.5(b) of Chapter 3, the conclusion follows. $\qquad\Box$

Problem 9.1.1. *Let $k = 2^{2^n} + 1$ for some positive integer n. Show that k is a prime if and only if k is a factor of $3^{(k-1)/2} + 1$.*

(1997 Taiwanese Mathematical Olympiad)

Solution. Suppose k is a factor of $3^{(k-1)/2} + 1$. This is equivalent to $3^{(k-1)/2} \equiv -1 \pmod{k}$. Hence $3^{k-1} \equiv 1 \pmod{k}$. Let d be the order of $3 \bmod k$. Then $d \nmid (k-1)/2 = 2^{2^n} - 1$, but $d \mid (k-1) = 2^{2^n}$. Hence $d = 2^{2^n} = k - 1$. Therefore k is prime.

Conversely, suppose k is prime. By the quadratic reciprocity law,

$$\left(\frac{3}{k}\right) = \left(\frac{k}{3}\right) = \left(\frac{2}{3}\right) = -1.$$

By Euler's criterion, $3^{(k-1)/2} \equiv \left(\frac{3}{k}\right) \equiv -1 \pmod{k}$, as claimed.

Problem 9.1.2. *Prove that if n is a positive integer such that the equation $x^3 - 3xy^2 + y^3 = n$ has an integer solution (x, y), then it has at least three such solutions.*

Show there are no solutions if $n = 2891$.

(23rd International Mathematical Olympiad)

Solution. The idea of the solution is to find a nonsingular change of coordinates with integer coefficients

$$(x, y) \rightarrow (ax + by, cx + dy)$$

such that the polynomial $x^3 - 3xy^2 + y^3$ does not change after changing coordinates. Such a transformation can be found after noticing the identity

$$x^3 - 3xy^2 + y^3 = (y - x)^3 - 3x^2y + 2x^3 = (y - x)^3 - 3(y - x)x^2 + (-x)^3.$$

Thus, such a transformation is $T(x, y) = (y - x, -x)$. It can be represented by the linear transformation

$$T\begin{pmatrix} x \\ y \end{pmatrix} = \begin{pmatrix} -1 & 1 \\ -1 & 0 \end{pmatrix}\begin{pmatrix} x \\ y \end{pmatrix} = \begin{pmatrix} -x + y \\ -x \end{pmatrix}.$$

We have

$$T^2 = \begin{pmatrix} -1 & 1 \\ -1 & 0 \end{pmatrix} \begin{pmatrix} -1 & 1 \\ -1 & 0 \end{pmatrix} = \begin{pmatrix} 0 & -1 \\ 1 & -1 \end{pmatrix}$$

and

$$T^3 = \begin{pmatrix} 0 & -1 \\ 1 & -1 \end{pmatrix} \begin{pmatrix} -1 & 1 \\ -1 & 0 \end{pmatrix} = \begin{pmatrix} 1 & 0 \\ 0 & 1 \end{pmatrix}.$$

Thus, $T^2(x, y) = (-y, x - y)$. Moreover, it is easy to see that if $x^3 - 3xy^2 + y^3 = n$, $n \geq 0$, then the pairs (x, y), $(-y, x - y)$, and $(y - x, -x)$ are distinct.

For the second part, observe that $2891 = 7^2 \cdot 59$. Suppose that x, y are integers such that $x^3 - 3xy^2 + y^3 = 2891$. Then x, y are relatively prime, because from $d = \gcd(x, y)$ we obtain $d^3 \mid 2891$. The numbers x, y are not divisible by 7; thus they are invertible modulo 7. Thus, from the equation we obtain

$$\left(\frac{y}{x}\right)^3 - 3\left(\frac{y}{x}\right)^2 + 1 \equiv 0 \pmod{7}.$$

This proves that the congruence

$$a^3 - 3a^2 + 1 \equiv 0 \pmod{7}$$

has a solution, $a \in \mathbb{Z}$. Since 7 is not a divisor of a, by Fermat's little theorem one has $a^6 \equiv 1 \pmod 7$. There are two possibilities: $a^3 \equiv 1 \pmod 7$ or $a^3 \equiv -1 \pmod 7$. When $a^3 \equiv 1 \pmod 7$ we obtain

$$a^3 - 3a^2 + 1 \equiv 0 \pmod 7 \Rightarrow 3a^2 \equiv 2 \pmod 7 \Rightarrow a^2 \equiv 3 \pmod 7.$$

Using the Legendre symbol and the quadratic reciprocity law,

$$\left(\tfrac{3}{7}\right) = (-1)^{\frac{3-1}{2} \cdot \frac{7-1}{2}} \left(\tfrac{7}{3}\right) = (-1)\left(\tfrac{1}{3}\right) = -1.$$

This proves that 3 is not a square modulo 7. When $a^3 \equiv -1 \pmod 7$ we obtain the contradiction from $3a^2 \equiv 0 \pmod 7$. Thus, the equation $x^3 - 3xy^2 + y^3 = 2891$ has no solution in integers (x, y).

Problem 9.1.3. *Let m, n be positive integers such that*

$$A = \frac{(m + 3)^n + 1}{3m}$$

is an integer. Prove that A is odd.

(1998 Bulgarian Mathematical Olympiad)

Solution. If m is odd, then $(m + 3)^n + 1$ is odd and A is odd. Now we suppose that m is even. Since A is an integer,

$$0 \equiv (m + 3)^n + 1 \equiv m^n + 1 \pmod 3,$$

so $n = 2k + 1$ is odd and $m \equiv -1 \pmod 3$. We consider the following cases:
   (a) $m = 8m'$ for some positive integer m'. Then

$$(m + 3)^n + 1 \equiv 3^{2k+1} + 1 \equiv 4 \pmod 8$$

and $3m \equiv 0 \pmod 8$. So A is not an integer.
   (b) $m = 2m'$ for some odd positive integer m', i.e., $m \equiv 2 \pmod 4$. Then

$$(m + 3)^n + 1 \equiv (2 + 3)^n + 1 \equiv 2 \pmod 4$$

and $3m \equiv 2 \pmod 4$. So A is odd.
   (c) $m = 4m'$ for some odd positive integer m'. Because $m \equiv -1 \pmod 3$, there exists an odd prime p such that $p \equiv -1 \pmod 3$ and $p \mid m$. Since A is an integer,
$$0 \equiv (m + 3)^n + 1 \equiv 3^{2k+1} + 1 \pmod m$$

and $3^{2k+1} \equiv -1 \pmod p$. Let a be a primitive root modulo p; let b be a positive integer such that $3 \equiv a^b \pmod p$. Thus $a^{(2k+1)b} \equiv -1 \pmod p$. Note that $(p/3) = (-1/3) = -1$. We consider the following cases.
   (i) $p \equiv 1 \pmod 4$. From the quadratic reciprocity law, $(-1/p) = 1$, so

$$a^{2c} \equiv -1 \equiv a^{(2k+1)b} \pmod p$$

for some positive integer c. Therefore b is even and $(3/p) = 1$. Again, from the quadratic reciprocity law,

$$-1 = (3/p)(p/3) = (-1)^{(3-1)(p-1)/4} = 1,$$

a contradiction.
   (ii) $p \equiv 3 \pmod 4$. From the quadratic reciprocity law, $(-1/p) = -1$, so

$$a^{2c+1} \equiv -1 \equiv a^{(2k+1)b} \pmod p$$

for some positive integer c. Therefore b is odd and $(3/p) = -1$. Again, from the quadratic reciprocity law,

$$1 = (3/p)(p/3) = (-1)^{(3-1)(p-1)/4} = -1,$$

a contradiction.
   Thus for $m = 4m'$ and m' odd, A is not an integer.
   From the above, we see that if A is an integer, A is odd.

Problem 9.1.4. *Prove that $2^n + 1$ has no prime factors of the form $8k + 7$.*

(2004 Vietnamese International Mathematical Olympiad Team Selection Test)

Solution. Assume that we have a prime p such that $p \mid 2^n + 1$ and $p \equiv -1$ (mod 8). If n is even, then $p \equiv 3$ (mod 4) and $\left(\frac{-1}{p}\right) = 1$, a contradiction. If n is odd, $2^n \equiv -1$ (mod p), and so $-2 \equiv (2^{\frac{n+1}{2}})^2$ (mod p); hence $\left(\frac{-2}{p}\right) = 1$. We get $(-1)^{\frac{p^2-1}{8}}(-1)^{\frac{p-1}{2}} = 1$, again a contradiction.

Problem 9.1.5. *Prove that $2^{3^n} + 1$ has at least n prime divisors of the form $8k+3$.*

Solution. Using the result of the previous problem, we deduce that $2^n + 1$ does not have prime divisors of the form $8k + 7$. We will prove that if n is odd, then it has no prime divisors of the form $8k + 5$ either. Indeed, let p be a prime divisor of $2^n + 1$. Again we have $2^n \equiv -1$ (mod p) and so $-2 \equiv (2^{\frac{n+1}{2}})^2$ (mod p). Using the same argument as the one in the previous problem, we deduce that $\frac{p^2-1}{8} + \frac{p-1}{2}$ is even, which cannot happen if p is of the form $8k + 5$.

Now, let us solve the additional problem. We will assume $n > 2$ (otherwise, the verification is trivial). The essential observation is the identity

$$2^{3^n} + 1 = (2 + 1)(2^2 - 2 + 1)(2^{2 \cdot 3} - 2^3 + 1) \cdots (2^{2 \cdot 3^{n-1}} - 2^{3^{n-1}} + 1).$$

Now we will prove that for all $1 \leq i < j \leq n - 1$,

$$\gcd(2^{2 \cdot 3^i} - 2^{3^i} + 1, 2^{2 \cdot 3^j} - 2^{3^j} + 1) = 3.$$

Indeed,

$$2^{3^j} = (2^{3^{i+1}})^{3^{j-i-1}} \equiv (-1)^{3^{j-i-1}} \equiv -1 \pmod{2^{3^{i+1}} + 1},$$

implying

$$2^{2 \cdot 3^j} - 2^{3^j} + 1 \equiv 3 \pmod{2^{3^{i+1}} + 1}.$$

Therefore the greatest common divisor is at most 3. Since

$$2^{2 \cdot 3^i} - 2^{3^i} + 1 \equiv 1 - (-1) + 1 = 3 \pmod 3,$$

both quantities are divisible by 3 and therefore the greatest common divisor is 3, as claimed.

It remains to show that each of the numbers $2^{2 \cdot 3^i} - 2^{3^i} + 1$, with $1 \leq i \leq n-1$ has at least one prime divisor of the form $8k+3$ different from 3. It would follow in this case that $2^{3^n} + 1$ has at least $n - 1$ distinct prime divisors of the form $8k + 3$ (from the previous remarks), and since it is also divisible by 3, the conclusion would follow. Fix $i \in \{1, 2, \ldots, n - 1\}$ and observe that any prime factor of $2^{2 \cdot 3^i} - 2^{3^i} + 1$, is also a prime factor of $2^{3^n} + 1$, and thus, from the first remark, it must be of the form $8k + 1$ or $8k + 3$. Because $v_3(2^{2 \cdot 3^i} - 2^{3^i} + 1) = 1$, it follows that if all prime divisors of $2^{2 \cdot 3^i} - 2^{3^i} + 1$ except for 3 are of the form $8k + 1$, then $2^{2 \cdot 3^i} - 2^{3^i} + 1 \equiv 3 \pmod 8$, which is clearly impossible. Thus at least one prime divisor of $2^{2 \cdot 3^i} - 2^{3^i} + 1$ is different from 3 and is of the form $8k + 3$, and so the claim is proved. The conclusion follows.

Problem 9.1.6. *Find a number n between* 100 *and* 1997 *such that* $n \mid 2^n + 2$.

(1997 Asian-Pacific Mathematical Olympiad)

Solution. The first step would be choosing $n = 2p$, for some prime number p. Unfortunately, this cannot work by Fermat's little theorem. So let us try setting $n = 2pq$, with p, q different prime numbers. We need $pq \mid 2^{2pq-1} + 1$, and so we must have $\left(\frac{-2}{p}\right) = \left(\frac{-2}{q}\right) = 1$. Also, using Fermat's little theorem, $p \mid 2^{2q-1} + 1$ and $q \mid 2^{2p-1} + 1$. A small verification shows that $q = 3, 5, 7$ are not good choices, so let us try $q = 11$. In this case we obtain $p = 43$, and so it suffices to show that $pq \mid 2^{2pq-1} + 1$ for $q = 11$ and $p = 43$. This is immediate, since the hard work has already been completed: we have shown that it suffices to have $p \mid 2^{2q-1}, q \mid 2^{2p-1} + 1$, and $\left(\frac{-2}{p}\right) = \left(\frac{-2}{q}\right) = 1$ in order to have $pq \mid 2^{2pq-1} + 1$. But as one can easily check, all these conditions are satisfied, and the number $2 \cdot 11 \cdot 43 = 946$ is a valid answer.

Additional Problems

Problem 9.1.7. Let $f, g : Z^+ \to Z^+$ functions with the following properties:

    (i) g is surjective;

    (ii) $2f^2(n) = n^2 + g^2(n)$ for all positive integers n.

    If, moreover, $|f(n) - n| \le 2004\sqrt{n}$ for all n, prove that f has infinitely many fixed points.

(2005 Moldavian International Mathematical Olympiad Team Selection Test)

Problem 9.1.8. Suppose that the positive integer a is not a perfect square. Then $\left(\frac{a}{p}\right) = -1$ for infinitely many primes p.

Problem 9.1.9. Suppose that $a_1, a_2, \ldots, a_{2004}$ are nonnegative integers such that $a_1^n + a_2^n + \cdots + a_{2004}^n$ is a perfect square for all positive integers n. What is the minimal number of such integers that must equal 0?

(2004 Mathlinks Contest)

Problem 9.1.10. Find all positive integers n such that $2^n - 1 \mid 3^n - 1$.

(American Mathematical Monthly)

Problem 9.1.11. Find the smallest prime factor of $12^{2^{15}} + 1$.

9.2 Special Numbers

9.2.1 Fermat Numbers

Trying to find all primes of the form $2^m + 1$, Fermat noticed that m must be a power of 2. Indeed, if m equaled $k \cdot h$ with k an odd integer greater than 1, then

$$2^m + 1 = (2^h)^k + 1 = (2^h + 1)(2^{h(k-1)} - 2^{h(k-2)} + \cdots - 2^h + 1),$$

and so $2^m + 1$ would not be a prime.

The integers $f_n = 2^{2^n} + 1$, $n \geq 0$, are called *Fermat numbers*. We have

$$f_0 = 3, \quad f_1 = 5, \quad f_2 = 17, \quad f_3 = 257, \quad f_4 = 65{,}573.$$

After checking that these five numbers are primes, Fermat conjectured that f_n is a prime for all n. But Euler proved that $641 \mid f_5$. His argument was the following:

$$f_5 = 2^{32} + 1 = 2^{28}(5^4 + 2^4) - (5 \cdot 2^7)^4 + 1 \quad = 2^{28} \cdot 641 - (640^4 - 1)$$
$$= 641(2^{28} - 639(640^2 + 1)).$$

It is still an open problem whether there are infinitely many Fermat primes. Also, the question whether there are any Fermat primes after f_4 is still open. The answer to this question is important, because Gauss proved that a regular polygon $Q_1 Q_2 \ldots Q_n$ can be constructed using only a straightedge and compass if and only if $n = 2^h p_1 \cdots p_k$, where $k \geq 0$ and $p_1, \ldots, p_k$ are distinct Fermat primes. Gauss was the first to construct such a polygon for $n = 17$.

Problem 9.2.1. *Prove that for f_n, the nth Fermat number,*
  (i) $f_n = f_0 \cdots f_{n-1} + 2$, $n \geq 1$;
  (ii) $\gcd(f_k, f_h) = 1$ if $k \neq h$;
  (iii) f_n ends in 7 for all $n \geq 2$.

Solution. (i) We have

$$f_k = 2^{2^k} + 1 = \left(2^{2^{k-1}}\right)^2 + 1 = (f_{k-1} - 1)^2 + 1 = f_{k-1}^2 - 2f_{k-1} + 2;$$

hence

$$f_k - 2 = f_{k-1}(f_{k-1} - 2), \quad k \geq 1. \tag{1}$$

Multiplying relations (1) for $k = 1, \ldots, n$ yields

$$f_n - 2 = f_0 \cdots f_{n-1}(f_0 - 2),$$

and the conclusion follows.

For a different proof we can use directly the identity

$$\frac{x^{2^n} - 1}{x - 1} = \prod_{k=0}^{n-1}(x^{2^k} + 1).$$

(ii) From (i) we have

$$\gcd(f_n, f_0) = \gcd(f_n, f_1) = \cdots = \gcd(f_n, f_{n-1}) = 1$$

for all $n \geq 1$; hence $\gcd(f_k, f_h) = 1$ for all $k \neq h$.

(iii) Because $f_1 = 5$ and $f_0 \cdots f_{n-1}$ is odd, using (i), it follows that f_n ends in $5 + 2 = 7$ for all $n \geq 2$.

Problem 9.2.2. *Find all Fermat numbers that can be written as a sum of two primes.*

Solution. All Fermat numbers are odd. If $f_n = p + q$ for some primes p and q, $p \leq q$, then $p = 2$ and $q > 2$. We obtain

$$q = 2^{2^n} - 1 = \left(2^{2^{n-1}}\right)^2 - 1 = \left(2^{2^{n-1}} - 1\right)\left(2^{2^{n-1}} + 1\right);$$

hence $2^{2^{n-1}} - 1$ must equal 1. That is, $n = 1$ and $f_1 = 2 + 3$ is the unique Fermat number with this property.

An alternative solution uses Problem 1 (iii): if $n \geq 2$, then f_n ends in 7, so q must end in 5. Hence $q = 5$ and $2 + 5 \neq f_n$ for $n \geq 2$. The only Fermat number with the given property is f_1.

Problem 9.2.3. *Show that for any $n \geq 2$ the prime divisors p of f_n are of the form $p = s \cdot 2^{n+2} + 1$.*

Solution. Because $p \mid f_n$, it follows that $2^{2^n} \equiv -1 \pmod{p}$. Hence squaring gives $2^{2^{n+1}} \equiv 1 \pmod{p}$. Thus $o_p(2) \mid 2^{n+1}$. Since $2^{2^n} \not\equiv 1 \pmod{p}$, we have $o_p(2) = 2^{n+1}$. Thus $2^{n+1} \mid p - 1$. Hence $p \equiv 1 \pmod 8$ and $\left(\frac{2}{p}\right) = 1$. So 2 is a quadratic residue mod p and there is some a with $a^2 \equiv 2 \pmod{p}$. Hence $o_p(a) = 2^{n+2}$ and $2^{n+2} \mid p - 1$, that is, $p = s \cdot 2^{n+2} + 1$ for some s.

Additional Problems

Problem 9.2.4. Find all positive integers n such that $2^n - 1$ is a multiple of 3 and $\frac{2^n - 1}{3}$ is a divisor of $4m^2 + 1$ for some integer m.

(1999 Korean Mathematical Olympiad)

Problem 9.2.5. Prove that the greatest prime factor of f_n, $n \geq 2$, is greater than $2^{n+2}(n + 1)$.

(2005 Chinese International Mathematical Olympiad Team Selection Test)

9.2.2 Mersenne Numbers

The integers $M_n = 2^n - 1$, $n \geq 1$, are called *Mersenne*[2] *numbers*. It is clear that if n is composite, then so is M_n. Moreover, if $n = ab$, where a and b are integers greater than 1, then M_a and M_b both divide M_n. But there are primes n for which M_n is composite, for example $47 \mid M_{23}$, $167 \mid M_{83}$, $263 \mid M_{131}$, and so on.

It is not known if there are infinitely many primes with this property. The largest known prime is

$$2^{32582657} - 1,$$

and it is a Mersenne number. Presently, we know 42 Mersenne numbers that are primes.

Theorem 9.2.1. *Let p be an odd prime and let q be a prime divisor of M_p. Then $q = 2kp + 1$ for some positive integer k.*

Proof. From the congruence $2^p \equiv 1 \pmod{q}$ and from the fact that p is a prime, it follows that p is the least positive integer satisfying this property. Using Fermat's little theorem, we have $2^{q-1} \equiv 1 \pmod{q}$, and hence $p \mid q - 1$. But $q - 1$ is an even integer, so $q - 1 = 2kp$, and the conclusion follows.

Problem 9.2.6. *Let p be a prime of the form $4k + 3$. Then $2p + 1$ is a prime if and only if $2p + 1$ divides M_p.*

Solution. Suppose that $q = 2p + 1$ is a prime. Then

$$\left(\frac{2}{q}\right) = (-1)^{\frac{q^2-1}{8}} = (-1)^{\frac{p(p+1)}{2}} = (-1)^{2(k+1)(4k+3)} = 1;$$

hence 2 is a quadratic residue mod q.

Using Euler's criterion, it follows that $2^{\frac{q-1}{2}} \equiv 1 \pmod{q}$, that is, $2^p \equiv 1 \pmod{q}$, and the conclusion follows.

If q is composite, then it has a prime divisor q_1 such that $q_1 \leq \sqrt{q}$. Using Fermat's little theorem, we have $2^{q_1-1} \equiv 1 \pmod{q_1}$. But $2^p \equiv 1 \pmod{q_1}$ with p prime implies that p is the least positive integer with the property. Hence $p \mid q_1 - 1$, and thus $q_1 \geq p + 1 > \sqrt{p}$, contradicting the choice of q_1. Therefore q must be a prime and the conclusion follows.

Additional Problems

Problem 9.2.7. Let P^* denote all the odd primes less than 10000, and suppose $p \in P^*$. For each subset $S = \{p_1, p_2, \ldots, p_k\}$ of P^*, with $k \geq 2$ and not including p, there exists a $q \in P^* \setminus S$ such that

$$(q + 1) \mid (p_1 + 1)(p_2 + 1) \cdots (p_k + 1).$$

---

[2]Marin Mersenne (1588–1648), French monk who is best known for his role as a clearing house for correspondence with eminent philosophers and scientists and for his work in number theory.

Find all such possible values of p.

(1999 Taiwanese Mathematical Olympiad)

9.2.3 Perfect Numbers

An integer $n \geq 2$ is called *perfect* if the sum of its divisors is equal to $2n$. That is, $\sigma(n) = 2n$. For example, the numbers $6, 28, 496$ are perfect. The even perfect numbers are closely related to Mersenne numbers. It is not known whether any odd perfect numbers exist.

Theorem 9.2.2. (Euclid) *If M_k is a prime, then $n = 2^{k-1}M_k$ is a perfect number.*

Proof. Because $\gcd(2^{k-1}, 2^k - 1) = 1$, and the fact that σ is a multiplicative function, it follows that

$$\sigma(n) = \sigma(2^{k-1})\sigma(2^k - 1) = (2^k - 1) \cdot 2^k = 2n. \qquad \square$$

There is also a partial converse, due to Euler.

Theorem 9.2.3. *If the even positive integer n is perfect, then $n = 2^{k-1}M_k$ for some positive integer k for which M_k is a prime.*

Proof. Let $n = 2^t u$, where $t \geq 1$ and u is odd. Because n is perfect, we have $\sigma(n) = 2n$; hence $\sigma(2^t u) = 2^{t+1}u$. Using again that σ is multiplicative, we get

$$\sigma(2^t u) = \sigma(2^t)\sigma(u) = (2^{t+1} - 1)\sigma(u).$$

This is equivalent to

$$(2^{t+1} - 1)\sigma(u) = 2^{t+1}u.$$

Because $\gcd(2^{t+1} - 1, 2^{t+1}) = 1$, it follows that $2^{t+1} \mid \sigma(u)$; hence $\sigma(u) = 2^{t+1}v$ for some positive integer v. We obtain $u = (2^{t+1} - 1)v$.

The next step is to show that $v = 1$. If $v > 1$, then

$$\sigma(u) \geq 1 + v + 2^{t+1} - 1 + v(2^{t+1} - 1) = (v+1)2^{t+1} > v \cdot 2^{t+1} = \sigma(u),$$

a contradiction. We get $v = 1$; hence $u = 2^{t+1} - 1 = M_{t+1}$ and $\sigma(u) = 2^{t+1}$. If M_{t+1} is not a prime, then $\sigma(u) > 2^{t+1}$, which is impossible. Finally, $n = 2^{k-1}M_k$, where $k = t + 1$. $\qquad \square$

Remark. Recall that M_k is a prime only if k is a prime. This fact reflects also in Theorem 9.2.2 and Theorem 9.2.3.

Problem 9.2.8. *Show that any even perfect number is triangular.*

Solution. Using Theorem 9.2.3, we have

$$n = 2^{k-1}M_k = \frac{2^k}{2}(2^k - 1) = \frac{m(m+1)}{2},$$

where $m = 2^k - 1$ and we are done.

Additional Problems

Problem 9.2.9. Prove that if n is an even perfect number, then $8n + 1$ is a perfect square.

Problem 9.2.10. Show that if k is an odd positive integer, then $2^{k-1} M_k$ can be written as the sum of the cubes of the first $2^{\frac{k-1}{2}}$ odd positive integers. In particular, any perfect number has this property.

9.3   Sequences of Integers

9.3.1   Fibonacci and Lucas Sequences

Leonardo Fibonacci[3] introduced in 1228 the sequence $F_1 = F_2 = 1$ and $F_{n+1} = F_n + F_{n-1}, n \geq 2$. It is not difficult to prove by induction that the closed form for F_n is given by Binet's formula

$$F_n = \frac{1}{\sqrt{5}} \left[\left(\frac{1+\sqrt{5}}{2}\right)^n - \left(\frac{1-\sqrt{5}}{2}\right)^n \right] \tag{1}$$

for all $n \geq 1$. As a consequence of the recursive definition or of formula above, it is a convention to define $F_0 = 0$. Identities for Fibonacci numbers are usually proved either by induction or from Binet's formula. It is also an useful matrix form for the Fibonacci numbers

$$\begin{pmatrix} 1 & 1 \\ 1 & 0 \end{pmatrix}^n = \begin{pmatrix} F_{n+1} & F_n \\ F_n & F_{n-1} \end{pmatrix}, \quad n \geq 1 \tag{2}$$

that easily follows by induction.

In what follows we give some arithmetical properties of the Fibonacci numbers.

(1) If $m \mid n$, then $F_m \mid F_n$. If $n \geq 5$ and F_n is a prime, then so is n.
(2) For any $m, n \geq 0$, $\gcd(F_m, F_n) = F_{\gcd(m,n)}$.
(3) If $\gcd(m, n) = 1$, then $F_m F_n \mid F_{mn}$.

In order to prove (1) suppose that $n = mk$ for some integer $k > 1$ and denote $\alpha = \frac{1+\sqrt{5}}{2}, \beta = \frac{1-\sqrt{5}}{2}$. Using (1), we have

$$\frac{F_n}{F_m} = \frac{\alpha^n - \beta^n}{\alpha^m - \beta^m} = \frac{(\alpha^m)^k - (\beta^m)^k}{\alpha^m - \beta^m} = \alpha^{m(k-1)} + \alpha^{m(k-2)}\beta^m + \cdots + \beta^{m(k-1)}.$$

---

[3]Leonardo Pisano Fibonacci (1170–1250) was among the first to introduce the Hindu-Arabic number system into Europe. His book on how to do arithmetic in the decimal system, called *Liber abbaci* (meaning *Book of the Abacus* or *Book of Calculating*), completed in 1202, persuaded many European mathematicians of his day to use this "new" system.

Because $\alpha + \beta = 1$ and $\alpha\beta = -1$ it follows by induction that $\alpha^i + \beta^i$ is an integer for all integers $i \geq 1$ and the conclusion follows.

It is now clear that if $n = kh$, $k \geq 3$, then F_k divides F_n hence F_n is not a prime.

For (2) let $d = \gcd(m, n)$ and suppose that $n > m$. Applying Euclid's Algorithm, we get

$$n = mq_1 + r_1$$
$$m = r_1 q_2 + r_2$$
$$r_1 = r_2 q_3 + r_3$$
$$\cdots$$
$$r_{i-1} = r_i q_{i+1}$$

and so $d = r_i$. It is not difficult to check that for any positive integers m, n,

$$F_{m+n} = F_{m-1} F_n + F_m F_{n+1}. \tag{3}$$

The standard proof of (3) is by induction on n after fixing m. Another argument follows from the matrix form (2). Indeed, we have

$$\begin{pmatrix} F_{m+n+1} & F_{m+n} \\ F_{m+n} & F_{m+n-1} \end{pmatrix} = \begin{pmatrix} 1 & 1 \\ 1 & 0 \end{pmatrix}^{m+n} = \begin{pmatrix} 1 & 1 \\ 1 & 0 \end{pmatrix}^n \begin{pmatrix} 1 & 1 \\ 1 & 0 \end{pmatrix}^m$$
$$= \begin{pmatrix} F_{n+1} & F_n \\ F_n & F_{n-1} \end{pmatrix} \begin{pmatrix} F_{m+1} & F_m \\ F_m & F_{m-1} \end{pmatrix}$$
$$= \begin{pmatrix} F_{n+1} F_{m+1} + F_n F_m & F_{n+1} F_m + F_n F_{m-1} \\ F_n F_{m+1} + F_{n-1} F_m & F_n F_m + F_{n-1} F_{m-1} \end{pmatrix}.$$

Suppose $n > m \geq 1$. From the identity $F_n = F_{m-1} F_{n-m} + F_m F_{n-m+1}$ we have

$$\gcd(F_m, F_n) = \gcd(F_m, F_{m-1} F_{n-m} + F_m F_{n-m+1}) = \gcd(F_m, F_{m-1} F_{n-m}).$$

By the inductive hypothesis,

$$\gcd(F_m, F_{m-1}) = F_1 = 1$$

and

$$\gcd(F_m, F_{n-m}) = F_{\gcd(m, n-m)} = F_{\gcd(m,n)}.$$

Therefore $\gcd(F_m, F_n) = F_{\gcd(m,n)}$.

Property (3) follows from (2) by observing that

$$\gcd(F_m, F_n) = F_{\gcd(m,n)} = F_1 = 1$$

and then using (1).

Lucas's sequence is defined by $L_0 = 2$, $L_1 = 1$, and $L_{n+1} = L_n + L_{n-1}$, $n \geq 1$. The Lucas numbers are the companions to the Fibonacci numbers because they satisfy the same recursion.

The analogue of Binet's Fibonacci number formula for Lucas numbers is

$$L_n = \left(\frac{1 + \sqrt{5}}{2}\right)^n + \left(\frac{1 - \sqrt{5}}{2}\right)^n, \quad n \geq 0, \tag{4}$$

and they are connected with Fibonacci numbers by $L_n = F_{2n}/F_n$, $L_n = 2F_{n+1} - F_n$, $n \geq 0$.

Problem 9.3.1. *Show that there is a positive number in the Fibonacci sequence that is divisible by 1000.*

(1999 Irish Mathematical Olympiad)

Solution. In fact, for any natural number n, there exist infinitely many positive Fibonacci numbers divisible by n.

Consider ordered pairs of consecutive Fibonacci numbers (F_0, F_1), (F_1, F_2), ... taken modulo n. Because the Fibonacci sequence is infinite and there are only n^2 possible ordered pairs of integers modulo n, two such pairs (F_j, F_{j+1}) must be congruent: $F_i \equiv F_{i+m}$ and $F_{i+1} \equiv F_{i+m+1}$ (mod n) for some i and m.

If $i \geq 1$ then $F_{i-1} \equiv F_{i+1} - F_i \equiv F_{i+m+1} - F_{i+m} \equiv F_{i+m-1}$ (mod n). Likewise, $F_{i+2} \equiv F_{i+1} + F_i \equiv F_{i+m+1} + F_{i+m} \equiv F_{i+2+m}$ (mod n). Continuing similarly, we have $F_j \equiv F_{j+m}$ (mod n) for all $j \geq 0$. In particular, $0 = F_0 \equiv F_m \equiv F_{2m} \equiv \cdots$ (mod n), so the numbers $F_m, F_{2m}, \ldots$ are all positive Fibonacci numbers divisible by n. Applying this to $n = 1000$, we are done.

Problem 9.3.2. *Prove that*

(i) *The statement "$F_{n+k} - F_n$ is divisible by 10 for all positive integers n" is true if $k = 60$ and false for any positive integer $k < 60$;*

(ii) *The statement "$F_{n+t} - F_n$ is divisible by 100 for all positive integers n" is true if $t = 300$ and false for any positive integer $t < 300$.*

(1996 Irish Mathematical Olympiad)

First solution. A direct computation shows that the Fibonacci sequence has period 3 modulo 2 and 20 modulo 5 (compute terms until the initial terms 0, 1 repeat, at which time the entire sequence repeats), yielding (a). As for (b), one computes that the period mod 4 is 6. The period mod 25 turns out to be 100, which is awfully many terms to compute by hand, but knowing that the period must be a multiple of 20 helps, and verifying the recursion $F_{n+8} = 7F_{n+4} - F_n$ shows that the period divides 100; finally, an explicit computation shows that the period is not 20.

Second solution. Expanding Binet's formula using the binomial theorem gives

$$2^{n-1} F_n = \sum_{k=0}^{n/2} 5^k \binom{n}{2k+1}.$$

Therefore $F_n \equiv 3^{n-1} \left(n + 5\binom{n}{3} \right)$ (mod 25). Modulo 25, 3^{n-1} has period 20, n has period 25, and $5\binom{n}{3}$ has period 5. Therefore F_n clearly has period dividing 100. The period cannot divide 50, since this formula gives $F_{n+50} \equiv 3^{10} F_n \equiv -F_n$ (mod 25) and the period cannot divide 20 since it gives $F_{n+20} \equiv F_n + 3^{n-1} \cdot 20$ (mod 25).

Problem 9.3.3. *Let $(a_n)_{n \geq 0}$ be the sequence defined by $a_0 = 0$, $a_1 = 1$, and*

$$\frac{a_{n+1} - 3a_n + a_{n-1}}{2} = (-1)^n$$

for all integers $n > 0$. Prove that a_n is a perfect square for all $n \geq 0$.

Solution. Note that $a_2 = 1$, $a_3 = 4$, $a_4 = 9$, $a_5 = 25$, so $a_0 = F_0^2$, $a_1 = F_1^2$, $a_2 = F_2^2$, $a_3 = F_3^2$, $a_4 = F_4^2$, $a_5 = F_5^2$, where $(F_n)_{n \geq 0}$ is the Fibonacci sequence.

We induct on n to prove that $a_n = F_n^2$ for all $n \geq 0$. Assume that $a_k = F_k^2$ for all $k \leq n$. Hence

$$a_n = F_n^2, \quad a_{n-1} = F_{n-1}^2, \quad a_{n-2} = F_{n-2}^2. \tag{1}$$

From the given relation we obtain

$$a_{n+1} - 3a_n + a_{n-1} = 2(-1)^n$$

and

$$a_n - 3a_{n-1} + a_{n-2} = 2(-1)^{n-1}, \quad n \geq 2.$$

Summing up these equalities yields

$$a_{n+1} - 2a_n - 2a_{n-1} + a_{n-2} = 0, \quad n \geq 2. \tag{2}$$

Using the relations (1) and (2) we obtain

$$a_{n+1} = 2F_n^2 + 2F_{n-1}^2 - F_{n-2}^2 = (F_n + F_{n-1})^2 + (F_n - F_{n-1})^2 - F_{n-2}^2$$
$$= F_{n+1}^2 + F_{n-2}^2 - F_{n-2}^2 = F_{n+1}^2,$$

as desired.

Problem 9.3.4. *Define the sequence $(a_n)_{n \geq 0}$ by $a_0 = 0$, $a_1 = 1$, $a_2 = 2$, $a_3 = 6$, and*

$$a_{n+4} = 2a_{n+3} + a_{n+2} - 2a_{n+1} - a_n, \quad n \geq 0.$$

Prove that n divides a_n for all $n > 0$.

Solution. From the hypothesis it follows that $a_4 = 12$, $a_5 = 25$, $a_6 = 48$. We have $\frac{a_1}{1}$, $\frac{a_2}{2} = 1$, $\frac{a_3}{3} = 2$, $\frac{a_4}{4} = 3$, $\frac{a_5}{5} = 5$, $\frac{a_6}{6} = 8$, so $\frac{a_n}{n} = F_n$ for all $n = 1, 2, 3, 4, 5, 6$, where $(F_n)_{n\geq 1}$ is the Fibonacci sequence.

We prove by induction that $a_n = n F_n$ for all n. Indeed, assuming that $a_k = k F_k$ for $k \leq n + 3$, we have

$$
\begin{aligned}
a_{n+4} &= 2(n + 3)F_{n+3} + (n + 2)F_{n+2} - 2(n + 1)F_{n+1} - n F_n \\
&= 2(n + 3)F_{n+3} + (n + 2)F_{n+2} - 2(n + 1)F_{n+1} - n(F_{n+2} - F_{n+1}) \\
&= 2(n + 3)F_{n+3} + 2F_{n+2} - (n + 2)F_{n+1} \\
&= 2(n + 3)F_{n+3} + 2F_{n+2} - (n + 2)(F_{n+3} - F_{n+2}) \\
&= (n + 4)(F_{n+3} + F_{n+2}) = (n + 4)F_{n+4},
\end{aligned}
$$

as desired.

Additional Problems

Problem 9.3.5. Determine the maximum value of $m^2 + n^2$, where m and n are integers satisfying $1 \leq m, n \leq 1981$ and $(n^2 - mn - m^2)^2 = 1$.

(22nd International Mathematical Olympiad)

Problem 9.3.6. Prove that for any integer $n \geq 4$, $F_n + 1$ is not a prime.

Problem 9.3.7. Let k be an integer greater than 1, $a_0 = 4$, $a_1 = a_2 = (k^2 - 2)^2$, and

$$a_{n+1} = a_n a_{n-1} - 2(a_n + a_{n-1}) - a_{n-2} + 8 \text{ for } n \geq 2.$$

Prove that $2 + \sqrt{a_n}$ is a perfect square for all n.

9.3.2   Problems Involving Linear Recursive Relations

A sequence $x_0, x_1, x_2, \ldots$ of complex numbers is defined recursively by a *linear recursion of order k* if

$$x_n = a_1 x_{n-1} + a_2 x_{n-2} + \cdots + a_k x_{n-k}, \quad n \geq k, \tag{1}$$

where $a_1, a_2, \ldots, a_k$ are given complex numbers and $x_0 = \alpha_0$, $x_1 = \alpha_1, \ldots,$ $x_{k-1} = \alpha_{k-1}$ are also given.

The main problem is to find a general formula for x_n in terms of $a_1, a_2, \ldots,$ a_k, $\alpha_0, \alpha_1, \ldots, \alpha_{k-1}$, and n. In order to solve this problem we attach to (1) the algebraic equation

$$t^k - a_1 t^{k-1} - a_2 t^{k-2} - \cdots - a_k = 0, \tag{2}$$

which is called the characteristic equation of (1).

Theorem 9.3.1. *If the characteristic equation (2) has distinct roots $t_1, t_2, \ldots, t_k$, then*

$$x_n = c_1 t_1^n + c_2 t_2^n + \cdots + c_k t_k^n, \tag{3}$$

where the constants $c_1, c_2, \ldots, c_k$ are determined by the initial conditions $x_0 = \alpha_0, x_1 = \alpha_1, \ldots, x_{k-1} = \alpha_{k-1}$.

Proof. Consider the sequence $y_0, y_1, y_2, \ldots$ given by

$$y_n = c_1 t_1^n + c_2 t_2^n + \cdots + c_n t_n^n.$$

It is not difficult to prove that the sequence $(y_n)_{n \geq 0}$ satisfies the linear recursion (1), since $t_1, t_2, \ldots, t_k$ are the roots of the characteristic equation (2). Consider the following system of linear equations:

$$c_1 + c_2 + \cdots + c_k = \alpha_0,$$
$$c_1 t_1 + c_2 t_2 + \cdots + c_k t_k = \alpha_1,$$
$$\cdots$$
$$c_1 t_1^{k-1} + c_2 t_2^{k-1} + \cdots + c_k t_k^{k-1} = \alpha_{k-1}, \tag{4}$$

whose determinant is the so-called Vandermonde determinant

$$V(t_1, t_2, \ldots, t_k) = \prod_{1 \leq i < j \leq k} (t_j - t_i).$$

This determinant is nonzero, because $t_1, t_2, \ldots, t_k$ are distinct.

Hence $c_1, c_2, \ldots, c_k$ are uniquely determined as a solution to system (4). Moreover, $y_0 = \alpha_0 = x_0, y_1 = \alpha_1 = x_1, \ldots, y_{k-1} = \alpha_{k-1} = x_{k-1}$. Using strong induction, from (1) it follows that $y_n = x_n$ for all n. $\square$

The case in which the roots of the characteristic equation (2) are not distinct is addressed in the following theorem.

Theorem 9.3.2. *Suppose that equation (2) has the distinct roots $t_1, \ldots, t_h$, with multiplicities $s_1, \ldots, s_h$, respectively. Then x_n is a linear combination of*

$$t_1^n, n t_1^n, \ldots n^{s_1 - 1} t_1^n$$
$$\cdots$$
$$t_h^n, n t_h^n, \ldots, n^{s_h - 1} t_h^n$$

One can also say that

$$x_n = f_1(n) t_1^n + f_2(n) t_2^n + \cdots + f_h(n) t_h^n,$$

where f_i is a polynomial of degree s_i, for each i.

The proof of this result uses the so-called Hermite interpolation polynomial or formal series.

The most frequent situation is with $k = 2$. Then the linear recursion becomes

$$x_n = a_1 x_{n-1} + a_2 x_{n-2}, \quad n \geq 2,$$

where a_1, a_2 are given complex numbers and $x_0 = \alpha_0$, $x_1 = \alpha_1$.

If the characteristic equation $t^2 - a_1 t - a_2 = 0$ has distinct roots t_1, t_2, then

$$x_n = c_1 t_1^n + c_2 t_2^n, \quad n \geq 0,$$

where c_1, c_2 are solutions to the system of linear equations

$$c_1 + c_2 = \alpha_0, \quad c_1 t_1 + c_2 t_2 = \alpha_1,$$

that is,

$$c_1 = \frac{\alpha_1 - \alpha_0 t_2}{t_1 - t_2}, \quad c_2 = \frac{\alpha_0 t_1 - \alpha_1}{t_1 - t_2}.$$

If the characteristic equation has the nonzero double root t_1, then

$$x_n = c_1 t_1^n + c_2 n t_1^n = (c_1 + c_2 n) t_1^n,$$

where c_1, c_2 are determined from the system of equations $x_0 = \alpha_0$, $x_1 = \alpha_1$, that is,

$$c_1 = \alpha_0, \quad c_2 = \frac{\alpha_1 - \alpha_0 t_1}{t_1}.$$

Example. Let us find the general term of the sequence

$$P_0 = 0, \quad P_1 = 1, \ldots, \quad P_n = 2P_{n-1} + P_{n-2}, \quad n \geq 2.$$

The characteristic equation is $t^2 - 2t - 1 = 0$, whose roots are $t_1 = 1 + \sqrt{2}$ and $t_2 = 1 - \sqrt{2}$. We have $P_n = c_1 t_1^n + c_2 t_2^n$, $n \geq 0$, where $c_1 + c_2 = 0$ and $c_1(1 + \sqrt{2}) + c_2(1 - \sqrt{2}) = 1$; hence

$$P_n = \frac{1}{2\sqrt{2}}[(1 + \sqrt{2})^n - (1 - \sqrt{2})^n], \quad n \geq 0.$$

This sequence is called *Pell's sequence*, and it plays an important part in Diophantine equations.

In some situations we encounter inhomogeneous recursions of order k of the form

$$x_n = a_1 x_{n-1} + a_2 x_{n-2} + \cdots + a_k x_{n-k} + b, \quad n \geq k,$$

where $a_1, a_2, \ldots, a_k, b$ are given complex numbers and $x_1 = \alpha_1$, $x_2 = \alpha_2, \ldots,$ $x_{k-1} = \alpha_{k-1}$. The method of attack consists in performing a translation $x_n = y_n + \beta$, where β is the solution to the equation $(1 - a_1 - a_2 - \cdots - a_k)\beta = b$ when $a_1 + a_2 + \cdots + a_k \neq 1$. The sequence $(y_n)_{n \geq 0}$ satisfies the linear recursion (1).

Example. Let us find x_n if $x_0 = \alpha$, $x_n = ax_{n-1} + b$, $n \geq 1$.

If $a = 1$, we have an arithmetic sequence whose first term is α and whose common difference is b. In this case $x_n = \alpha + nb$.

If $a \neq 1$, we perform the translation $x_n = y_n + \beta$, where $\beta = \frac{b}{1-a}$. In this case $(y_n)_{n\geq 0}$ satisfies the recursion $y_0 = \alpha - \beta$, $y_n = ay_{n-1}$, $n \geq 1$, which is a geometric sequence whose first term is $\alpha - \beta$ and whose ratio is a. We obtain $y_n = (\alpha - \beta)a^n$; hence

$$x_n = \left(\alpha - \frac{b}{1-a}\right)a^n + \frac{b}{1-a}, \quad n \geq 0.$$

Problem 9.3.8. *Let a and b be positive integers and let the sequence $(x_n)_{n\geq 0}$ be defined by $x_0 = 1$ and $x_{n+1} = ax_n + b$ for all nonnegative integers n. Prove that for any choice of a and b, the sequence $(x_n)_{n\geq 0}$ contains infinitely many composite numbers.*

(1995 German Mathematical Olympiad)

Solution. The case $a = 1$ gives $x_n = 1 + b + \cdots + b^{n-1} = \frac{b^n-1}{b-1}$, $n \geq 0$. If n is even, $n = 2k$, then

$$x_n = (b^k + 1)\frac{b^k - 1}{b - 1} = (b^k + 1)x_k, \quad k \geq 0$$

and we are done.

Let $a \neq 1$.

Assume to the contrary that x_n is composite for only finitely many n. Take N larger than all such n, so that x_m is prime for all $n > N$. Choose such a prime $x_m = p$ not dividing $a - 1$ (this excludes only finitely many candidates). Let t be such that $t(1 - a) \equiv b \pmod{p}$; then

$$x_{n+1} - t \equiv ax_n + b - b = a(x_n - t) \pmod{p}.$$

In particular,

$$x_{m+p-1} = t + (x_{m+p-1} - t) \equiv t + a^{p-1}(x_m - t) \equiv (1 - a^{p-1})t \equiv 0 \pmod{p}.$$

However, x_{m+p-1} is a prime greater than p, yielding a contradiction. Hence infinitely many of the x_n are composite.

Problem 9.3.9. *Find a_n if $a_0 = 1$ and $a_{n+1} = 2a_n + \sqrt{3a_n^2 - 2}$, $n \geq 0$.*

Solution. We have $(a_{n+1} - 2a_n)^2 = 3a_n^2 - 2$, so

$$a_{n+1}^2 - 4a_{n+1}a_n + a_n^2 + 2 = 0, \quad n \geq 0.$$

Then

$$a_n^2 - 4a_na_{n-1} + a_{n-1}^2 + 2 = 0, \quad n \geq 1;$$

hence, by subtraction,

$$a_{n+1}^2 - a_{n-1}^2 - 4a_n(a_{n+1} - a_{n-1}) = 0$$

for all $n \geq 1$. Because it is clear that $(a_n)_{n \geq 0}$ is increasing, we have $a_{n+1} - a_{n-1} \neq 0$, for all $n \geq 1$, so

$$a_{n+1} + a_{n-1} - 4a_n = 0, \quad n \geq 1,$$

that is, $a_{n+1} = 4a_n - a_{n-1}, n \geq 1$. Moreover, $a_0 = 1$ and $a_1 = 3$. The characteristic equation is $t^2 - 4t + 1 = 0$, whose roots are $t_1 = 2 + \sqrt{3}$ and $t_2 = 2 - \sqrt{3}$. We obtain

$$a_n = \frac{1}{2\sqrt{3}}\left[(1 + \sqrt{3})(2 + \sqrt{3})^n - (1 - \sqrt{3})(2 - \sqrt{3})^n\right], \quad n \geq 0.$$

We can also write a_n as follows:

$$a_n = \frac{1}{\sqrt{3}}\left[\left(\frac{1 + \sqrt{3}}{2}\right)^{2n+1} - \left(\frac{1 - \sqrt{3}}{2}\right)^{2n+1}\right], \quad n \geq 0.$$

Note that from $a_0 = 1$, $a_1 = 3$, and $a_{n+1} = 4a_n - a_{n-1}$ it follows by strong induction that a_n is a positive integer for all n.

Problem 9.3.10. *Consider the sequence $\{a_n\}$ such that $a_0 = 4$, $a_1 = 22$, and $a_n - 6a_{n-1} + a_{n-2} = 0$ for $n \geq 2$. Prove that there exist sequences $\{x_n\}$ and $\{y_n\}$ of positive integers such that*

$$a_n = \frac{y_n^2 + 7}{x_n - y_n}$$

for any $n \geq 0$.

(2001 Bulgarian Mathematical Olympiad)

Solution. Consider the sequence $\{c_n\}$ of positive integers such that $c_0 = 2$, $c_1 = 1$, and $c_n = 2c_{n-1} + c_{n-2}$ for $n \geq 2$.

We prove by induction that $a_n = c_{2n+2}$ for $n \geq 0$. We check the base cases of $a_0 = 4 = c_2$ and $a_1 = 9 = c_4$. Then, for any $k \geq 2$, assuming that the claim holds for $n = k - 2$ and $n = k - 1$,

$$\begin{aligned}
c_{2k+2} &= 2c_{2k+1} + c_{2k} \\
&= 2(2c_{2k} + c_{2k-1}) + a_{k-1} \\
&= 4c_{2k} + (c_{2k} - c_{2k-2}) + a_{k-1} \\
&= 6a_{k-1} - a_{k-2} \\
&= a_k,
\end{aligned}$$

so the claim holds for $n = k$ as well, and the induction is complete.

For $n \geq 1$,

$$\begin{pmatrix} a_{n+1} & a_n \\ a_{n+2} & a_{n+1} \end{pmatrix} = \begin{pmatrix} 0 & 1 \\ 1 & 2 \end{pmatrix} \begin{pmatrix} a_n & a_{n-1} \\ a_{n+1} & a_n \end{pmatrix}$$

and

$$\begin{vmatrix} a_{n+1} & a_n \\ a_{n+2} & a_{n+1} \end{vmatrix} = \begin{vmatrix} 0 & 1 \\ 1 & 2 \end{vmatrix} \begin{vmatrix} a_n & a_{n-1} \\ a_{n+1} & a_n \end{vmatrix} = - \begin{vmatrix} a_n & a_{n-1} \\ a_{n+1} & a_n \end{vmatrix}.$$

Thus, for $n \geq 0$,

$$c_{n+1}^2 - c_n c_{n+2} = (-1)^n (c_1^2 - c_0 c_2) = (-1)^n (1^2 - 2 \cdot 4) = (-1)^n (-7).$$

In particular, for all $n \geq 0$,

$$c_{2n+1}^2 - c_{2n} a_n = c_{2n+1}^2 - c_{2n} c_{2n+2} = (-1)^{2n} (-7) = -7$$

and

$$a_n = \frac{c_{2n+1}^2 + 7}{c_{2n}}.$$

We may therefore take $y_n = c_{2n+1}$ and $x_n = c_{2n} + y_n$.

Problem 9.3.11. *The sequence* $a_1, a_2, \ldots$ *is defined by the initial conditions* $a_1 = 20$, $a_2 = 30$ *and the recursion* $a_{n+2} = 3a_{n+1} - a_n$ *for* $n \geq 1$. *Find all positive integers* n *for which* $1 + 5a_n a_{n+1}$ *is a perfect square.*

(2002 Balkan Mathematical Olympiad)

Solution. The only solution is $n = 3$. We can check that $20 \cdot 30 \cdot 5 + 1 = 3001$ and $30 \cdot 70 \cdot 5 + 1 = 10501$ are not perfect squares, while $70 \cdot 180 \cdot 5 + 1 = 63001 = 251^2$ is a perfect square. Then we have only to prove that $1 + 5a_n a_{n+1}$ is not a perfect square for $n \geq 4$. First, we will prove a lemma.

Lemma. *For any integer* $n \geq 2$,

$$a_n^2 + 500 = a_{n-1} a_{n+1}.$$

Proof. We will prove this by induction on n. In the base case, $30^2 + 500 = 1400 = 20 \cdot 70$. Now assume that $a_n^2 + 500 = a_{n-1} a_{n+1}$. Then

$$a_n a_{n+2} = (3a_{n+1} - a_n)(a_n) = 3a_{n+1} a_n - a_n^2$$
$$= 3a_{n+1} a_n - (a_{n-1} a_{n+1} - 500)$$
$$= 500 + a_{n+1}(3a_n - a_{n-1}) = 500 + a_{n+1}^2,$$

proving the inductive step. Therefore the desired statement is true from induction. $\square$

Now, for $n \geq 4$, $(a_n + a_{n+1})^2 = a_n^2 + a_{n+1}^2 + 2a_n a_{n+1}$. But

$$a_{n+1}^2 = 9a_n^2 + a_{n-1}^2 - 6a_{n-1}a_n,$$

so

$$
\begin{aligned}
(a_n + a_{n+1})^2 &= 2a_n a_{n+1} + 3a_n(3a_n - a_{n-1}) + a_{n-1}^2 + a_n^2 - 3a_n a_{n-1} \\
&= 5a_n a_{n+1} + a_{n-1}^2 - a_n a_{n-2} \\
&= 5a_n a_{n+1} + a_{n-1}^2 - (a_{n-1}^2 + 500) = 5a_n a_{n+1} - 500,
\end{aligned}
$$

by the lemma and the definition of a.

Therefore $(a_n + a_{n+1})^2 = 5a_n a_{n+1} - 500 < 5a_n a_{n+1} + 1$. Since a_n is increasing and $n \geq 4$,

$$a_n + a_{n+1} \geq 180 + 470 = 650,$$

so

$$
\begin{aligned}
(a_n + a_{n+1} + 1)^2 &= (a_n + a_{n+1})^2 + 2(a_n + a_{n+1}) + 1 \\
&> (a_n + a_{n+1})^2 + 501 = 5a_n a_{n+1} + 1.
\end{aligned}
$$

Because two adjacent integers have squares above and below $5a_n a_{n+1} + 1$, that value is not a perfect square for $n \geq 4$.

Additional Problems

Problem 9.3.12. Let a, b be integers greater than 1. The sequence $x_1, x_2, \ldots$ is defined by the initial conditions $x_0 = 0$, $x_1 = 1$ and the recursion

$$x_{2n} = ax_{2n-1} - x_{2n-2}, \quad x_{2n+1} = bx_{2n} - x_{2n-1}$$

for $n \geq 1$. Prove that for any natural numbers m and n, the product $x_{n+m}x_{n+m-1} \cdots x_{n+1}$ is divisible by $x_m x_{m-1}$.

(2001 St. Petersburg City Mathematical Olympiad)

Problem 9.3.13. Let m be a positive integer. Define the sequence $\{a_n\}_{n \geq 0}$ by $a_0 = 0$, $a_1 = m$, and $a_{n+1} = m^2 a_n - a_{n-1}$ for $n \geq 1$. Prove that an ordered pair (a, b) of nonnegative integers, with $a \leq b$, is a solution of the equation

$$\frac{a^2 + b^2}{ab + 1} = m^2$$

if and only if $(a, b) = (a_n, a_{n+1})$ for some $n \geq 0$.

(1998 Canadian Mathematical Olympiad)

Problem 9.3.14. Let b, c be positive integers, and define the sequence $a_1, a_2, \ldots$ by $a_1 = b, a_2 = c$, and

$$a_{n+2} = |3a_{n+1} - 2a_n|$$

for $n \geq 1$. Find all such (b, c) for which the sequence $a_1, a_2, \ldots$ has only a finite number of composite terms.

(2002 Bulgarian Mathematical Olympiad)

9.3.3   Nonstandard Sequences of Integers

Problem 9.3.15. *Let k be a positive integer. The sequence a_n is defined by $a_1 = 1$, and for $n \geq 2$, a_n is the nth positive integer greater than a_{n-1} which is congruent to n modulo k. Find a_n is closed form.*

(1997 Austrian Mathematical Olympiad)

Solution. We have $a_n = \frac{n(2+(n-1)k)}{2}$. If $k = 2$, then $a_n = n^2$. First, observe that $a_1 \equiv 1 \pmod{k}$. Thus, for all n, $a_n \equiv n \pmod{k}$, and the first positive integer greater than a_{n-1} which is congruent to n modulo k must be $a_{n-1}+1$. The nth positive integer greater than a_{n-1} that is congruent to n modulo k is simply $(n-1)k$ more than the first positive integer greater than a_{n-1} which satisfies that condition. Therefore, $a_n = a_{n-1} + 1 + (n-1)k$. Solving this recursion gives

$$a_n = n + \frac{(n-1)n}{2}k.$$

Problem 9.3.16. *Let $a_1 = 19$, $a_2 = 98$. For $n \geq 1$, define a_{n+2} to be the remainder of $a_n + a_{n+1}$ when it is divided by 100. What is the remainder when*

$$a_1^2 + a_2^2 + \cdots + a_{1998}^2$$

is divided by 8?

(1998 United Kingdom Mathematical Olympiad)

Solution. The answer is 0. Consider $a_n \pmod 4$ which is not changed by taking the remainder divided by 100, there's the cycle 3, 2, 1, 3, 0, 3 which repeats 333 times. Then

$$a_1^2 + a_2^2 + \cdots + a_{1998}^2 \equiv 333(1 + 4 + 1 + 1 + 0 + 1) \equiv 0 \pmod 8,$$

as claimed.

Problem 9.3.17. *A sequence of integers $\{a_n\}_{n \geq 1}$ satisfies the following recursion relation:*

$$a_{n+1} = a_n^3 + 1999 \quad \text{for} \quad n = 1, 2, \ldots.$$

Prove that there exists at most one n for which a_n is a perfect square.

(1999 Austrian–Polish Mathematics Competition)

Solution. Consider the possible values of (a_n, a_{n+1}) modulo 4:

$$
\begin{array}{c|cccc}
a_n & 0 & 1 & 2 & 3 \\
\hline
a_{n+1} & 3 & 0 & 3 & 2
\end{array}
$$

No matter what a_1 is, the terms $a_3, a_4, \ldots$ are all 2 or 3 (mod 4). However, all perfect squares are 0 or 1 (mod 4), so at most two terms (a_1 and a_2) can be perfect squares. If a_1 and a_2 are both perfect squares, then writing $a_1 = a^2$, $a_2 = b^2$ we have $a^6 + 1999 = b^2$ or $1999 = b^2 - (a^3)^2 = (b + a^3)(b - a^3)$. Because 1999 is prime, $b - a^3 = 1$ and $b + a^3 = 1999$. Thus $a^3 = \frac{1999-1}{2} = 999$, which is impossible. Hence at most one term of the sequence is a perfect square.

Problem 9.3.18. *Determine whether there exists an infinite sequence of positive integers such that*

(i) no term divides any other term;

(ii) every pair of terms has a common divisor greater than 1, but no integer greater than 1 divides all the terms.

(1999 Hungarian Mathematical Olympiad)

Solution. The desired sequence exists. Let $p_0, p_1, \ldots$ be the primes greater than 5 in order, and let $q_{3i} = 6$, $q_{3i+1} = 10$, $q_{3i+2} = 15$ for each nonnegative integer i. Then let $s_i = p_i q_i$ for all $i \geq 0$. The sequence $s_0, s_1, s_2, \ldots$ clearly satisfies (i) because s_i is not even divisible by p_j for $i \neq j$. For the first part of (ii), any two terms have their indices both in $\{0, 1\}$, both in $\{0, 2\}$, or both in $\{1, 2\}$ (mod 3), so they have a common divisor of 2, 3, or 5, respectively. For the second part, we just need to check that no prime divides all the s_i. Indeed, $2 \nmid s_2$, $3 \nmid s_1$, $5 \nmid s_0$, and no prime greater than 5 divides more than one s_i.

Problem 9.3.19. *Let $a_1, a_2, \ldots$ be a sequence satisfying $a_1 = 2$, $a_2 = 5$, and*

$$a_{n+2} = (2 - n^2)a_{n+1} + (2 + n^2)a_n$$

for all $n \geq 1$. Do there exist indices p, q, and r such that $a_p a_q = a_r$?

(1995 Czech–Slovak Match)

Solution. No such p, q, r exist. We show that for all n, $a_n \equiv 2$ (mod 3). This holds for $n = 1$ and $n = 2$ by assumption and follows for all n by induction:

$$
\begin{aligned}
a_{n+2} &= (2 - n^2)a_{n+1} + (2 + n^2)a_n \\
&\equiv 2(2 - n^2) + 2(2 + n^2) = 8 \equiv 2 \pmod{3}.
\end{aligned}
$$

Let p, q, r be positive integers. We have $a_p a_q \equiv 1$ (mod 3), so $a_p a_q$ is different from a_r, which is congruent to 2 (mod 3).

Problem 9.3.20. *Is there a sequence of natural numbers in which every natural number occurs just once and moreover, for any $k = 1, 2, 3, \ldots$ the sum of the first k terms is divisible by k?*

(1995 Russian Mathematical Olympiad)

Solution. We recursively construct such a sequence. Suppose $a_1, \ldots, a_m$ have been chosen, with $s = a_1 + \cdots + a_m$, and let n be the smallest number not yet appearing. By the Chinese remainder theorem, there exists t such that $t \equiv -s$ (mod $m + 1$) and $t \equiv -s - n$ (mod $m + 2$). We can increase t by a suitably large multiple of $(m + 1)(m + 2)$ to ensure that it does not equal any of $a_1, \ldots, a_m$. Then $a_1, \ldots, a_m, t, n$ also has the desired property, and the construction ensures that $1, \ldots, m$ all occur among the first $2m$ terms.

Additional Problems

Problem 9.3.21. Let $\{a_n\}$ be a sequence of integers such that for $n \geq 1$,

$$(n - 1)a_{n+1} = (n + 1)a_n - 2(n - 1).$$

If 2000 divides a_{1999}, find the smallest $n \geq 2$ such that 2000 divides a_n.

(1999 Bulgarian Mathematical Olympiad)

Problem 9.3.22. The sequence $(a_n)_{n \geq 0}$ is defined by $a_0 = 1$, $a_1 = 3$, and

$$a_{n+2} = \begin{cases} a_{n+1} + 9a_n & \text{if } n \text{ is even,} \\ 9a_{n+1} + 5a_n & \text{if } n \text{ is odd.} \end{cases}$$

Prove that
(a) $\sum_{k=1995}^{2000} a_k^2$ is divisible by 20,
(b) a_{2n+1} is not a perfect square for any $n = 0, 1, 2, \ldots$.

(1995 Vietnamese Mathematical Olympiad)

Problem 9.3.23. Prove that for any natural number $a_1 > 1$, there exists an increasing sequence of natural numbers $a_1, a_2, \ldots$ such that $a_1^2 + a_2^2 + \cdots + a_k^2$ is divisible by $a_1 + a_2 + \cdots + a_k$ for all $k \geq 1$.

(1995 Russian Mathematical Olympiad)

Problem 9.3.24. The sequence $a_0, a_1, a_2, \ldots$ satisfies

$$a_{m+n} + a_{m-n} = \tfrac{1}{2}(a_{2m} + a_{2n})$$

for all nonnegative integers m and n with $m \geq n$. If $a_1 = 1$, determine a_n.

(1995 Russian Mathematical Olympiad)

Problem 9.3.25. The sequence of real numbers $a_1, a_2, a_3, \ldots$ satisfies the initial conditions $a_1 = 2$, $a_2 = 500$, $a_3 = 2000$ as well as the relation

$$\frac{a_{n+2} + a_{n+1}}{a_{n+1} + a_{n-1}} = \frac{a_{n+1}}{a_{n-1}}$$

for $n = 2, 3, 4, \ldots$. Prove that all the terms of this sequence are positive integers and that 2^{2000} divides the number a_{2000}.

(1999 Slovenian Mathematical Olympiad)

Problem 9.3.26. Let k be a fixed positive integer. We define the sequence $a_1, a_2, \ldots$ by $a_1 = k + 1$ and the recursion $a_{n+1} = a_n^2 - ka_n + k$ for $n \geq 1$. Prove that a_m and a_n are relatively prime for distinct positive integers m and n.

Problem 9.3.27. Suppose the sequence of nonnegative integers $a_1, a_2, \ldots, a_{1997}$ satisfies

$$a_i + a_j \leq a_{i+j} \leq a_i + a_j + 1$$

for all $i, j \geq 1$ with $i + j \leq 1997$. Show that there exists a real number x such that $a_n = \lfloor nx \rfloor$ for all $1 \leq n \leq 1997$.

(1997 USA Mathematical Olympiad)

Problem 9.3.28. The sequence $\{a_n\}$ is given by the following relation:

$$a_{n+1} = \begin{cases} \frac{a_n - 1}{2}, & \text{if } a_n \geq 1, \\ \frac{2a_n}{1 - a_n}, & \text{if } a_n < 1. \end{cases}$$

Given that a_0 is a positive integer, $a_n \neq 2$ for each $n = 1, 2, \ldots, 2001$, and $a_{2002} = 2$, find a_0.

(2002 St. Petersburg City Mathematical Olympiad)

Problem 9.3.29. Let $x_1 = x_2 = x_3 = 1$ and $x_{n+3} = x_n + x_{n+1}x_{n+2}$ for all positive integers n. Prove that for any positive integer m there is an integer $k > 0$ such that m divides x_k.

Problem 9.3.30. Find all infinite bounded sequences $a_1, a_2, \ldots$ of positive integers such that for all $n > 2$,

$$a_n = \frac{a_{n-1} + a_{n-2}}{\gcd(a_{n-1}, a_{n-2})}.$$

(1999 Russian Mathematical Olympiad)

Problem 9.3.31. Let $a_1, a_2, \ldots$ be a sequence of positive integers satisfying the condition $0 < a_{n+1} - a_n \leq 2001$ for all integers $n \geq 1$. Prove that there exists an infinite number of ordered pairs (p, q) of distinct positive integers such that a_p is a divisor of a_q.

(2001 Vietnamese Mathematical Olympiad)

Problem 9.3.32. Define the sequence $\{x_n\}_{n\geq 0}$ by $x_0 = 0$ and

$$x_n = \begin{cases} x_{n-1} + \frac{3^{r+1}-1}{2}, & \text{if } n = 3^r(3k+1), \\[2mm] x_{n-1} - \frac{3^{r+1}+1}{2}, & \text{if } n = 3^r(3k+2), \end{cases}$$

where k and r are nonnegative integers. Prove that every integer appears exactly once in this sequence.

(1999 Iranian Mathematical Olympiad)

Problem 9.3.33. Suppose that $a_1, a_2, \ldots$ is a sequence of natural numbers such that for all natural numbers m and n, $\gcd(a_m, a_n) = a_{\gcd(m,n)}$. Prove that there exists a sequence $b_1, b_2, \ldots$ of natural numbers such that $a_n = \prod_{d\mid n} b_d$ for all integers $n \geq 1$.

(2001 Iranian Mathematical Olympiad)

10

Problems Involving Binomial Coefficients

10.1 Binomial Coefficients

One of the main problems leading to the consideration of binomial coefficients is the expansion of $(a + b)^n$, where a, b are complex numbers and n is a positive integer. It is well known that

$$(a + b)^n = \binom{n}{0}a^n + \binom{n}{1}a^{n-1}b + \cdots + \binom{n}{n-1}ab^{n-1} + \binom{n}{n}b^n,$$

where $\binom{n}{k} = \frac{n!}{k!(n-k)!}$, $k = 0, 1, \ldots, n$ with the convention $0! = 1$. The integers $\binom{n}{0}, \binom{n}{1}, \ldots, \binom{n}{n}$ are called *binomial coefficients*. They can be obtained recursively by using *Pascal's*[1] *triangle*,

$$
\begin{array}{ccccccccccc}
 & & & & & 1 & & & & & \\
 & & & & 1 & & 1 & & & & \\
 & & & 1 & & 2 & & 1 & & & \\
 & & 1 & & 3 & & 3 & & 1 & & \\
 & 1 & & 4 & & 6 & & 4 & & 1 & \\
1 & & 5 & & 10 & & 10 & & 5 & & 1
\end{array}
$$

$$\cdots \cdots \cdots \cdots \cdots \cdots \cdots \cdots \cdots \cdots \cdots$$

in which every entry different from 1 is the sum of the two entries above and adjacent to it.

The fundamental properties of the binomial coefficients are the following:

---

[1] Blaise Pascal (1623–1662) was a very influential French mathematician and philosopher who contributed to many areas of mathematics.

(1) (symmetry) $\binom{n}{k} = \binom{n}{n-k}$;

(2) (Pascal's triangle property) $\binom{n}{k+1} = \binom{n-1}{k+1} + \binom{n-1}{k}$;

(3) (monotonicity) $\binom{n}{0} < \binom{n}{1} < \cdots < \binom{n}{\lfloor \frac{n-1}{2} \rfloor + 1} = \binom{n}{\lfloor \frac{n}{2} \rfloor}$;

(4) (sum of binomial coefficients) $\binom{n}{0} + \binom{n}{1} + \cdots + \binom{n}{n} = 2^n$;

(5) (alternating sum) $\binom{n}{0} - \binom{n}{1} + \cdots + (-1)^n \binom{n}{n} = 0, n \geq 1$;

(6) (Vandermonde property) $\sum_{i=0}^{k} \binom{m}{i} \binom{n}{k-i} = \binom{m+n}{k}$;

(7) If p is a prime, then $p \mid \binom{p}{k}, k = 1, \ldots, p-1$.

Problem 10.1.1. *Let n be an odd positive integer. Prove that the set*

$$\left\{ \binom{n}{1}, \binom{n}{2}, \ldots, \binom{n}{\frac{n-1}{2}} \right\}$$

contains an odd number of odd numbers.

Solution. For $n = 1$ the claim is clear, so let $n \geq 3$.

Define $S_n = \binom{n}{1} + \binom{n}{2} + \cdots + \binom{n}{\frac{n-1}{2}}$. Then

$$2S_n = \binom{n}{1} + \binom{n}{2} + \cdots + \binom{n}{n-1} = 2^n - 2,$$

or $S_n = 2^{n-1} - 1$. Because S_n is odd it follows that the sum S_n contains an odd number of odd terms, as desired.

Problem 10.1.2. *Determine all positive integers $n \geq 3$ such that 2^{2000} is divisible by*

$$1 + \binom{n}{1} + \binom{n}{2} + \binom{n}{3}.$$

(1998 Chinese Mathematical Olympiad)

Solution. The solutions are $n = 3, 7, 23$. Since 2 is a prime,

$$1 + \binom{n}{1} + \binom{n}{2} + \binom{n}{3} = 2^k$$

for some positive integer $k \leq 2000$. We have

$$1 + \binom{n}{1} + \binom{n}{2} + \binom{n}{3} = \frac{(n+1)(n^2 - n + 6)}{6},$$

i.e., $(n + 1)(n^2 - n + 6) = 3 \times 2^{k+1}$. Let $m = n + 1$; then $m \geq 4$ and $m(m^2 - 3m + 8) = 3 \times 2^{k+1}$. We consider the following two cases:

(a) $m = 2^s$. Since $m \geq 4$, $s \geq 2$. We have

$$2^{2s} - 3 \times 2^s + 8 = m^2 - 3m + 8 = 3 \times 2^t$$

for some positive integer t. If $s \geq 4$, then

$$8 \equiv 3 \times 2^t \pmod{16} \Rightarrow 2^t = 8 \Rightarrow m^2 - 3m + 8 = 24 \Rightarrow m(m - 3) = 16,$$

which is impossible. Thus either $s = 3$, $m = 8$, $t = 4$, $n = 7$, or $s = 2$, $m = 4$, $t = 2$, $n = 3$.

(b) $m = 3 \times 2^u$. Since $m \geq 4$, $m > 4$ and $u \geq 1$. We have

$$9 \times 2^{2u} - 9 \times 2^u + 8 = m^2 - 3m + 8 = 2^v$$

for some positive integer v. It is easy to check that there is no solution for v when $u = 1, 2$. If $u \geq 4$, we have $8 \equiv 2^v \pmod{16} \Rightarrow v = 3$ and $m(m - 3) = 0$, which is impossible. So $u = 3$, $m = 3 \times 2^3 = 24$, $v = 9$, $n = 23$.

Problem 10.1.3. *Let m and n be integers such that $1 \leq m \leq n$. Prove that m is a divisor of*

$$n \sum_{k=0}^{m-1} (-1)^k \binom{n}{k}.$$

(2001 Hungarian Mathematical Olympiad)

Solution. We can write the given expression as follows:

$$n \sum_{k=0}^{m-1} (-1)^k \binom{n}{k} = n \sum_{k=0}^{m-1} (-1)^k \left(\binom{n-1}{k} + \binom{n-1}{k-1} \right)$$

$$= n \sum_{k=0}^{m-1} (-1)^k \binom{n-1}{k} + n \sum_{k=1}^{m-1} (-1)^k \binom{n-1}{k-1}$$

$$= n \sum_{k=0}^{m-1} (-1)^k \binom{n-1}{k} - n \sum_{k=0}^{m-2} (-1)^k \binom{n-1}{k}$$

$$= n(-1)^{m-1} \binom{n-1}{m-1}$$

$$= m(-1)^{m-1} \binom{n}{m}.$$

The final expression is clearly divisible by m.

Problem 10.1.4. *Show that for any positive integer n, the number*

$$S_n = \binom{2n+1}{0} \cdot 2^{2n} + \binom{2n+1}{2} \cdot 2^{2n-2} \cdot 3 + \cdots + \binom{2n+1}{2n} \cdot 3^n$$

is the sum of two consecutive perfect squares.

(1999 Romanian International Mathematical Olympiad Team Selection Test)

Solution. It is easy to see that:

$$S_n = \tfrac{1}{4}\left[(2+\sqrt{3})^{2n+1} + (2-\sqrt{3})^{2n+1}\right].$$

The required property says that there exists $k > 0$ such that $S_n = (k-1)^2 + k^2$, or, equivalently,

$$2k^2 - 2k + 1 - S_n = 0.$$

The discriminant of this equation is $\Delta = 4(2S_n - 1)$, and, using $\left(\frac{1+\sqrt{3}}{\sqrt{2}}\right)^2 = 2 + \sqrt{3}$, after the usual computations, we obtain

$$\Delta = \left(\frac{(1+\sqrt{3})^{2n+1} + (1-\sqrt{3})^{2n+1}}{2^n}\right)^2.$$

After solving the equation, we find that

$$k = \frac{2^{n+1} + (1+\sqrt{3})^{2n+1} + (1-\sqrt{3})^{2n+1}}{2^{n+2}}.$$

Therefore, it is sufficient to prove that k is an integer. Let us set $E_m = (1 + \sqrt{3})^m + (1 - \sqrt{3})^m$, where m is a positive integer. Clearly, E_m is an integer. We shall prove that $2^{\lceil \frac{m}{2} \rceil}$ divides E_m. For $E_0 = 2$, $E_1 = 2$, $E_2 = 8$, the assertion is true. Moreover, the numbers E_m satisfy the relation

$$E_m = 2E_{m-1} + 2E_{m-2}.$$

The property now follows by induction.

Problem 10.1.5. *Prove that for every pair m, k of natural numbers, m has a unique representation in the form*

$$m = \binom{a_k}{k} + \binom{a_{k-1}}{k-1} + \cdots + \binom{a_t}{t},$$

where

$$a_k > a_{k-1} > \cdots > a_t \geq t \geq 1.$$

(1996 Iranian Mathematical Olympiad)

First solution. We first show uniqueness. Suppose m is represented by two sequences $a_k, \ldots, a_t$ and $b_k, \ldots, b_t$. Find the first position in which they differ; without loss of generality, assume that this position is k and that $a_k > b_k$. Then

$$m \le \binom{b_k}{k} + \binom{b_k - 1}{k - 1} + \cdots + \binom{b_k - k + 1}{1} < \binom{b_k + 1}{k} \le m,$$

a contradiction.

To show existence, apply the greedy algorithm: find the largest a_k such that $\binom{a_k}{k} \le m$, and apply the same algorithm with m and k replaced by $m - \binom{a_k}{k}$ and $k - 1$. We need only make sure that the sequence obtained is indeed decreasing, but this follows because by assumption, $m < \binom{a_k+1}{m}$, and so $m - \binom{a_k}{k} < \binom{a_k}{k-1}$.

Second solution. Sort all unordered k-tuples of distinct nonnegative integers lexicographically. Then the k-tuple $\{a_k, a_{k-1}, \ldots, a_t, t - 1, t - 3, \ldots, 0\}$ is preceded by exactly $\binom{a_k}{k} + \binom{a_{k-1}}{k-1} + \cdots + \binom{a_t}{t}$ other k-tuples. (The first term counts the number of k-tuples whose largest element is smaller than a_k. The second term counts k-tuples that begin with a_k but whose second-largest element is smaller than a_{k-1}, etc.) Since there is necessarily a unique k-tuple preceded by m other k-tuples, every m has a unique representation in this form.

Problem 10.1.6. *Show that for every positive integer $n \ge 3$, the least common multiple of the numbers $1, 2, \ldots, n$ is greater than 2^{n-1}.*

(1999 Czech–Slovak Match)

Solution. For any $n \ge 3$ we have

$$2^{n-1} = \sum_{k=0}^{n-1} \binom{n-1}{k} < \sum_{k=0}^{n-1} \binom{n-1}{\lfloor \frac{n-1}{2} \rfloor} = n \binom{n-1}{\lfloor \frac{n-1}{2} \rfloor}.$$

Hence it suffices to show that $n\binom{n-1}{\lfloor \frac{n-1}{2} \rfloor}$ divides $\operatorname{lcm}(1, 2, \ldots, n)$. Using an argument involving prime factorizations, we will prove the more general assertion that for each $k < n$, $\operatorname{lcm}(n, n - 1, \ldots, n - k)$ is divisible by $n\binom{n-1}{k}$.

Let k and n be fixed natural numbers with $k < n$, and let $p \le n$ be an arbitrary prime. Let p^α be the highest power of p that divides $\operatorname{lcm}(n, n - 1, \ldots, n - k)$, where $p^\alpha \mid n - l$ for some l. Then for each $i \le \alpha$, we know that $p^i \mid n - l$. Thus exactly $\lfloor \frac{l}{p^i} \rfloor$ of $\{n - l + 1, n - l + 2, \ldots, n\}$ and exactly $\lfloor \frac{k-l}{p^i} \rfloor$ of $\{n - l - 1, n - l - 2, \ldots, n - k\}$ are multiples of p^i, so p^i divides $\lfloor \frac{l}{p^i} \rfloor + \lfloor \frac{k-l}{p^i} \rfloor \le \lfloor \frac{k}{p^i} \rfloor$ of the remaining k numbers, that is, at most the number of multiples of p^i between 1 and k. It follows that p divides

$$n \binom{n-1}{k} = \frac{n(n-1)\cdots(n-l+1)(n-l-1)\cdots(n-k)}{k!}(n-l)$$

at most α times, so that indeed $n\binom{n-1}{k} \mid \operatorname{lcm}(n, n - 1, \ldots, n - k)$.

Additional Problems

Problem 10.1.7. Show that the sequence

$$\binom{2002}{2002}, \binom{2003}{2002}, \binom{2004}{2002}, \ldots,$$

considered modulo 2002, is periodic.

(2002 Baltic Mathematical Competition)

Problem 10.1.8. Prove that

$$\binom{2p}{p} \equiv 2 \pmod{p^2}$$

for any prime number p.

Problem 10.1.9. Let k, m, n be positive integers such that $m + k + 1$ is a prime number greater than $n + 1$. Let us set $C_s = s(s + 1)$. Show that the product

$$(C_{m+1} - C_k)(C_{m+2} - C_k) \cdots (C_{m+n} - C_k)$$

is divisible by $C_1 C_2 \cdots C_n$.

(18th International Mathematical Olympiad)

Problem 10.1.10. Let n, k be arbitrary positive integers. Show that there exist positive integers $a_1 > a_2 > a_3 > a_4 > a_5 > k$ such that

$$n = \pm\binom{a_1}{3} \pm \binom{a_2}{3} \pm \binom{a_3}{3} \pm \binom{a_4}{3} \pm \binom{a_5}{3}.$$

(2000 Romanian International Mathematical Olympiad Team Selection Test)

Problem 10.1.11. Prove that if n and m are integers, and m is odd, then

$$\frac{1}{3^m n} \sum_{k=0}^{m} \binom{3m}{3k} (3n - 1)^k$$

is an integer.

(2004 Romanian International Mathematical Olympiad Team Selection Test)

Problem 10.1.12. Show that for any positive integer n the number

$$\sum_{k=0}^{n} \binom{2n + 1}{2k + 1} 2^{3k}$$

is not divisible by 5.

(16th International Mathematical Olympiad)

Problem 10.1.13. Prove that for a positive integer k there is an integer $n \geq 2$ such that $\binom{n}{1}, \ldots, \binom{n}{n-1}$ are all divisible by k if and only if k is a prime.

10.2 Lucas's and Kummer's Theorems

The following theorems of E. Lucas[2] (1878) and E. Kummer[3] (1852) are very useful in number theory. Let n be a positive integer, and let p be a prime. Let $\overline{n_m n_{m-1} \cdots n_0}_p$ denote the base-p representation of n; that is,

$$n = \overline{n_m n_{m-1} \cdots n_0}_p = n_0 + n_1 p + \cdots + n_m p^m,$$

where $0 \leq n_0, n_1, \ldots, n_m \leq p - 1$ and $n_m \neq 0$.

Theorem 10.2.1. (Lucas) *Let p be a prime, and let n be a positive integer with $n = \overline{n_m n_{m-1} \cdots n_0}_p$. Let i be a positive integer less than n. If $i = i_0 + i_1 p + \cdots + i_m p^m$, where $0 \leq i_0, i_1, \ldots, i_m \leq p - 1$, then*

$$\binom{n}{i} \equiv \prod_{j=0}^{m} \binom{n_j}{i_j} \quad (\text{mod } p). \tag{1}$$

Here $\binom{0}{0} = 1$ and $\binom{n_j}{i_j} = 0$ if $n_j < i_j$.

To prove this theorem, we need some additional techniques. Let p be a prime, and let $f(x)$ and $g(x)$ be two polynomials with integer coefficients. We say that $f(x)$ is congruent to $g(x)$ modulo p, and write $f(x) \equiv g(x)$ (mod p), if all of the coefficients of $f(x) - g(x)$ are divisible by p. (Note that the congruence of polynomials is different from the congruence of the values of polynomials. For example, $x(x + 1) \not\equiv 0$ (mod 2) even though $x(x + 1)$ is divisible by 2 for all integers x.) The following properties can be easily verified:

(a) $f(x) \equiv f(x)$ (mod p);

(b) if $f(x) \equiv g(x)$ (mod p), then $g(x) \equiv f(x)$ (mod p);

(c) if $f(x) \equiv g(x)$ (mod p) and $g(x) \equiv h(x)$ (mod p), then

$$f(x) \equiv h(x) \quad (\text{mod } p);$$

(d) if $f(x) \equiv g(x)$ (mod p) and $f_1(x) \equiv g_1(x)$ (mod p), then

$$f(x) \pm f_1(x) \equiv g(x) \pm g_1(x) \quad (\text{mod } p)$$

and

$$f(x) f_1(x) \equiv g(x) g_1(x) \quad (\text{mod } p).$$

Proof of Theorem 10.2.1. By property (7) (Part I, Section 10.1), the binomial coefficients $\binom{p}{k}$, where $1 \leq k \leq p - 1$, are divisible by p. Thus,

$$(1 + x)^p \equiv 1 + x^p \quad (\text{mod } p)$$

---

[2]François Edouard Anatole Lucas (1842–1891), French mathematician best known for his results in number theory. He studied the Fibonacci sequence and devised the test for Mersenne primes.

[3]Ernst Eduard Kummer (1810–1893), German mathematician whose main achievement was the extension of results about integers to other integral domains by introducing the concept of an ideal.

and
$$(1+x)^{p^2} = [(1+x)^p]^p \equiv [1+x^p]^p \equiv 1 + x^{p^2} \pmod{p},$$
and so on, so that for any positive integer r,
$$(1+x)^{p^r} \equiv 1 + x^{p^r} \pmod{p}$$
by induction.

We have
$$(1+x)^n = (1+x)^{n_0 + n_1 p + \cdots + n_m p^m}$$
$$= (1+x)^{n_0}[(1+x)^p]^{n_1} \cdots [(1+x)^{p^m}]^{n_m}$$
$$\equiv (1+x)^{n_0}(1+x^p)^{n_1} \cdots (1+x^{p^m})^{n_m} \pmod{p}.$$

The coefficient of x^i in the expansion of $(1+x)^n$ is $\binom{n}{i}$. On the other hand, because $i = i_0 + i_1 p + \cdots + i_m p^m$, the coefficient of x^i is the coefficient of $x^{i_0}(x^p)^{i_1} \cdots (x^{p^m})^{i_m}$, which is equal to $\binom{n_0}{i_0}\binom{n_1}{i_1} \cdots \binom{n_m}{i_m}$. Hence
$$\binom{n}{i} \equiv \binom{n_0}{i_0}\binom{n_1}{i_1} \cdots \binom{n_m}{i_m} \pmod{p},$$
as desired.   □

Theorem 10.2.2. (Kummer) *Let n and i be positive integers with $i \leq n$, and let p be a prime. Then p^t divides $\binom{n}{i}$ if and only if t is less than or equal to the number of carries in the addition $(n-i) + i$ in base p.*

Proof. We will use the formula
$$e_p(n) = \frac{n - S_p(n)}{p - 1}, \tag{2}$$
where e_p is the Legendre function and $S_p(n)$ is the sum of the digits of n in base p (see Section 6.5). We actually prove that the largest nonnegative integer t such that p^t divides $\binom{n}{i}$ is exactly the number of carries in the addition $(n-i) + i$ in base p.

Let $n = \overline{a_m a_{m-1} \cdots a_0}_p$, $i = \overline{b_k b_{k-1} \ldots b_0}_p$, $(n-i) = \overline{(c_l c_{l-1} \ldots c_0)}_p$. Because $1 \leq i \leq n$, it follows that $k, l \leq m$. Without loss of generality, we assume that $k \leq l$. Let a, b, c, and t' be integers such that $p^a \| n!$, $p^b \| i!$, $p^c \| (n-i)!$, and $p^{t'} \| \binom{n}{i}$. Then $t' = a - b - c$.

From formula (2) we have
$$a = \frac{n - (a_m + a_{m-1} + \cdots + a_0)}{p - 1},$$
$$b = \frac{i - (b_k + b_{k-1} + \cdots + b_0)}{p - 1},$$
$$c = \frac{(n-i) - (c_l + c_{l-1} + \cdots + c_0)}{p - 1}.$$

Thus

$$t' = \frac{-(a_m + \cdots + a_0) + (b_k + \cdots + b_0) + (c_l + \cdots + c_0)}{p - 1}. \tag{3}$$

On the other hand, if we add $n - i$ and i in base p, we have

$$
\begin{array}{ccccccccc}
 & & & b_k & b_{k-1} & \ldots & b_1 & b_0 \\
 & c_l & c_{l-1} & \ldots & c_k & c_{k-1} & \ldots & c_1 & c_0 \\
\hline
a_m & a_{m-1} & \ldots & a_l & a_{l-1} & \ldots & a_k & a_{k-1} & \ldots & a_1 & a_0
\end{array}
$$

Then we have either $b_0 + c_0 = a_0$ (with no carry) or $b_0 + c_0 = a_0 + p$ (with a carry of 1). More generally, we have

$$b_0 + c_0 = a_0 + \alpha_1 p,$$
$$b_1 + c_1 + \alpha_1 = a_1 + \alpha_2 p,$$
$$b_2 + c_2 + \alpha_2 = a_2 + \alpha_3 p,$$

$$\cdots$$

$$b_m + c_m + \alpha_m = a_m,$$

where α_i denotes the carry at the $(i - 1)$th digit from the right. (Note also that $b_j = 0$ for $j > k$ and that $c_j = 0$ for $j > l$.) Adding the above equations together yields

$$(b_0 + \cdots + b_k) + (c_0 + \cdots + c_l) = (a_0 + \cdots + a_m) + (p - 1)(\alpha_1 + \cdots + \alpha_m).$$

Thus, equation (3) becomes

$$t' = \alpha_1 + \cdots + \alpha_m,$$

as desired. $\qquad \square$

Problem 10.2.1. *Let n be a positive integer. Prove that the number of $k \in \{0, 1, \ldots, n\}$ for which $\binom{n}{k}$ is odd is a power of 2.*

Solution. Let the base-2 expansion of n be $2^0 n_0 + 2^1 n_1 + \cdots + 2^a n_a$, where $n_i \in \{0, 1\}$ for each i. Then for any $k = 2^0 k_0 + 2^1 k_1 + \cdots + 2^a k_a$, we have

$$\binom{n}{k} \equiv \binom{n_0}{k_0}\binom{n_1}{k_1} \cdots \binom{n_a}{k_a} \pmod 2$$

by Lucas's theorem. Thus $\binom{n}{k}$ is odd if and only if $k_i \le n_i$ for each i. Let m be the number of n_i's equal to 1. Then the values of $k \in \{0, 1, \ldots, 2^{a+1} - 1\}$ for which $\binom{n}{k}$ is odd are obtained by setting $k_i = 0$ or 1 for each of the m values of i such that $n_i = 1$, and $k_i = 0$ for the other values of i. Thus there are 2^m values of k in $\{0, 1, \ldots, 2^{a+1} - 1\}$ for which $\binom{n}{k}$ is odd. Finally, note that for $k > n$, $\binom{n}{k} = 0$ is never odd, so the number of $k \in \{0, 1, \ldots, n\}$ for which $\binom{n}{k}$ is odd is 2^m, a power of 2.

Problem 10.2.2. *Determine all positive integers n, n ≥ 2, such that $\binom{n-k}{k}$ is even for $k = 1, 2, \ldots, \lfloor \frac{n}{2} \rfloor$.*

(1999 Belarusian Mathematical Olympiad)

Solution. Suppose that $p = 2$, $a = 2^s - 1$, and $a_{s-1} = a_{s-2} = \cdots = a_0 = 1$. For any b with $0 \le b \le 2^s - 1$, each term $\binom{a_i}{b_i}$ in the above equation equals 1. Therefore, $\binom{a}{b} \equiv 1 \pmod 2$.

This implies that $n + 1$ is a power of two. Otherwise, let $s = \lfloor \log_2 n \rfloor$ and let

$$k = n - (2^s - 1) = n - \frac{2^{s+1} - 2}{2} \le n - \frac{n}{2} = \frac{n}{2}.$$

Then $\binom{n-k}{k} = \binom{2^s-1}{k}$ is odd, a contradiction.

Conversely, suppose that $n = 2^s - 1$ for some positive integer s. For $k = 1, 2, \ldots, \lfloor \frac{n}{2} \rfloor$, there is at least one 0 in the binary representation of $a = n - k$ (not counting leading zeros, of course). Whenever there is a 0 in the binary representation of $n - k$, there is a 1 in the corresponding digit of $b = k$. Then the corresponding $\binom{a_i}{b_i}$ equals 0, and by Lucas's theorem, $\binom{n-k}{k}$ is even.

Therefore, $n = 2^s - 1$ for integers $s \ge 2$.

Problem 10.2.3. *Prove that $\binom{2^n}{k}$, $k = 1, 2, \ldots, 2^n - 1$, are all even and that exactly one of them is not divisible by 4.*

Solution. All these numbers are even, since

$$\binom{2^n}{k} = \frac{2^n}{k}\binom{2^n - 1}{k - 1}$$

and $2^n/k$ is different from 1 for all $k = 1, 2, \ldots, 2^n - 1$.

From the same relation it follows that $\binom{2^n}{k}$ is a multiple of 4 for all k different from 2^{n-1}. For $k = 2^{n-1}$ we have

$$\binom{2^n}{2^{n-1}} = 2\binom{2^n - 1}{2^{n-1} - 1}.$$

But from Lucas's theorem it follows that $\binom{2^n-1}{2^{n-1}-1}$ is odd, since $2^n - 1$ contains only 1's in its binary representation and $\binom{1}{k} = 1$ if $k = 0$ or 1. This solves the problem.

Additional Problems

Problem 10.2.4. Let p be an odd prime. Find all positive integers n such that $\binom{n}{1}, \binom{n}{2}, \ldots, \binom{n}{n-1}$ are all divisible by p.

Problem 10.2.5. Let p be a prime. Prove that p does not divide any of $\binom{n}{1}, \ldots, \binom{n}{n-1}$ if and only if $n = sp^k - 1$ for some positive integer k and some integer s with $1 \le s \le p - 1$.

11

Miscellaneous Problems

Problem 11.1. *Find all positive integers x, y, z that satisfy the conditions $x + y \geq 2z$ and $x^2 + y^2 - 2z^2 = 8$.*

(2003 "Alexandru Myller" Romanian Regional Contest)

First solution. There are two possible cases:

Case I. $x \geq y \geq z$.

We set $x - z = a \geq 0$, $y - z = b \geq 0$, $a \geq b$. We then obtain the equation $2z(a + b) + a^2 + b^2 = 8$. When $z \geq 3$, there are no solutions. For $z = 2$, we get $(a + 2)^2 + (b + 2)^2 = 16$, which again has no solution. When $z = 1$ we obtain solutions $(x, y, z) = (3, 1, 1)$ and $(x, y, z) = (1, 3, 1)$. When $z = 0$, $a^2 + b^2 = 8$ and we get the solution $(x, y, z) = (2, 2, 0)$.

Case II. $x \geq z \geq y$.

Note again that $x - z = a$, $y - z = b$ and obtain the solution $(x, y, z) = (n + 2, n - 2, n)$ or $(x, y, z) = (n - 2, n + 2, n)$.

Second solution. Let $x = z + a \geq y = z + b$, where $a + b \geq 0$ (b may be negative). Then the equation becomes $2(a + b)z + a^2 + b^2 = 8$. Note that this implies that $a + b$ is even and $a + b < 4$. If $a + b = 0$, then we get $a = 2$ and $b = -2$; hence $(x, y, z) = (n + 2, n - 2, n)$ or $(n - 2, n + 2, n)$. If $a + b = 2$, then $z = 1$, $a = 2$ and $b = 0$; hence $(x, y, z) = (3, 1, 1)$ or $(1, 3, 1)$.

Problem 11.2. *Let n be a positive integer. Find all integers that can be written as*

$$\frac{1}{a_1} + \frac{2}{a_2} + \cdots + \frac{n}{a_n},$$

for some positive integers $a_1, a_2, \ldots, a_n$.

Solution. First, observe that $k = \frac{1}{a_1} + \frac{2}{a_2} + \cdots + \frac{n}{a_n}$. Then

$$k \leq 1 + 2 + 3 + \cdots + n = \frac{n(n + 1)}{2}.$$

T. Andreescu and D. Andrica, *Number Theory*, DOI: 10.1007/b11856_11,
© Birkhäuser Boston, a part of Springer Science + Business Media, LLC 2009

We prove that any integer $k \in \left\{1, 2, \ldots, \frac{n(n+1)}{2}\right\}$ can be written as requested.

For $k = 1$, put $a_1 = a_2 = \cdots = a_n = \frac{n(n+1)}{2}$.

For $k = n$, set $a_1 = 1, a_2 = 2, \ldots, a_n = n$.

For $1 < k < n$, let $a_{k-1} = 1$ and $a_i = \frac{n(n+1)}{2} - k + 1$ for $i \neq k - 1$.

Thus

$$\frac{1}{a_1} + \frac{2}{a_2} + \cdots + \frac{n}{a_n} = \frac{k-1}{1} + \sum_{\substack{i=1 \\ i \neq k-1}} \frac{i}{a_i} = k - 1 + \frac{\frac{n(n+1)}{2} - k + 1}{\frac{n(n+1)}{2} - k + 1} = k.$$

For $n < k < \frac{n(n+1)}{2}$, write k as

$$k = n + p_1 + p_2 + \cdots + p_i,$$

with $1 \leq p_i < \cdots < p_2 < p_1 \leq n - 1$.

Setting $a_{p_1+1} = a_{p_2+1} = \cdots = a_{p_i+1} = 1$ and $a_j = j$ otherwise, we are done.

Problem 11.3. *Find all positive integers $a < b < c < d$ with the property that each of them divides the sum of the other three.*

Solution. Since $d \mid (a + b + c)$ and $a + b + c < 3d$, it follows that $a + b + c = d$ or $a + b + c = 2d$.

Case (i). If $a + b + c = d$, since $a \mid (b + c + d)$, we have $a \mid 2d$ and similarly $b \mid 2d$, $c \mid 2d$.

Let $2d = ax = by = cz$, where $2 < z < y < x$. Thus $\frac{1}{x} + \frac{1}{y} + \frac{1}{z} = \frac{1}{2}$.

(a) If $z = 3$, then $\frac{1}{x} + \frac{1}{y} = \frac{1}{6}$. The solutions are

$$(x, y) = \{(42, 7), (24, 8), (18, 9), (15, 10)\};$$

hence

$$(a, b, c, d) \in \big\{(k, 6k, 14k, 21k), (k, 3k, 8k, 12k),$$
$$(k, 2k, 6k, 9k), (2k, 3k, 10k, 15k)\big\},$$

for $k > 0$.

(b) If $z = 4$, then $\frac{1}{x} + \frac{1}{y} = \frac{1}{4}$, and

$$(x, y) = \{(20, 5), (12, 6)\}.$$

The solutions are

$$(a, b, c, d) = (k, 4k, 5k, 10k) \text{ and } (a, b, c, d) = (k, 2k, 3k, 6k),$$

for $k > 0$.

(c) If $z = 5$, then $\frac{1}{x} + \frac{1}{y} = \frac{3}{10}$, and $(3x - 10)(3y - 10) = 100$.

Since $3x - 10 \equiv 2 \pmod{3}$, it follows that $3x - 10 = 20$ and $3y - 10 = 5$. Thus $y = 5$, false.

(d) If $z \geq 6$ then $\frac{1}{x} + \frac{1}{y} + \frac{1}{z} < \frac{1}{6} + \frac{1}{6} + \frac{1}{6} = \frac{1}{2}$, so there are no solutions.

Case (ii). If $a + b + c = 2d$, we obtain $a \mid 3d, b \mid 3d, c \mid 3d$.

Then $3d = ax = by = cz$, with $x > y > z > 3$ and $\frac{1}{x} + \frac{1}{y} + \frac{1}{z} = \frac{2}{3}$. Since $x \geq 4, y \geq 5, z \geq 6$ we have $\frac{1}{x} + \frac{1}{y} + \frac{1}{z} \leq \frac{1}{6} + \frac{1}{5} + \frac{1}{4} = \frac{37}{60} < \frac{2}{3}$, so there are no solutions in this case.

Problem 11.4. *Find the greatest number that can be written as a product of some positive integers whose sum is* 1976.

(18th International Mathematical Olympiad)

Solution. Let $x_1, x_2, \ldots, x_n$ be the numbers having the sum $x_1 + x_2 + \cdots + x_n = 1976$ and the maximum value of the product $x_1 \cdot x_2 \cdots x_n = p$.

If one of the numbers, say x_1, is equal to 1, then $x_1 + x_2 = 1 + x_2 > x_2 = x_1 x_2$. Hence the product $(x_1 + x_2) \cdot x_3 \cdots x_n$ is greater than $x_1 \cdot x_2 \cdots x_n = p$, false. Therefore $x_k \geq 2$ for all k.

If one of the numbers is equal to 4, we can replace it with two numbers 2 without changing the sum or the product.

Suppose that $x_k \geq 5$ for some k. Then $x_k < 3(x_k - 3)$, so replacing the number x_k with the numbers 3 and $x_k - 3$, the sum remains constant while the product increases, contradiction.

Therefore all the numbers are equal to 2 or 3. If there are more than three numbers equal to 2, we can replace them by two numbers equal to 3, preserving the sum and increasing the product (since $2 \cdot 2 \cdot 2 < 3 \cdot 3$). Hence at most two terms equal to 2 are allowed. Since $1976 = 3 \cdot 658 + 2$, the maximum product is equal to $2 \cdot 3^{658}$.

Problem 11.5. *Prove that there exist infinitely many positive integers that cannot be written in the form*

$$x_1^3 + x_2^5 + x_3^7 + x_4^9 + x_5^{11}$$

for some positive integers x_1, x_2, x_3, x_4, x_5.

(2002 Belarusian Mathematical Olympiad)

Solution. For each integer N, we consider the number of integers in $[1, N]$ that can be written in the above form. Because $x_1 \leq N^{\frac{1}{3}}$, there are at most $N^{\frac{1}{3}}$ ways to choose x_1. A similar argument applies to the other x_i's. Therefore, there are at most $N^{\frac{1}{3}} N^{\frac{1}{5}} N^{\frac{1}{7}} N^{\frac{1}{9}} N^{\frac{1}{11}} = N^{\frac{3043}{3465}}$ combinations. So there are at least $N - N^{\frac{3043}{3465}}$ integers not covered. It is easy to see that this value can be arbitrarily large as N approaches infinity. Therefore, there exist infinitely many positive integers that cannot be written in the form $x_1^3 + x_2^5 + x_3^7 + x_4^9 + x_5^{11}$.

Additional Problems

Problem 11.6. Let a, b be positive integers. By integer division of $a^2 + b^2$ by $a + b$ we obtain the quotient q and the remainder r. Find all pairs (a, b) such that $q^2 + r = 1977$.

(19th International Mathematical Olympiad)

Problem 11.7. Let m, n be positive integers. Show that $25^n - 7^m$ is divisible by 3 and find the least positive integer of the form $|25^n - 7^m - 3^m|$, where m, n run over the set of positive integers.

(2004 Romanian Mathematical Regional Contest)

Problem 11.8. Given an integer d, let

$$S = \{m^2 + dn^2 \mid m, n \in \mathbb{Z}\}.$$

Let $p, q \in S$ be such that p is a prime and $r = \frac{q}{p}$ is an integer. Prove that $r \in S$.

(1999 Hungary–Israel Mathematical Competition)

Problem 11.9. Prove that every positive rational number can be represented in the form
$$\frac{a^3 + b^3}{c^3 + d^3},$$
where a, b, c, d are positive integers.

(1999 International Mathematical Olympiad Shortlist)

Problem 11.10. Two positive integers are written on the board. The following operation is repeated: if $a < b$ are the numbers on the board, then a is erased and $ab/(b - a)$ is written in its place. At some point the numbers on the board are equal. Prove that again they are positive integers.

(1998 Russian Mathematical Olympiad)

Problem 11.11. Let $f(x) + a_0 + a_1x + \cdots + a_mx^m$, with $m \geq 2$ and $a_m \neq 0$, be a polynomial with integer coefficients. Let n be a positive integer, and suppose that:

(i) $a_2, a_3, \ldots, a_m$ are divisible by all the prime factors of n;
(ii) a_1 and n are relatively prime.

Prove that for any positive integer k, there exists a positive integer c such that $f(c)$ is divisible by n^k.

(2001 Romanian International Mathematical Olympiad Team Selection Test)

Problem 11.12. Let x, a, b be positive integers such that $x^{a+b} = a^b b$. Prove that $a = x$ and $b = x^x$.

(1998 Iranian Mathematical Olympiad)

Problem 11.13. Let m, n be integers with $1 \leq m < n$. In their decimal representations, the last three digits of 1978^m are equal, respectively, to the last three digits of 1978^n. Find m and n such that $m + n$ is minimal.

(20th International Mathematical Olympiad)

Part II

Solutions to Additional Problems

1

Divisibility

1.1 Divisibility

Problem 1.1.10. *Show that for any natural number n, one can find three distinct natural numbers a, b, c between n^2 and $(n + 1)^2$ such that $a^2 + b^2$ is divisible by c.*

(1998 St. Petersburg City Mathematical Olympiad)

Solution. (We must assume $n > 1$.) Take

$$a = n^2 + 2, \quad b = n^2 + n + 1, \quad c = n^2 + 1.$$

Then $a^2 + b^2 = (2n^2 + 2n + 5)c$.

Problem 1.1.11. *Find all odd integers n greater than 1 such that for any relatively prime divisors a and b of n, the number $a + b - 1$ is also a divisor of n.*

(2001 Russian Mathematical Olympiad)

Solution. We will call a number "good" if it satisfies the given conditions. It is not difficult to see that all prime powers are good. Suppose n is a good number that has at least two distinct prime factors. Let $n = p^r s$, where p is the smallest prime dividing n, and s is not divisible by p. Because n is good, $p + s - 1$ must divide n. For any prime q dividing s, $s < p + s - 1 < s + q$, so q does not divide $p + s - 1$. Therefore, the only prime factor of $p + s - 1$ is p. Then $s = p^c - p + 1$ for some $c > 1$. Because p^c must also divide n, $p^c + s - 1 = 2p^c - p$ divides n. Because $2p^{c-1} - 1$ has no factors of p, it must divide s. Since every prime divisor of s is larger than p, we must have either $s > p(2p^{c-1} - 1)$ or $s = 2p^{c-1} - 1$. In the first case, rearranging gives $1 > p^c$, a contradiction. In the second case, rearranging gives $(p - 2)(p^{c-1} - 1) = 0$. Hence $p = 2$, contrary to the assumption that n is odd.

T. Andreescu and D. Andrica, *Number Theory*, DOI: 10.1007/b11856_12,
© Birkhäuser Boston, a part of Springer Science + Business Media, LLC 2009

Problem 1.1.12. *Find all positive integers n such that $3^{n-1}+5^{n-1}$ divides 3^n+5^n.*
So only prime powers are good.

(1996 St. Petersburg City Mathematical Olympiad)

First solution. This occurs only for $n = 1$. Let $s_n = 3^n + 5^n$ and note that

$$s_n = (3+5)s_{n-1} - 3 \cdot 5 \cdot s_{n-2},$$

so s_{n-1} must also divide $3 \cdot 5 \cdot s_{n-2}$. If $n > 1$, then s_{n-1} is coprime to 3 and 5, so s_{n-1} must divide s_{n-2}, which is impossible since $s_{n-1} > s_{n-2}$.

Second solution. Note that $1 < \frac{3^n+5^n}{3^{n-1}+5^{n-1}} < 5$, so we can have only $\frac{3^n+5^n}{3^{n-1}+5^{n-1}} \in \{2, 3, 4\}$ cases, which are easily checked.

Problem 1.1.13. *Find all positive integers n such that the set*

$$\{n, n+1, n+2, n+3, n+4, n+5\}$$

can be split into two disjoint subsets such that the products of elements in these subsets are the same.

(12th International Mathematical Olympiad)

Solution. At least one of six consecutive numbers is divisible by 5. From the given condition it follows that two numbers must be divisible by 5. These two numbers are necessarily n and $n+5$. Therefore n and $n+5$ are in distinct subsets. Since $n(n + 1) > n + 5$ for $n \geq 3$, it follows that a required partition cannot be considered with subsets of different cardinality. Thus each subset must contain three numbers. The following possibilities have to be considered:

(a) $\{n, n+2, n+4\} \cup \{n+1, n+3, n+5\}$,
(b) $\{n, n+3, n+4\} \cup \{n+1, n+2, n+5\}$.

In case (a), $n < n+1$, $n+2 < n+3$, and $n+4 < n+5$.
In case (b), the condition of the problem gives

$$n(n+3)(n+4) = (n+1)(n+3)(n+5).$$

We obtain $n^2 + 5n + 10 = 0$, and this equation has no real solution.

Remark. One can prove that if p is a prime of the form $4k + 3$, then one cannot partition $p - 1$ consecutive integers into two classes with equal product. This problem is the particular case $p = 7$.

Problem 1.1.14. *The positive integers $d_1, d_2, \ldots, d_n$ are distinct divisors of 1995. Prove that there exist d_i and d_j among them such that the numerator of the reduced fraction d_i/d_j is at least n.*

(1995 Israeli Mathematical Olympiad)

Solution. Note that $3 \cdot 5 \cdot 7 \cdot 19 = 1995$. If the chosen divisors include one divisible by 19 and another not divisible by 19, the quotient of the two has numerator divisible by 19, solving the problem since $n \leq 16$. If this is not the case, either all divisors are divisible by 19 or none of them has this property, and in particular $n \leq 8$. Without loss of generality, assume that the divisors are all not divisible by 19.

Under this assumption, we are done if the divisors include one divisible by 7 and another not divisible by 7, unless $n = 8$. In the latter case all of the divisors not divisible by 19 occur, including 1 and $3 \cdot 5 \cdot 7$, so this case also follows. We now assume that none of the chosen divisors is divisible by 4, so that in particular $n \leq 4$.

Again, we are done if the divisors include one divisible by 5 and another not divisible by 5. But this can fail to occur only if $n = 1$ or $n = 2$. The former case is trivial, while in the latter case we simply divide the larger divisor by the smaller one, and the resulting numerator has at least one prime divisor and so is at least 3. Hence the problem is solved in all cases.

Problem 1.1.15. *Determine all pairs (a, b) of positive integers such that $ab^2 + b + 7$ divides $a^2 b + a + b$.*

(39th International Mathematical Olympiad)

Solution. From the divisibility $ab^2 + b + 7 \mid a^2 b + a + b$ we obtain

$$ab^2 + b + 7 \mid b(a^2 b + a + b) - a(ab^2 + b + 7) \Rightarrow ab^2 + b + 7 \mid b^2 - 7a.$$

When $b^2 - 7a = 0$, it follows that $b^2 = 7k$, $a = 7k^2$. Observe that all pairs $(7k^2, 7k)$, $k \geq 1$, are solutions to the problem.

Suppose $b^2 - 7a > 0$. Then $ab^2 + b + 7 \leq b^2 - 7a$, and we get a contradiction:

$$b^2 - 7a < b^2 < ab^2 + b + 7.$$

Suppose $b^2 - 7a < 0$. Then $ab^2 + b + 7 \leq 7a - b^2$. This is possible only for $b^2 < 7$, i.e., either $b = 1$ or $b = 2$. If $b = 1$, we obtain $a + 8 \mid a^2 + a + 1$, that is, $a + 8 \mid a(a + 8) - 7(a + 8) + 57$. Hence $a + 8 \mid 57$ and we get $a + 8 = 19$ or $a + 8 = 49$, so $a = 11$ or $a = 49$.

If $b = 2$, we obtain $4a + 9 \mid a + 22 \Rightarrow 4a + 9 \leq a + 22 \Rightarrow 3a \leq 13$. This case cannot give a solution.

Hence, the solutions of the problem are $(7k^2, 7k)$, $(11, 1)$, and $(49, 1)$.

Problem 1.1.16. *Find all integers a, b, c with $1 < a < b < c$ such that $(a - 1)(b - 1)(c - 1)$ is a divisor of $abc - 1$.*

(33rd International Mathematical Olympiad)

Solution. It is convenient to define $a - 1 = x$, $b - 1 = y$, and $c - 1 = z$. Then we have the conditions $1 \leq x < y < z$ and $xyz \mid xy + yz + zx + x + y + z$.

The idea of the solution is to point out that we cannot have $xyz \leq xy + yz + zx + x + y + z$ for infinitely many triples (x, y, z) of positive integers. Let $f(x, y, z)$ be the quotient of the required divisibility.

From the algebraic form

$$f(x, y, z) = \frac{1}{x} + \frac{1}{y} + \frac{1}{z} + \frac{1}{xy} + \frac{1}{yz} + \frac{1}{zx}$$

we can see that f is a decreasing function in one of the variables x, y, z. By symmetry and because x, y, z are distinct numbers,

$$f(x, y, z) \leq f(1, 2, 3) = 2 + \tfrac{5}{6} < 3.$$

Thus, if the divisibility is fulfilled we can have either $f(x, y, z) = 1$ or $f(x, y, z) = 2$. So, we have to solve in positive integers the equations

$$xy + yz + zx + x + y + z = kxyz, \tag{1}$$

where $k = 1$ or $k = 2$.

Observe that $f(3, 4, 5) = \frac{59}{60} < 1$. Thus $x \in \{1, 2\}$. Also $f(2, 3, 4) = \frac{35}{24} < 2$. Thus, for $x = 2$, we necessarily have $k = 1$. The conclusion is that only three equations have to be considered in (1).

Case 1. $x = 1$ and $k = 1$. We obtain the equation

$$1 + 2(y + z) + yz = yz.$$

It has no solutions.

Case 2. $x = 1$ and $k = 2$. We obtain the equation

$$1 + 2(y + z) = yz.$$

Write it in the form $(y - 2)(z - 2) = 5$ and obtain $y - 2 = 1$, $z - 2 = 5$. It has a unique solution: $y = 3$, $z = 7$.

Case 3. $x = 2$ and $k = 1$. We obtain the equation

$$2 + 3(y + z) = yz.$$

By writing it in the form $(y - 3)(z - 3) = 11$, we obtain $y - 3 = 1$, $z - 3 = 11$. Thus, it has a unique solution: $y = 4$, $z = 15$.

From Case 2 and Case 3 we obtain respectively $a = 2$, $b = 4$, $c = 8$ and $a = 3$, $b = 5$, $c = 16$. These are the solutions of the problem.

Problem 1.1.17. *Find all pairs of positive integers (x, y) for which*

$$\frac{x^2 + y^2}{x - y}$$

is an integer that divides 1995.

(1995 Bulgarian Mathematical Olympiad)

Solution. It is enough to find all pairs (x, y) for which $x > y$ and $x^2 + y^2 = k(x - y)$, where k divides $1995 = 3 \cdot 5 \cdot 7 \cdot 19$. We shall use the following well-known fact: if p is prime of the form $4q + 3$ and if it divides $x^2 + y^2$ then p divides x and y. (For $p = 3, 7, 19$ the last statement can be proved directly.) If k is divisible by 3, then x and y are divisible by 3 too. Simplifying by 9 we get an equality of the form $x_1^2 + y_1^2 = k_1(x_1 - y_1)$, where k_1 divides $5 \cdot 7 \cdot 19$. Considering 7 and 19 analogously we get an equality of the form $a^2 + b^2 = 5(a - b)$, where $a > b$. (It is not possible to get an equality of the form $a^2 + b^2 = a - b$.) From here $(2a - 5)^2 + (2b + 5)^2 = 50$, i.e., $a = 3, b = 1$, or $a = 2, b = 1$. The above consideration implies that the pairs we are looking for are of the form $(3c, c)$, $(2c, c)$, $(c, 3c)$, $(c, 2c)$, where $c = 1, 3, 7, 19, 3 \cdot 7, 3 \cdot 19, 7 \cdot 19, 3 \cdot 7 \cdot 19$.

Problem 1.1.18. *Find all positive integers (x, n) such that $x^n + 2^n + 1$ is a divisor of $x^{n+1} + 2^{n+1} + 1$.*

(1998 Romanian International Mathematical Olympiad Team Selection Test)

Solution. The solutions are $(x, n) = (4, 1)$ and $(11, 1)$. If $n = 1$, we need $x + 3 = x + 2 + 1 \mid x^2 + 4 + 1 = x^2 + 5 = (x + 3)(x - 3) + 14$, so $x + 3$ divides 14 and $x = 4$ or 11. Suppose $n \geq 2$. For $x \in \{1, 2, 3\}$ we have

$$1 + 2^n + 1 < 1 + 2^{n+1} + 1 < 2(1 + 2^n + 1),$$

$$2^n + 2^n + 1 < 2^{n+1} + 2^{n+1} + 1 < 2(2^n + 2^n + 1),$$

$$2(3^n + 2^n + 1) < 3^{n+1} + 2^{n+1} + 1 < 3(3^n + 2^n + 1),$$

so $x^n + 2^n + 1$ does not divide $x^{n+1} + 2^{n+1} + 1$. For $x \geq 4$, $x^n = x^n/2 + x^n/2 \geq 2^{2n}/2 + x^2/2$, so

$$(2^n + 1)x \leq ((2^n + 1)^2 + x^2)/2$$
$$= (2^{2n} + 2^{n+1} + 1 + x^2)/2 < 2^{n+1} + x^n + 2^n + 2.$$

Therefore

$$(x - 1)(x^n + 2^n + 1) = x^{n+1} + 2^n x + x - x^n - 2^n - 1$$
$$< x^{n+1} + 2^{n+1} + 1 < x(x^n + 2^n + 1);$$

again $x^n + 2^n + 1$ does not divide $x^{n+1} + 2^{n+1} + 1$. So the only solutions are $(4, 1)$ and $(11, 1)$.

Problem 1.1.19. *Find the smallest positive integer k such that every k-element subset of* $\{1, 2, \ldots, 50\}$ *contains two distinct elements a, b such that* $a + b$ *divides* ab.

(1996 Chinese Mathematical Olympiad)

Solution. The minimal value is $k = 39$. Suppose $a, b \in S$ are such that $a + b$ divides ab. Let c be the greatest common divisor of a and b, and put $a = ca_1$, $b = cb_1$, so that a_1 and b_1 are relatively prime. Then $c(a_1 + b_1)$ divides $c^2 a_1 b_1$, so $a_1 + b_1$ divides $ca_1 b_1$. Since a_1 and b_1 have no common factor, neither do a_1 and $a_1 + b_1$, or b_1 and $a_1 + b_1$. In short, $a_1 + b_1$ divides c.

Since $S \subseteq \{1, \ldots, 50\}$, we have $a + b \leq 99$, so $c(a_1 + b_1) \leq 99$, which implies $a_1 + b_1 \leq 9$; on the other hand, of course $a_1 + b_1 \geq 3$. An exhaustive search produces 23 pairs a, b satisfying the condition:

$$a_1 + b_1 = 3 \qquad (6, 3), (12, 6), (18, 9), (24, 12),$$
$$(30, 15), (36, 18), (42, 21), (48, 24)$$
$$a_1 + b_1 = 4 \qquad (12, 4), (24, 8), (36, 12), (48, 16)$$
$$a_1 + b_1 = 5 \; (20, 5), (40, 10), (15, 10), (30, 20), (45, 30)$$
$$a_1 + b_1 = 6 \qquad\qquad\qquad\qquad\qquad\qquad (30, 6)$$
$$a_1 + b_1 = 7 \qquad\qquad\qquad (42, 7), (35, 14), (28, 21)$$
$$a_1 + b_1 = 8 \qquad\qquad\qquad\qquad\qquad\qquad (40, 24)$$
$$a_1 + b_1 = 9 \qquad\qquad\qquad\qquad\qquad\qquad (45, 36)$$

The twelve pairs $(3, 6), (4, 12), (5, 20), (7, 42), (8, 24), (9, 18), (10, 40),$ $(14, 35), (16, 48), (15, 30), (21, 28)$ and $(36, 45)$ are disjoint. Hence any 39-element subset must contain one of these pairs and hence two elements a and b with $a + b \mid ab$. Conversely, the 12-element set

$$\{6, 10, 12, 18, 20, 21, 24, 30, 35, 42, 45, 48\}$$

meets every pair on the list, so

$$\{1, 2, \ldots, 50\} \setminus \{6, 10, 12, 18, 20, 21, 24, 30, 35, 42, 45, 48\}$$

is a 38-element set without this property.

1.2 Prime Numbers

Problem 1.2.10. *For each integer n such that* $n = p_1 p_2 p_3 p_4$, *where* p_1, p_2, p_3, p_4 *are distinct primes, let*

$$d_1 = 1 < d_2 < d_3 < \cdots < d_{16} = n$$

be the sixteen positive integers that divide n. Prove that if $n < 1995$, *then* $d_9 - d_8 \neq 22$.

(1995 Irish Mathematical Olympiad)

Solution. Note that $35 \cdot 57 = 1995 = 2 \cdot 3 \cdot 7 \cdot 19$. Suppose that $n < 1995$ and $d_9 - d_8 = 22$; then $d_8 d_9 = n$, so $d_8 < 35$. Moreover, d_8 cannot be even, since that would make n divisible by 4, whereas n has distinct prime factors. Hence d_8, d_9, and n are odd.

The divisors $d_1, \ldots, d_8$ each are the product of distinct odd primes, since they divide n. Since $3 \cdot 5 \cdot 7 > 35$, none of $d_1, \ldots, d_8$ is large enough to have three odd prime factors, so each is either prime or the product of two primes. Since n has only four prime factors, four of the d_i must be the product of two odd primes. But the smallest such numbers are

$$15, 21, 33, 35, \ldots .$$

Assume that $p_1 < p_2 < p_3 < p_4$. If $d_8 = p_1 p_4$, then clearly $d_8 \neq 15$ and $d_8 \neq 21$. Moreover, if $d_8 = 33$, then $p_1 = 3$ and $p_4 = 11$; hence $p_2 = 5$ and $p_3 = 7$, and we get $d_9 = p_2 p_3 = 35$, giving the difference $d_9 - d_8 = 2$, which is not possible. If $d_8 = p_3 p_4$, then $d_8 \neq 15$, $d_8 \neq 21$, and $d_8 \neq 33$, since $p_3 > 3$.

In both situations we must have $d_8 \geq 35$, contrary to assumption.

Problem 1.2.11. *Prove that there are infinitely many positive integers a such that the sequence $(z_n)_{n \geq 1}$, $z_n = n^4 + a$, does not contain any prime number.*

(11th International Mathematical Olympiad)

Solution. To consider all positive integers of the form $n^4 + a$, $n \geq 1$, means to consider all values of the polynomial $P(X) = X^4 + a$ in the positive integers. A decomposition of the polynomial $P(X)$ gives us decompositions of the numbers $n^4 + a$, except in the case of factors taking the value 1.

The polynomial $P(X)$ can have a decomposition in integer polynomials only into quadratic factors:

$$P(X) = (X^2 + mX + n)(X^2 + m'X + n').$$

Such a decomposition is possible if and only if

$$m + m' = 0, \quad mm' + n + n' = 0, \quad mn' + m'n = 0 \text{ and } nn' = a.$$

We obtain $m' = -m$, $n = n'$, $m^2 - 2n = 0$, and $n^2 = a$.

The third equation forces m to be even. Taking $m = -m' = 2k$ gives $n = n' = 2k^2$ and $a = 4k^4$. The corresponding factorization is $X^4 + 4k^4 = (X^2 - 2kX + 2k^2)(X^2 + 2kX + 2k^2)$. For $k \geq 2$, these factors are

$$X^2 \pm 2kX + 2k^2 = (X \pm k)^2 + k^2 \geq k^2,$$

hence are nontrivial and $X^4 + 4k^4$ is composite.

For the record, this is the third problem to use this factorization.

Problem 1.2.12. *Let p, q, r be distinct prime numbers and let A be the set*

$$A = \{p^a q^b r^c : 0 \le a, b, c \le 5\}.$$

Find the smallest integer n such that any n-element subset of A contains two distinct elements x, y such that x divides y.

(1997 Romanian Mathematical Olympiad)

Solution. Define an order relation on A by setting $p^a q^b r^c \le p^{a_1} q^{b_1} r^{c_1}$ iff $a \le a_1$, $b \le b_1$, $c \le c_1$. Thus, we must find the longest antichain with respect to this relation, that is, the maximal number m such that there is $B \subset A$ with $|B| = m$ and no two elements of B are comparable. The answer will then be $n = m + 1$.

From now on, identify $p^a q^b r^c$ with (a, b, c) and regard it as a lattice point in $\mathbb{R}^3$. One can easily check that the set

$$B = \{(a, b, c) \mid a, b, c \in \{0, 1, \dots, 5\},\ a + b + c = 8\}$$

has 27 elements and that it is an antichain. We will prove that any set with 28 elements contains two comparable elements. Of course, it suffices to find 27 chains that partition $\{(a, b, c) \mid 0 \le a, b, c \le 5\}$ and such that each chain has a unique representation from B. Take $A = \{(a, b) \mid 0 \le a, b \le 5\}$ and partition it into six chains (draw a picture!)

$$
\begin{aligned}
A_1 &= \{(0, 0), (0, 1), \dots, (0, 5), (1, 5), \dots, (5, 5)\},\\
A_2 &= \{(1, 0), (1, 1), \dots, (1, 4), (2, 4), \dots, (5, 4)\},\\
A_3 &= \{(2, 0), (2, 1), \dots, (2, 3), (3, 3), \dots, (5, 3)\},\\
A_4 &= \{(3, 0), (3, 1), (3, 2), (4, 2), (5, 2)\},\\
A_5 &= \{(4, 0), (4, 1), (5, 1)\},\\
A_6 &= \{(5, 0)\}.
\end{aligned}
$$

Next define $A_{1j} = \{(a, b, j) \mid (a, b) \in A_1\}$ and similarly for A_2, A_3. We have found 18 chains so far.

For $(a, b) \in A_4 \cup A_5 \cup A_6$ we define the chain $A_{(a,b)} = \{(a, b, j) \mid 0 \le j \le 5\}$, and we have 9 chains, for a total of 27 chains.

Problem 1.2.13. *Prove Bonse's inequality:*

$$p_1 p_2 \cdots p_n > p_{n+1}^2$$

for $n \ge 4$, where $p_1 = 2, p_2 = 3, \dots$ is the increasing sequence of prime numbers.

Solution. Let us define $A_{n-1} = p_1 p_2 \cdots p_{n-1}$ and $a_k = kA_{n-1} - p_n$ for $1 \le k \le p_n - 1$. Observe that these numbers are relatively prime. Indeed, a prime common divisor of a_{k_1} and a_{k_2} would divide $(k_1 - k_2)A_{n-1}$, and since $\gcd(a_{k_1}, p_n) = 1$, this divisor would be one of $p_1, \ldots, p_{n-1}$, which is clearly impossible. Of course, this implies that $a_k \ge p_{n+k}$ (since a_k is relatively prime to $p_1, \ldots, p_{n-1}$). Thus for $k = p_n - 1$ we have $A_n - A_{n-1} - p_n > p_{p_n+n-1}$, and so $p_1 p_2 \cdots p_n > p_{p_n+n-1} > p_{3n-1}$ for $n \ge 5$. From here we find that for $n \ge 6$ we have $p_1 \cdots p_n > \left(p_1 \cdots p_{\lfloor \frac{n}{2} \rfloor}\right)^2 > p_{3\lfloor \frac{n}{2} \rfloor -1}^2 > p_{n+1}^2$. In the last inequality it is necessary to have $\lfloor \frac{n}{2} \rfloor \ge 5$, that is, $n \ge 10$. Let us remark that checking cases shows that this inequality holds for $n \ge 6$. For $n = 5$ one can easily check the inequality.

Problem 1.2.14. *Show that there exists a set A of positive integers with the following property: for any infinite set S of primes, there exist two positive integers $m \in A$ and $n \notin A$ each of which is a product of k distinct elements of S for some $k \ge 2$.*

(35th International Mathematical Olympiad)

Solution. There are several constructions for such A, involving different ideas about the decomposition of integers.

First example. Let $p_1 < p_2 < \cdots < p_n < \cdots$ be the increasing sequence of all prime numbers. Define A as the set of numbers of the form $p_{i_1} p_{i_2} \cdots p_{i_k}$ where $i_1 < i_2 < \cdots < i_k$ and $k = p_{i_1}$. For example, $3 \cdot 5 \cdot 7 \in A$; $3 \cdot 11 \cdot 13 \in A$ and $5 \cdot 7 \cdot 11 \notin A$; $3 \cdot 5 \cdot 7 \cdot 11 \notin A$.

We will see that A satisfies the required condition. Let S be an infinite set of prime numbers, say $q_1 < q_2 < \cdots < q_n < \cdots$. Take $m = q_1 q_2 \cdots q_{q_1}$ and $n = q_1 q_2 \cdots q_{q_1+1}$. Then $m \in A$ and $n \notin A$.

Second example. Define $A = \bigcup_{i=1}^{\infty} A_i$, where A_i is the set of numbers that are the product of $i + 1$ distinct primes that are different from p_i. For example, $3 \cdot 5 \cdot 7 \in A_2, 2 \cdot 3 \cdot 7 \cdot 11 \in A_3$ and $2 \cdot 3 \cdot 7 \notin A_2, 3 \cdot 5 \cdot 7 \cdot 13 \notin A_3$.

Let S be an infinite set of prime numbers, say $q_1 < q_2 < \cdots < q_n < \cdots$. Suppose that $q_1 = p_{i_1}$. If $i_1 > 1$, note that $i_1 = k$. Then $n = q_1 q_2 \cdots q_{k+1} \notin A$, because it contains a prime factor $q_1 = p_{i_1} = p_k$. The number $m = q_2 q_3 \cdots q_{k+2}$ contains $k + 1$ factors, all different from $p_k = q_1$. Thus $m \in A$. If $i_1 = 1$, take $k = i_2$, and the same construction will answer the question.

Third example. Let P be the set of all positive primes and let $P_1 \subset P_2 \subset \cdots \subset P_n \subset \cdots$ be a nested sequence of finite distinct subsets of P such that $P = \bigcup_{i=1}^{\infty} P_i$. Define A to be the set of elements of the form

$$a = p_1 p_2 \cdots p_k,$$

where $k = i_1 < i_2 < \cdots < i_k$ and $p_1 \in P_{i_1} \setminus P_{i_1-1}, p_2 \in P_{i_2}, \ldots, p_k \in P_{i_k}$.

Let S be an infinite set of prime numbers and let $S_i = S \cap P_i$. Since $S = \bigcup_{i=1}^{\infty} S_i$, there must be infinitely many indices $i > 1$ such that $S_{i-1} \neq S_i$. Let i_m be the mth such index. Then since $S_{i_m} \subset S_{i_m+1} \neq S_{i_m+1}$, we see that $S_{i_1} \subset S_{i_2} \subset \cdots \subset S_{i_m} \subset \cdots$ is an infinite nested subsequence of distinct sets.

Suppose that $S_{i_n} = S_{i_n+1} = \cdots = S_{i_{n+1}-1} \subset S_{i_{n+1}}$. Set $i_1 = k > 1$ and choose $p_1 \in S_{i_1} \setminus S_{i_1-1}$, $p_2 \in S_{i_2} \setminus S_{i_2-1}$, ..., $p_k \in S_{i_k} \setminus S_{i_k-1}$, and $p_{k+1} \in S_{i_{k+1}} \setminus S_{i_k}$. Then $m = p_1 p_2 \cdots p_k \in A$ and $n = p_2 p_3 \cdots p_{k+1} \notin A$ because $p_2 \notin S_{i_1} = S_k$.

Problem 1.2.15. *Let n be an integer, $n \geq 2$. Show that if $k^2 + k + n$ is a prime number for every integer k, $0 \leq k \leq \sqrt{n/3}$, then $k^2 + k + n$ is a prime number for any k, $0 \leq k \leq n - 2$.*

(28th International Mathematical Olympiad)

Solution. It is not difficult to verify the property for $n = 2, 3$, so we may suppose $n \geq 5$. Assume the contrary. Then there is some number $\sqrt{n/3} < m \leq n - 2$ such that $m^2 + m + n$ is composite and $k^2 + k + n$ is prime for $k < m$. Note that $(-k-1)^2 + (-k-1) + n = k^2 + k + n$. Therefore $k^2 + k + n$ is prime for $-m \leq k < m$. Let $m^2 + m + n = ab$ be a nontrivial decomposition such that $1 < a \leq b$. Since $n < 3m^2$, $ab = m^2 + m + n < 4m^2 + m < (2m+1)^2$. Therefore $a < 2m + 1$ and $-m \leq m - a < m$. Therefore $(m-a)^2 + (m-a) + n$ is a prime number. However,

$$(m-a)^2 + (m-a) + n = m^2 + m + n + a(a - 2m - 1) = a(b + a - 2m - 1).$$

It follows that $b + a - 2m - 1 = 1$ or $a + b = 2(m + 1)$. By the AM–GM inequality,

$$m^2 + m + n = ab \leq \frac{(a+b)^2}{4} = (m+1)^2 = m^2 + 2m + 1;$$

hence $n \leq m + 1$, contradicting the choice of $m \leq n - 2$.

Remark. The problem is related to the famous example of Euler of a polynomial generator of primes: $x^2 + x + 41$ produces primes for $0 \leq x \leq 39$. The problem shows that it suffices to check the primality only for the first four values of x.

Problem 1.2.16. *A sequence $q_1, q_2, \ldots$ of primes satisfies the following condition: for $n \geq 3$, q_n is the greatest prime divisor of $q_{n-1} + q_{n-2} + 2000$. Prove that the sequence is bounded.*

(2000 Polish Mathematical Olympiad)

Solution. Let $b_n = \max\{q_n, q_{n+1}\}$ for $n \geq 1$. We first prove that $b_{n+1} \leq b_n + 2002$ for all such n. Certainly $q_{n+1} \leq b_n$, so it suffices to show that $q_{n+2} \leq b_n + 2002$. If either q_n or q_{n+1} equals 2, then we have $q_{n+2} \leq q_n + q_{n+1} + 2000 = b_n + 2002$.

Otherwise, q_n and q_{n+1} are both odd, so $q_n + q_{n+1} + 2000$ is even. In this case $q_{n+2} \neq 2$ divides this number; hence we have

$$q_{n+2} \leq \tfrac{1}{2}(q_n + q_{n+1} + 2000) = \tfrac{1}{2}(q_n + q_{n+1}) + 1000 \leq b_n + 1000.$$

This proves the claim.

Choose k large enough that $b_1 \leq k \cdot 2003! + 1$. We prove by induction that $b_n \leq k \cdot 2003! + 1$ for all n. If this statement holds for some n, then $b_{n+1} \leq b_n + 2003 \leq k \cdot 2003! + 2003$. However, the numbers $k \cdot 2003! + m$ for $2 \leq m \leq 2003$ are all composite (since m is a factor). Since b_{n+1} is prime, it follows that $b_{n+1} \leq k \cdot 2003! + 1$. Thus, $q_n \leq b_n \leq k \cdot 2003! + 1$ for all n.

Problem 1.2.17. *Let $a > b > c > d$ be positive integers and suppose*

$$ac + bd = (b + d + a - c)(b + d - a + c).$$

Prove that $ab + cd$ is not prime.

(42nd International Mathematical Olympiad)

Solution. The given equality is equivalent to $a^2 - ac + c^2 = b^2 + bd + d^2$. Hence

$$(ab + cd)(ad + bc) = ac(b^2 + bd + d^2) + bd(a^2 - ac + c^2),$$

that is,

$$(ab + cd)(ad + bc) = (ac + bd)(a^2 - ac + c^2). \tag{1}$$

Now suppose that $ab + cd$ is prime. It follows from $a > b > c > d$ that

$$ab + cd > ac + bd > ad + bc; \tag{2}$$

hence $ac + bd$ is relatively prime to $ab + cd$. But then (1) implies that $ac + bd$ divides $ad + bc$, which is impossible by (2).

Problem 1.2.18. Find the least odd positive integer n such that for each prime p, $\frac{n^2-1}{4} + np^4 + p^8$ is divisible by at least four primes.

(Mathematical Reflections)

First solution. Let $n = 2k + 1$ with k a nonnegative integer. For $k = 0, 1, 2, 3$ it is easy to see that when $p = 2$ there are fewer than four prime divisors:

$$
\begin{aligned}
M &= p^8 + np^4 + \frac{n^2 - 1}{4} \\
&= \left(p^4 + \frac{n}{2}\right)^2 - \frac{1}{4} \\
&= \left(p^4 + \frac{n-1}{2}\right)\left(p^4 + \frac{n+1}{2}\right) \\
&= (p^4 + k)(p^4 + k + 1).
\end{aligned}
$$

Let $k = 4$. Then

$$M = (p^4 + 4)(p^4 + 5) = (p^2 + 2p + 2)(p^2 - 2p + 2)(p^4 + 5).$$

If $p = 2$, then m is divisible by 2, 3, 5, 7. If p is odd we have

$$\gcd(p^2 + 2p + 2, \, p^2 - 2p + 2)$$
$$= \gcd(p^2 + 2p + 2, 4p) = 1,$$
$$\gcd(p^2 + 2p + 2, \, p^4 + 5)$$
$$= \gcd(p^2 + 2p + 2, \, p^4 + 5 - p^4 - 4 - 4p^3 - 4p)$$
$$= \gcd(p^2 + 2p + 2, \, 4p^3 + 8p^2 + 4p + 1)$$
$$= \gcd(p^2 + 2p + 2, \, 4p^3 + 8p^2 + 4p + 1 - 4p^3 - 8p^2 - 4p)$$
$$= \gcd(p^2 + 2p + 2, 1) = 1,$$

and

$$\gcd(p^2 - 2p + 2, \, p^4 + 5) = \gcd(p^2 - 2p + 2, \, 4p^3 - 8p^2 + 4p + 1)$$
$$= \gcd(p^2 - 2p + 2, 1) = 1.$$

Thus $p^2 + 2p + 2$, $p^2 - 2p + 2$, and $p^4 + 5$ are pairwise coprime. Since $p^4 + 5 \equiv 2 \pmod 4$ for all odd p, 2^1 is the greatest power of 2 dividing $p^4 + 5$. Since both $p^2 + 2p + 2$ and $p^2 - 2p + 2$ are odd, there is another prime different from 2 and from the divisors of $p^2 + 2p + 2$ and $p^2 - 2p + 2$ that divides $p^4 + 5$, and so $n = 9$ is the least desired number.

Second solution. Let $n = 2k + 1$. Then

$$\frac{n^2 - 1}{4} + np^4 + p^8 = k(k+1) + (2k+1)p^4 + p^8 = (p^4 + k)(p^4 + k + 1).$$

Note that for $k = 0, 1, 2, 3$ the result does not hold for $p = 2$. We prove that $k = 4$ is the least integer that satisfies the condition. For $k = 4$ we have

$$(p^4 + 4)(p^4 + 5) = (p^2 + 2p + 2)(p^2 - 2p + 2)(p^4 + 5).$$

Since $(p^2 + 2p + 2)(p^2 - 2p + 2) = (p^4 + 5) - 1$, we have that

$$\gcd(p^2 + 2p + 2, \, p^4 + 5) = \gcd(p^2 - 2p + 2, \, p^4 + 5) = 1.$$

This implies that any prime that divides $(p^2 + 2p + 2)(p^2 - 2p + 2)$ does not divide $p^4 + 5$ and vice versa. Then, it is enough to prove that two primes divide $(p^2 + 2p + 2)(p^2 - 2p + 2)$ and another two divide $p^4 + 5$.

For $p = 2$ the result holds. Assume that p is an odd prime. Note that $2 \mid p^4+5$. To prove that another prime divides $p^4 + 5$ it is enough to prove that $4 \nmid p^4 + 5$. This results follows from the fact that $4 \mid p^4 + 3$.

In order to prove that two primes divide $(p^2+2p+2)(p^2-2p+2)$ it is enough to prove that $(p^2+2p+2, p^2-2p+2) = 1$. Let $\gcd(p^2+2p+2, p^2-2p+2) = d$. Note that d is odd and that $d \mid 4p$. This implies that $d \mid p$. If $d = p$ then $p \mid p^2 + 2p + 2$, which is a contradiction. Therefore, $d = 1$, as we wanted to prove. This implies that $k = 4$ is the least integer value that satisfies the condition of the problem, from which we conclude that $n = 9$ is the least odd positive integer that satisfies the condition.

1.3 The Greatest Common Divisor and the Least Common Multiple

Problem 1.3.9. *A sequence* $a_1, a_2, \ldots$ *of natural numbers satisfies*

$$\gcd(a_i, a_j) = \gcd(i, j) \quad \text{for all} \quad i \neq j.$$

Prove that $a_i = i$ *for all* i.

(1995 Russian Mathematical Olympiad)

Solution. For any integer m, we have $\gcd(a_m, a_{2m}) = \gcd(2m, m)$, and so $m \mid a_m$. This means that for any other integer n, m divides a_n if and only if m divides $\gcd(a_m, a_n) = \gcd(m, n)$; hence if and only if $m \mid n$. Therefore a_n has exactly the same divisors as n and so must equal n for all n.

Problem 1.3.10. *The natural numbers* a *and* b *are such that*

$$\frac{a+1}{b} + \frac{b+1}{a}$$

is an integer. Show that the greatest common divisor of a *and* b *is not greater than* $\sqrt{a+b}$.

(1996 Spanish Mathematical Olympiad)

Solution. Let $d = \gcd(a, b)$. Adding 2, we see that

$$\frac{a+1}{b} + \frac{b+1}{a} + 2 = \frac{(a+b)(a+b+1)}{ab}$$

is an integer. Since d^2 divides the denominator and $\gcd(d, a+b+1) = 1$, we must have $d^2 \mid a+b$; hence $d \leq \sqrt{a+b}$.

Problem 1.3.11. *The positive integers* m, n, m, n *are written on a blackboard. A generalized Euclidean algorithm is applied to this quadruple as follows: if the numbers* x, y, u, v *appear on the board and* $x > y$, *then* $x - y$, y, $u + v$, v *are written instead; otherwise,* x, $y - x$, u, $v + u$ *are written instead. The algorithm stops when the numbers in the first pair become equal (they will equal the greatest common divisor of* m *and* n). *Prove that the arithmetic mean of the numbers in the second pair at that moment equals the least common multiple of* m *and* n.

(1996 St. Petersburg City Mathematical Olympiad)

Solution. Note that $xv + yu$ does not change under the operation, so it remains equal to $2mn$ throughout. Thus when the first two numbers both equal $\gcd(m, n)$, the sum of the latter two is $2mn/\gcd(m, n) = 2 \operatorname{lcm}(m, n)$.

Problem 1.3.12. *How many pairs* (x, y) *of positive integers with* $x \le y$ *satisfy* $\gcd(x, y) = 5!$ *and* $\operatorname{lcm}(x, y) = 50!$?

(1997 Canadian Mathematical Olympiad)

Solution. First, note that there are 15 primes from 1 to 50:

$$(2, 3, 5, 7, 11, 13, 17, 19, 23, 29, 31, 37, 41, 43, 47).$$

To make this easier, let us define $f(a, b)$ to be greatest power of b dividing a. (Note that $g(50!, b) > g(5!, b)$ for all $b < 50$.) Therefore, for each prime p, we have either $f(x, p) = f(5!, p)$ and $f(y, p) = f(50!, p)$ or $f(y, p) = f(5!, p)$ and $f(x, p) = f(50!, p)$. Since we have 15 primes, this gives 2^{15} pairs, and clearly $x \ne y$ in any such pair (since the greatest common divisor and least common multiple are different), so there are 2^{14} pairs with $x \le y$.

Problem 1.3.13. *Several positive integers are written on a blackboard. One can erase any two distinct integers and write their greatest common divisor and least common multiple instead. Prove that eventually the numbers will stop changing.*

(1996 St. Petersburg City Mathematical Olympiad)

Solution. If a, b are erased and $c < d$ are written instead, we have $c \le \min(a, b)$ and $d \ge \max(a, b)$; moreover, $ab = cd$. From this we may conclude that $a + b \le c + d$. Indeed, $ab + a^2 = cd + a^2 \le ac + ad$ (the latter since $(d - a)(c - a) \le 0$) and divide both sides by a. Thus the sum of the numbers never decreases, and it is obviously bounded (e.g., by n times the product of the numbers, where n is the number of numbers on the board); hence it eventually stops changing, at which time the numbers never change.

Problem 1.3.14. *(a) For which positive integers* n *do there exist positive integers* x, y *such that*

$$\operatorname{lcm}(x, y) = n!, \quad \gcd(x, y) = 1998?$$

(b) For which n *is the number of such pairs* x, y *with* $x \le y$ *less than 1998?*

(1998 Hungarian Mathematical Olympiad)

Solution. (a) Let $x = 1998a$, $y = 1998b$. So a, b are positive integers such that $a < b$, $\gcd(a, b) = 1$. We have $\text{lcm}(x, y) = 1998ab = 2 \cdot 3^3 \cdot 37ab = n!$. Thus $n \geq 37$ and it is easy to see that this condition is also sufficient.

(b) The answers are $n = 37, 38, 39, 40$. We need to consider only positive integers $n \geq 37$. For $37 \leq n < 41$, let $k = ab = n!/1998$. Since $\gcd(a, b) = 1$, any prime factor of k that occurs in a cannot occur in b, and vice versa. There are 11 prime factors of k, namely 2, 3, 5, 7, 11, 13, 17, 19, 23, 29, 31. For each of those prime factors, one must decide only whether it occurs in a or in b. These 11 decisions can be made in a total of $2^{11} = 2048$ ways. However, only half of these ways will satisfy the condition $a < b$. Thus there will be a total of 1024 such pairs of (x, y) for $n = 37, 38, 39, 40$. Since 41 is a prime, we can see by a similar argument that there will be at least 2048 such pairs of (x, y) for $n \geq 41$.

Problem 1.3.15. *Determine all integers k for which there exists a function $f : \mathbb{N} \to \mathbb{Z}$ such that*
(a) $f(1997) = 1998$;
(b) for all $a, b \in \mathbb{N}$, $f(ab) = f(a) + f(b) + kf(\gcd(a, b))$.

(1997 Taiwanese Mathematical Olympiad)

Solution. Such an f exists for $k = 0$ and $k = -1$. First take $a = b$ in (b) to get $f(a^2) = (k + 2)f(a)$. Applying this twice, we get

$$f(a^4) = (k + 2)f(a^2) = (k + 2)^2 f(a).$$

On the other hand,

$$f(a^4) = f(a) + f(a^3) + kf(a) = (k + 1)f(a) + f(a^3)$$
$$= (k + 1)f(a) + f(a) + f(a^2) + kf(a)$$
$$= (2k + 2)f(a) + f(a^2) = (3k + 4)f(a).$$

Setting $a = 1997$, so that $f(a) \neq 0$, we deduce that $(k + 2)^2 = 3k + 4$, which has roots $k = 0, -1$. For $k = 0$, an example is given by

$$f(p_1^{e_1} \cdots p_n^{e_n}) = e_1 g(p_1) + \cdots + e_n g(p_n),$$

$g(1, 97) = 1998$, and $g(p) = 0$ for all primes $p \neq 1997$. For $k = -1$, an example is given by

$$f(p_1^{e_1} \cdots p_n^{e_n}) = g(p_1) + \cdots + g(p_n).$$

Problem 1.3.16. *Find all triples (x, y, n) of positive integers such that*

$$\gcd(x, n + 1) = 1 \text{ and } x^n + 1 = y^{n+1}.$$

(1998 Indian Mathematical Olympiad)

Solution. All solutions are of the form $(a^2 - 1, a, 1)$ with a even. We have $x^n = y^{n+1} - 1 = (y - 1)m$ with $m = y^n + y^{n-1} + \cdots + y + 1$. Thus $m \mid x^n$ and $\gcd(m, n + 1) = 1$. Rewrite m as

$$m = (y - 1)(y^{n-1} + 2y^{n-2} + 3y^{n-3} + \cdots + (n - 1)y + n) + (n + 1).$$

Thus we have $\gcd(m, y - 1) \mid n + 1$. But $\gcd(m, n + 1) = 1$, so $\gcd(m, y - 1) = 1$. Since $x^n = (y - 1)m$, m must be a perfect nth power. But

$$(y + 1)^n = y^n + \binom{n}{1} y^{n-1} + \cdots + \binom{n}{n-1} y + 1 > m > y^n,$$

for $n > 1$. So m can be a perfect nth power only if $n = 1$ and $x = y^2 - 1$. Since x and $n + 1 = 2$ are relatively prime, y must be even, yielding the presented solutions.

Problem 1.3.17. *Find all triples (m, n, l) of positive integers such that*

$$m + n = \gcd(m, n)^2, \quad m + l = \gcd(m, l)^2, \quad n + l = \gcd(n, l)^2.$$

(1997 Russian Mathematical Olympiad)

Solution. The only solution is $l = m = n = 2$. Let $d = \gcd(l, m, n)$, and put $l = dl_1, m = dm_1, n = dn_1$. Then $d(m_1 + n_1) = d^2 d_{mn}^2$, where $d_{mn} = \gcd(m_1, n_1)$, so $m_1 + n_1 = dd_{mn}^2$. Defining d_{ln} and d_{lm} likewise, we get

$$2(l_1 + m_1 + n_1) = d(d_{lm}^2 + d_{ln}^2 + d_{mn}^2).$$

Since $d/\gcd(d, 2)$ divides $l_1 + m_1 + n_1$ as well as $m_1 + n_1$, it divides l_1 and likewise m_1 and n_1. Since these three numbers are relatively prime, we have $d/\gcd(d, 2) = 1$, and so $d \leq 2$.

Note that d_{lm}, d_{ln}, d_{mn} are pairwise relatively prime; therefore we can write $l_1 = l_2 d_{lm} d_{ln}, m_1 = m_2 d_{lm} d_{mn}, n_1 = n_2 d_{ln} d_{mn}$. Then we have

$$d_{lm} d_{mn} m_2 + d_{ln} d_{mn} n_2 = dd_{mn}^2,$$

and so $m_2 d_{lm} + n_2 d_{ln} = dd_{mn}$, and so forth. Assuming without loss of generality that d_{mn} is no larger than d_{lm}, d_{ln}, we get

$$2d_{mn} \geq dd_{mn} = d_{lm} m_2 + d_{ln} n_2 \geq d_{lm} + d_{ln} \geq 2d_{mn}.$$

Thus we have equality throughout: $d = 2$, $m_2 = n_2 = 1$, and $d_{lm} = d_{ln} = d_{mn}$. But these three numbers are pairwise relatively prime, so they are all 1. Then $m_1 = n_1 = 1$ and from $l_1 + m_1 = dd_{lm}^2$, $l_1 = 1$ as well. Therefore $l = m = n = 2$.

Problem 1.3.18. *Let a, b be positive integers such that $\gcd(a, b) = 1$. Find all pairs (m, n) of positive integers such that $a^m + b^m$ divides $a^n + b^n$.*

(Mathematical Reflections)

Solution. The solution is any pair of the form $(m, (2k + 1)m)$, where k is any nonnegative integer, i.e., n must be an odd multiple of m.

Call $x_i = (-1)^i a^{(2k-i)m} b^{im} = a^{2km} r^i$, where k is any nonnegative integer, and $r = -\left(\frac{b}{a}\right)^m$. Clearly, the sum of all x_i for $i = 0, 1, 2, \ldots, 2k$ is an integer, and

$$\sum_{i=0}^{2k} x_i = a^{2km} \sum_{i=0}^{2k} r^i = a^{2km} \frac{1 - r^{2k+1}}{1 - r} = \frac{a^{(2k+1)m} + b^{(2k+1)m}}{a^m + b^m}.$$

Thus $a^m + b^m$ divides $a^n + b^n$ for all $n = (2k+1)m$, where k is any nonnegative integer. We prove that these are the only possible values of n.

Note that if a and b are relatively prime, then so are $a^m + b^m$ and ab. Let us assume that for some integer n such that $m < n \le 2m$, $a^m + b^m$ divides $a^n + b^n$. Now,

$$(a^m + b^m)(a^{n-m} + b^{n-m}) - (a^n + b^n) = (ab)^{n-m}(a^{2m-n} + b^{2m-n}),$$

so $a^m + b^m$ must divide $a^{2m-n} + b^{2m-n}$, since it is relatively prime to $(ab)^{n-m}$. But this is absurd, since $2m - n < m$. So the only n such that $0 \le m \le 2m - 1$ and $a^m + b^m$ divides $a^n + b^n$ is $n = m$. Let us complete our proof by showing by induction that for all nonnegative integers k, if $n = 2mk + d$, where $0 \le d \le 2m - 1$, then $a^m + b^m$ divides $d = m$. The result is already proved for $k = 0$. Let us assume it true for some $k - 1$. Then

$$(a^m + b^m)\left(a^{(2k-1)m+d} + b^{(2k-1)m+d}\right) - (ab)^m \left(a^{2(k-1)m+d} + b^{2(k-1)m+d}\right)$$
$$= a^{2km+d} + b^{2km+d} = a^n + b^n.$$

If $a^m + b^m$ divides $a^n + b^n$, since $a^m + b^m$ is prime to $(ab)^m$, then $a^m + b^m$ must also divide $a^{2(k-1)m+d} + b^{2(k-1)m+d}$. But by the induction hypothesis, $d = m$, and we are done.

1.4 Odd and Even

Problem 1.4.5. *We are given three integers a, b, c such that $a, b, c, a + b - c$, $a + c - b$, $b + c - a$, and $a + b + c$ are seven distinct primes. Let d be the difference between the largest and smallest of these seven primes. Suppose that $800 \in \{a + b, b + c, c + a\}$. Determine the maximum possible value of d.*

Solution. Answer: 1594.

First, observe that a, b, c must all be odd primes; this follows from the assumption that the seven quantities listed are distinct primes and the fact that there is only one even prime, 2. Therefore, the smallest of the seven primes is at least 3. Next, assume without loss of generality that $a + b = 800$. Because $a + b - c > 0$, we must have $c < 800$. We also know that c is prime; therefore, since $799 = 17 \cdot 47$, we have $c \leq 797$. It follows that the largest prime, $a+b+c$, is no more than 1597. Combining these two bounds, we can bound d by $d \leq 1597 - 3 = 1594$. It remains to observe that we can choose $a = 13$, $b = 787$, $c = 797$ to achieve this bound. The other four primes are then 3, 23, 1571, and 1597.

Problem 1.4.6. *Determine the number of functions $f : \{1, 2, \ldots, n\} \to \{1995, 1996\}$ that satisfy the condition that $f(1) + f(2) + \cdots + f(1996)$ is odd.*

(1996 Greek Mathematical Olympiad)

Solution. We can send $1, 2, \ldots, n - 1$ anywhere, and the value of $f(n)$ will then be uniquely determined. Hence there are 2^{n-1} such functions.

Problem 1.4.7. *Is it possible to place 1995 different natural numbers around a circle so that for any two adjacent numbers, the ratio of the greatest to the least is a prime?*

(1995 Russian Mathematical Olympiad)

Solution. No, this is impossible. Let $a_0, \ldots, a_{1995} = a_0$ be the integers. Then for $i = 1, \ldots, 1995$, a_{k-1}/a_k is either a prime or the reciprocal of a prime; suppose the former occurs m times and the latter $1995 - m$ times. The product of all of these ratios is $a_0/a_{1995} = 1$, but this means that the product of some m primes equals the product of some $1995 - m$ primes. This can occur only when the primes are the same (by unique factorization), and in particular there has to be the same number on both sides. But $m = 1995 - m$ is impossible, since 1995 is odd, contradiction.

Problem 1.4.8. *Let a, b, c, d be odd integers such that $0 < a < b < c < d$ and $ad = bc$. Prove that if $a + d = 2^k$ and $b + c = 2^m$ for some integers k and m, then $a = 1$.*

(25th International Mathematical Olympiad)

Solution. Since $ad = bc$, we have

$$a((a + d) - (b + c)) = (a - b)(a - c) > 0.$$

Thus $a + d > b + c$, $2^k > 2^m$, and $k > m$. Since $ad = a(2^k - a) = bc = b(2^m - b)$, we obtain

$$2^m b - 2^k a = b^2 - a^2 = (b - a)(b + a).$$

By the equality $2^m(b - 2^{k-m}a) = (b - a)(b + a)$, we infer that $2^m \mid (b - a)(b + a)$. But $b - a$ and $b + a$ differ by $2a$, an odd multiple of 2, so either $b - a$

or $b + a$ is not divisible by 4. Hence, either $2^{m-1} \mid b - a$ or $2^{m-1} \mid b + a$. But $0 < b - a < b < 2^{m-1}$, so it must be that $2^{m-1} \mid b + a$.

Since $0 < b + a < b + c = 2^m$, it follows that $b + a = 2^{m-1}$ and $b = 2^{m-1} - a$. Then $c = 2^{m-1}$ and $ad = bc = (2^{m-1} - a)(2^{m-1} + a)$.

From this equality we obtain $a(a + d) = 2^{2m-2}$; hence $a = 1$.

1.5 Modular Arithmetic

Problem 1.5.7. *Find all integers $n > 1$ such that every prime divisor of $n^6 - 1$ is a divisor of $(n^3 - 1)(n^2 - 1)$.*

(2002 Baltic Mathematics Competition)

Solution. We show that $n = 2$ is the only such integer. It is clear that $n = 2$ satisfies the conditions. For $n > 2$, write

$$n^6 - 1 = (n^3 - 1)(n^3 + 1) = (n^3 - 1)(n + 1)(n^2 - n + 1);$$

hence, all prime factors of $n^2 - n + 1$ must divide $n^3 - 1$ or $n^2 - 1 = (n-1)(n+1)$. Note, however, that $(n^2 - n + 1, n^3 - 1) \leq (n^3 + 1, n^3 - 1) \leq 2$; on the other hand, $n^2 - n + 1 = n(n - 1) + 1$ is odd, so all prime factors of $n^2 - n + 1$ must divide $n + 1$. But $n^2 - n + 1 = (n + 1)(n - 2) + 3$, so we must have $n^2 - n + 1 = 3^k$ for some k. Because $n > 2$, we have $k \geq 2$. Now $3 \mid (n^2 - n + 1)$ gives $n \equiv 2$ (mod 3); but for each of the cases $n \equiv 2, 5, 8$ (mod 9), we have $n^2 - n + 1 \equiv 3$ (mod 9), a contradiction.

Problem 1.5.8. *Let $f(n)$ be the number of permutations $a_1, \ldots, a_n$ of the integers $1, \ldots, n$ such that*

  (i) $a_1 = 1$;

  (ii) $|a_i - a_{i+1}| \leq 2$, $i = 1, \ldots, n - 1$.

  Determine whether $f(1996)$ is divisible by 3.

(1996 Canadian Mathematical Olympiad)

Solution. We will prove the recursion $f(n) = f(n - 1) + f(n - 3) + 1$ for $n \geq 4$ as follows:

Call such a permutation "special." Suppose $a_2 = 2$. Then the sequence $b_i = a_{i+1} - 1$, $1 \leq i \leq n - 1$, is a special permutation of $1, \ldots, n - 1$. Conversely, if b_i is special permutation of $1, \ldots, n - 1$, then defining $a_1 = 1$ and $a_i = b_{i-1} + 1$ for $2 \leq i \leq n$ gives a special permutation of $1, \ldots, n$. Thus the number of these is $f(n - 1)$.

If $a_2 \neq 2$, then $a_2 = 3$. Suppose $a_3 = 2$; hence $a_4 = 4$. Then the sequence $b_i = a_{i+3} - 3$ is a special permutation of $1, \ldots, n - 3$. As above, the converse also holds. Hence the number of these is $f(n - 3)$.

If $a_2 = 3$ and $a_3 \neq 2$, look at which i, $4 \leq i \leq n$, has $a_i = 2$. Since $|a_{i-1} - a_i| \leq 2$ and 1 and 3 are already used as a_1 and a_2, we must have $a_{i-1} = 4$. However, if $i \neq n$, the same argument shows that $a_{i+1} = 4$, a contradiction. Thus $a_n = 2$ and $a_{n-1} = 4$. Hence $a_3 = 5$, and iterating this argument shows that the only such permutation is $1, 3, 5, \ldots, 6, 4, 2$ with all the odd numbers in order followed by the even numbers in reverse order. Thus there is exactly one special permutation of this form.

Combining these three cases, we see that $f(n) = f(n-1) + f(n-3) + 1$ for $n \geq 4$. Calculating shows that $f(n) \pmod 3$ is $f(1) = 1, 1, 2, 1, 0, 0, 2, 0, 1, 1, 2, 1, \ldots$, repeating with period 8. Since $1996 \equiv 4 \pmod 8$, we have $f(1996) \equiv f(4) = 4 \pmod 3$, so $f(1996)$ is not divisible by 3.

Problem 1.5.9. *For natural numbers m, n, show that $2^n - 1$ is divisible by $(2^m - 1)^2$ if and only if n is divisible by $m(2^m - 1)$.*

(1997 Russian Mathematical Olympiad)

Solution. Since
$$2^{kn+d} - 1 \equiv 2^d - 1 \pmod{2^n - 1},$$
we have that $2^m - 1$ divides $2^n - 1$ if and only if m divides n. Thus in either case, we must have $n = km$, in which case
$$\frac{2^{km} - 1}{2^m - 1} = 1 + 2^m + \cdots + 2^{m(k-1)} \equiv k \pmod{2^m - 1}.$$

The two conditions are now that k is divisible by $2^m - 1$ and that n is divisible by $m(2^m - 1)$, which are equivalent.

Problem 1.5.10. *Suppose that n is a positive integer and let*
$$d_1 < d_2 < d_3 < d_4$$
be the four smallest positive integer divisors of n. Find all integers n such that
$$n = d_1^2 + d_2^2 + d_3^2 + d_4^2.$$

(1999 Iranian Mathematical Olympiad)

Solution. The answer is $n = 130$. Note that $x^2 \equiv 0 \pmod 4$ when x is even and that $x^2 \equiv 1 \pmod 4$ when x is odd.

If n is odd, then all the d_i are odd and $n \equiv d_1^2 + d_2^2 + d_3^2 + d_4^2 \equiv 1+1+1+1 \equiv 0 \pmod 4$, a contradiction. Thus, $2 \mid n$.

If $4 \mid n$ then $d_1 = 1$ and $d_2 = 2$, and $n \equiv 1 + 0 + d_3^2 + d_4^2 \not\equiv 0 \pmod 4$, a contradiction. Thus, $4 \nmid n$.

Therefore $\{d_1, d_2, d_3, d_4\} = \{1, 2, p, q\}$ or $\{1, 2, p, 2p\}$ for some odd primes p, q. In the first case, $n \equiv 3 \pmod 4$, a contradiction. Thus $n = 5(1 + p^2)$ and $5 \mid n$, so $p = d_3 = 5$ and $n = 130$.

Problem 1.5.11. *Let p be an odd prime. For each $i = 1, 2, \ldots, p - 1$ denote by r_i the remainder when i^p is divided by p^2. Evaluate the sum*

$$r_1 + r_2 + \cdots + r_{p-1}.$$

(Kvant)

Solution. Denote the sum in question by S. Combine the first summand with the last, the second one with the next-to-last, and so on, to get

$$2S = (r_1 + r_{p-1}) + (r_2 + r_{p-2}) + \cdots + (r_{p-1} + r_1). \tag{1}$$

We have $r_i + r_{p-i} \equiv i^p + (p - i)^p \pmod{p^2}$ by the definition of the numbers $r_1, r_2, \ldots, r_{p-1}$. Furthermore, because p is odd,

$$i^p + (p - i)^p = p^p - \binom{p}{1} p^{p-1} i + \binom{p}{2} p^{p-2} i^2 - \cdots + \binom{p}{p-1} p i^{p-1}.$$

Since p is a prime, each binomial coefficient above is divisible by p, which yields the conclusion that $r_i + r_{p-i}$ is divisible by p^2. But $0 < r_i < p^2$, $0 < r_{p-i} < p^2$, because p is a prime (so neither one equals 0), and now we may claim that

$$r_i + r_{p-i} = p^2 \text{ for } i = 1, 2, \ldots, p - 1. \tag{2}$$

The equalities (1) and (2) show that

$$S = \frac{p - 1}{2} p^2 = \frac{p^3 - p^2}{2}.$$

Problem 1.5.12. *Find the number of integers x with $|x| \le 1997$ such that 1997 divides $x^2 + (x + 1)^2$.*

(1998 Indian Mathematical Olympiad)

Solution. There are four such integers. With congruences all taken modulo 1997, we have

$$x^2 + (x + 1)^2 \equiv 2x^2 + 2x + 1 \equiv 4x^2 + 4x + 2 \equiv 0,$$

i.e., $(2x+1)^2 \equiv -1$. Since 1997 is a prime of the form $4k+1$, there are exactly two distinct solutions to $u^2 \equiv -1$ (see Section 9.1 for more details). Each corresponds to a different solution to $(2x + 1)^2 \equiv -1$.

Also, the two solutions to $(2x + 1)^2 \equiv -1$ are nonzero, since 0 does not satisfy the equation. Therefore, there are exactly two satisfactory integers x from -1997 to -1 and two more from 1 to 1997, for a total of four integer solutions, as claimed.

Problem 1.5.13. *Find the greatest common divisor of the numbers*

$$A_n = 2^{3n} + 3^{6n+2} + 5^{6n+2}$$

when $n = 0, 1, \ldots, 1999$.

(1999 Junior Balkan Mathematical Olympiad)

Solution. We have

$$A_0 = 1 + 9 + 25 = 35 = 5 \cdot 7.$$

Using congruence mod 5, it follows that

$$A_n \equiv 2^{3n} + 3^{6n+2} \equiv 2^{3n} + 9^{3n+1} \equiv 2^{3n} + (-1)^{3n+1} \quad (\text{mod } 5).$$

For $n = 1$, $A_1 \equiv 9 \not\equiv 0$ (mod 5); hence 5 is not a common divisor. On the other hand,

$$\begin{aligned}
A_n &= 8^n + 9 \cdot 9^{3n} + 25 \cdot 25^{3n} \\
&\equiv 1 + 2 \cdot 2^{3n} + 4 \cdot 4^{3n} \\
&\equiv 1 + 2 \cdot 8^n + 4 \cdot 64^n \\
&\equiv 1 + 2 \cdot 1^n + 4 \cdot 1^n \\
&\equiv 0 \quad (\text{mod } 7).
\end{aligned}$$

Therefore 7 divides A_n for all integers $n \geq 0$.

Consequently, the greatest common divisor of the numbers $A_0, A_1, \ldots, A_{1999}$ is equal to 7.

1.6   Chinese Remainder Theorem

Problem 1.6.3. *Let $P(x)$ be a polynomial with integer coefficients. Suppose that the integers $a_1, a_2, \ldots, a_n$ have the following property: For any integer x there exists an $i \in \{1, 2, \ldots, n\}$ such that $P(x)$ is divisible by a_i. Prove that there is an $i_0 \in \{1, 2, \ldots, n\}$ such that a_{i_0} divides $P(x)$ for any integer x.*

(St. Petersburg City Mathematical Olympiad)

Solution. Suppose that the claim is false. Then for each $i = 1, 2, \ldots, n$ there exists an integer x_i such that $P(x_i)$ is not divisible by a_i. Hence, there is a prime power $p_i^{k_i}$ that divides a_i and does not divide $P(x_i)$. Some of the powers $p_1^{k_1}, p_2^{k_2}, \ldots, p_n^{k_n}$ may have the same base. If so, ignore all but the one with the least exponent. To simplify notation, assume that the sequence obtained this way is $p_1^{k_1}, p_2^{k_2}, \ldots, p_m^{k_m}$, $m \leq n$ ($p_1, p_2, \ldots, p_m$ are distinct primes). Note that each a_i is divisible by some term of this sequence.

Since $p_1^{k_1}, p_2^{k_2}, \ldots, p_m^{k_m}$ are pairwise relatively prime, the Chinese remainder theorem yields a solution of the simultaneous congruences

$$x \equiv x_1 \pmod{p_1^{k_1}}, \quad x \equiv x_2 \pmod{p_2^{k_2}}, \quad \ldots, \quad x \equiv x_m \pmod{p_m^{k_m}}.$$

Now, since $P(x)$ is a polynomial with integer coefficients, the congruence $x \equiv x_j \pmod{p_j^{k_j}}$ implies $P(x) \equiv P(x_j) \pmod{p_j^{k_j}}$ for each index $j = 1, 2, \ldots, m$. By the definition of $p_j^{k_j}$, the number $P(x_j)$ is never divisible by $p_j^{k_j}$, $j = 1, 2, \ldots, m$. Thus, for the solution x given by the Chinese remainder theorem, $P(x)$ is not divisible by any of the powers $p_j^{k_j}$. And because each a_i is divisible by some $p_j^{k_j}$, $j = 1, 2, \ldots, m$, it follows that no a_i divides $P(x)$ either, a contradiction.

Problem 1.6.4. *For any set $\{a_1, a_2, \ldots, a_n\}$ of positive integers there exists a positive integer b such that the set $\{ba_1, ba_2, \ldots, ba_n\}$ consists of perfect powers.*

Solution. There is a finite number of primes $p_1, p_2, \ldots, p_k$ that participate in the prime factorization of $a_1, a_2, \ldots, a_n$. Let

$$a_i = p_1^{\alpha_{i1}} p_2^{\alpha_{i2}} \cdots p_k^{\alpha_{ik}} \quad \text{for } i = 1, 2, \ldots, n;$$

some of the exponents α_{ij} may be zeros. A positive integer with prime factorization $p_1^{u_1} p_2^{u_2} \cdots p_k^{u_k}$ is a perfect qth power if and only if all the exponents u_j are divisible by q. Thus it suffices to find positive integers $q_1, q_2, \ldots, q_n$ greater than 1 and nonnegative integers $l_1, l_2, \ldots, l_k$ such that

$$l_1 + \alpha_{11}, l_2 + \alpha_{12}, \ldots, l_k + \alpha_{1k} \text{ are divisible by } q_1,$$

$$l_1 + \alpha_{21}, l_2 + \alpha_{22}, \ldots, l_k + \alpha_{2k} \text{ are divisible by } q_2,$$

$$\cdots$$

$$l_1 + \alpha_{n1}, l_2 + \alpha_{n2}, \ldots, l_k + \alpha_{nk} \text{ are divisible by } q_n.$$

Now it is clear that we have many choices; let, for example, q_i be the ith prime number. As far as l_1 is concerned, the above conditions translate into

$$l_1 \equiv -\alpha_{j1} \pmod{q_j}, \quad j = 1, 2, \ldots, n.$$

This system of simultaneous congruences has a solution by the Chinese remainder theorem, because $q_1, q_2, \ldots, q_n$ are pairwise relatively prime. Analogously, each of the systems of congruences

$$l_2 \equiv -\alpha_{j2} \pmod{q_j}, \quad j = 1, 2, \ldots, n,$$

$$l_3 \equiv -\alpha_{j3} \pmod{q_j}, \quad j = 1, 2, \ldots, n,$$

$$\cdots$$

$$l_k \equiv -\alpha_{jk} \pmod{q_j}, \quad j = 1, 2, \ldots, n,$$

is solvable for the same reason. Take $l_1, l_2, \ldots, l_k$ such that all these congruences are satisfied. Multiplying each a_i by $b = p_1^{l_1} p_2^{l_2} \cdots p_k^{l_k}$ yields a set $\{ba_1, ba_2, \ldots, ba_n\}$ consisting of perfect powers (more exactly, ba_i is a perfect q_ith power).

Remarks. (1) The following problem is a direct consequence of the above result:

Prove that for every positive integer n there exists a set of n positive integers such that the sum of the elements of each of its subsets is a perfect power.

(Korean proposal for the 33rd International Mathematical Olympiad)

Indeed, let $\{x_1, x_2, \ldots, x_m\}$ be a finite set of positive integers and $S_1, S_2, \ldots, S_r$ the element sums of its nonempty subsets ($r = 2^m - 1$). Choose a b such that $bS_1, bS_2, \ldots, bS_r$ are all perfect powers. Then the set $\{bx_1, bx_2, \ldots, bx_m\}$ yields the desired example.

(2) Another consequence is the following: *There are arithmetic progressions of arbitrary finite length consisting only of powers.* Yet, no such infinite progression exists.

1.7   Numerical Systems

Problem 1.7.12. *The natural number A has the following property: the sum of the integers from 1 to A, inclusive, has decimal expansion equal to that of A followed by three digits. Find A.*

(1999 Russian Mathematical Olympiad)

Solution. We know that

$$k = (1 + 2 + \cdots + A) - 1000A$$
$$= \frac{A(A+1)}{2} - 1000A = A\left(\frac{A+1}{2} - 1000\right)$$

is between 0 and 999, inclusive. If $A < 1999$ then k is negative. If $A \geq 2000$, then $\frac{A+1}{2} - 1000 \geq \frac{1}{2}$ and $k \geq 1000$. Therefore $A = 1999$, and indeed $1 + 2 + \cdots + 1999 = 1999000$.

Problem 1.7.13. *A positive integer is said to be balanced if the number of its decimal digits equals the number of its distinct prime factors. For instance, 15 is balanced, while 49 is not. Prove that there are only finitely many balanced numbers.*

(1999 Italian Mathematical Olympiad)

Solution. Let $p_1 = 2$, $p_2 = 3, \ldots$ be the sequence of primes. If x is balanced and it has n prime factors, then

$$10^n \geq p_1 p_2 \cdots p_n \geq 2 \cdot 3 \cdot 5 \cdots (2n - 1) > 2 \cdot 2 \cdot 4 \cdots (2n - 2) > (n - 1)!,$$

which implies that n is bounded and so is x, since $x \leq 10^n$.

Problem 1.7.14. *Let $p \geq 5$ be a prime and choose $k \in \{0, \ldots, p-1\}$. Find the maximum length of an arithmetic progression none of whose elements contain the digit k when written in base p.*

(1997 Romanian Mathematical Olympiad)

Solution. We show that the maximum length is $p - 1$ if $k \neq 0$ and p if $k = 0$. In a p-term arithmetic progression, the lowest nonconstant digit takes all values from 0 to $p - 1$. This proves the upper bound for $k \neq 0$, which is also a lower bound because of the sequence $1, \ldots, p - 1$. However, for $k = 0$, it is possible that when 0 occurs, it is not actually a digit in the expansion but rather a leading zero. This can occur only for the first term in the progression, so extending the progression to $p + 1$ terms would cause an honest zero to appear. Thus the upper bound for $k = 0$ is p, and the sequence $1, p + 1, \ldots, (p - 1)p + 1$ shows that it is also a lower bound.

Problem 1.7.15. *How many 10-digit numbers divisible by 66667 are there whose decimal representation contains only the digits 3, 4, 5, and 6?*

(1999 St. Petersburg City Mathematical Olympiad)

First solution. Suppose that $66667n$ had 10 digits, all of which were 3, 4, 5, and 6. Then

$$3333333333 \leq 66667n \leq 6666666666 \implies 50000 \leq n \leq 99999.$$

Now consider the following cases:
(i) $n \equiv 0 \pmod 3$. Then

$$66667n = \tfrac{2}{3}n \cdot 10^5 + \tfrac{1}{3}n,$$

the five digits of $3 \cdot \frac{n}{3}$ followed by the five digits of $\frac{n}{3}$. These digits are all 3, 4, 5, or 6 if and only if $\frac{n}{3} = 33333$ and $n = 99999$.
(ii) $n \equiv 1 \pmod 3$. Then

$$66667n = \tfrac{2}{3}(n - 1) \cdot 10^5 + \tfrac{1}{3}(n + 2) + 66666,$$

the five digits of $\frac{2}{3}(n-1)$ followed by the five digits of $\frac{1}{3}(n+2)+66666$. Because $\frac{1}{3}(n + 2) + 66666$ must be between 66667 and 99999, its digits cannot be 3, 4, 5, or 6. Hence there are no satisfactory $n \equiv 1 \pmod 3$.
(iii) $n \equiv 2 \pmod 3$. Let $a = \frac{1}{3}(n - 2)$. Then

$$66667n = \left(\tfrac{2}{3}(n - 2) + 1\right) \cdot 10^5 + \tfrac{1}{3}(n - 2) + 33334,$$

the five digits of $x = 2a + 1$ followed by the five digits of $y = a + 33334$. The units digits in x and y are between 3 and 6 if and only if the units digit in a is 1 or 2. In this case the other digits in x and y are all between 3 and 6 if and only if the other digits in a are 2 or 3. Thus there are thirty-two satisfactory a's (we can choose each of its five digits from two options), and each a corresponds to a satisfactory $n = 3a + 2$.

Therefore there is exactly one satisfactory $n \equiv 0 \pmod 3$, and thirty-two satisfactory $n \equiv 2 \pmod 3$, making a total of thirty-three values of n and thirty-three ten-digit numbers.

Second solution. Write $66667n = 10^5 A + B$, where A and B are five digit numbers. Since $66667 \mid 2 \cdot 10^5 + 1$ and $66667 \mid 2 \cdot 10^5 A + 2B$, we have $66667 \mid 2B - A$. Since $-66666 \le 2B - A \le 166662$, this leaves only the two possibilities $2B - A = 0$ and $2B - A = 66667$.

If $A = 2B$, then working up from the least-significant digit, we see that $B = 33333$ and $A = 66666$ is the only solution. If $2B = A + 66667 = 10^6 + (A - 33333)$, then $2B$ must have six digits, with leading digit 1 and the remaining digits 0, 1, 2, or 3. Hence working up from the least significant digit we see that B has only 5's and 6's as digits. Conversely, a B with all digits 5 or 6 gives a $2B$ of the desired form and a corresponding A. There is 1 solution in the first case and 32 in the second, so 33 solutions total.

Problem 1.7.16. *Call positive integers similar if they are written using the same digits. For example, for the digits 1, 1, 2, the similar numbers are 112, 121, and 211. Prove that there exist three similar 1995-digit numbers containing no zero digit such that the sum of two them equals the third.*

(1995 Russian Mathematical Olympiad)

Solution. Noting that 1995 is a multiple of 3, we might first try to find three similar 3-digit numbers such that the sum of two of them equals the third. There are various digit arrangements to try, one of which is $\overline{abc} + \overline{acb} = \overline{cba}$. Since c, as a leading digit, cannot be zero, the middle column implies $c = 9$, and there are carries into and out of this column. Hence $2a + 1 = c$ and $b + c = a + 10$. The first equation gives $a = 3$, and then the second gives $b = 5$, and we discover that $459 + 495 = 954$.

Problem 1.7.17. *Let k and n be positive integers such that*

$$(n + 2)^{n+2}, \quad (n + 4)^{n+4}, (n + 6)^{n+6}, \quad \ldots, \quad (n + 2k)^{n+2k}$$

end in the same digit in decimal representation. At most how large is k?

(1995 Hungarian Mathematical Olympiad)

Solution. We cannot have $k \ge 5$, since then one of the terms would be divisible by 5 and so would end in a different digit from those not divisible by 5. Hence $k \le 4$. In fact, we will see that $k = 3$ is the best possible.

Since $x^5 \equiv x \pmod{10}$ for all x, $x^x \pmod{10}$ depends only on $x \pmod{20}$. Hence it suffices to tabulate the last digit of x^x for $x = 0, \ldots, 19$ and look for the longest run. For the evens, we get

$$0, 4, 6, 6, 6, 0, 6, 6, 6, 4,$$

while for the odds, we get

$$1, 7, 5, 3, 9, 1, 3, 5, 7, 9.$$

Clearly a run of 3 is the best possible.

Problem 1.7.18. *Let*

$$\prod_{n=1}^{1996} (1 + nx^{3^n}) = 1 + a_1 x^{k_1} + a_2 x^{k_2} + \cdots + a_m x^{k_m},$$

where $a_1, a_2, \ldots, a_m$ are nonzero and $k_1 < k_2 < \cdots < k_m$, Find a_{1996}.

(1996 Turkish Mathematical Olympiad)

Solution. Note that $k_i/3$ is the number obtained by writing i in base 2 and reading the result as a number in base 3, and a_i is the product of the exponents of the powers of 3 used in k_i. Thus

$$k_{1996} = 3^{11} + 3^{10} + 3^9 + 3^8 + 3^7 + 3^4 + 3^3$$

and

$$a_{1996} = 3 \cdot 4 \cdot 7 \cdot 8 \cdot 9 \cdot 10 \cdot 11 = 665280.$$

Problem 1.7.19. *For any positive integer k, let $f(k)$ be the number of elements in the set $\{k+1, k+2, \ldots, 2k\}$ whose base-2 representation has precisely three 1's.*

(a) Prove that for each positive integer m, there exists at least one positive integer k such that $f(k) = m$.

(b) Determine all positive integers m for which there exists exactly one k with $f(k) = m$.

(35th International Mathematical Olympiad)

Solution. (a) Let $g : \mathbb{N} \to \mathbb{N}$ be the function defined as follows: $g(k)$ is the number of elements in the set $\{1, 2, \ldots, k\}$ having three digits 1 in their binary representation. The following equalities are obvious:

$$f(k) = g(2k) - g(k)$$

and

$$f(k+1) - f(k) = g(2k+2) - g(2k) - (g(k+1) - g(k)).$$

The binary representation of $2k+2$ is obtained by adding a final 0 in the binary representation of $k + 1$. Thus, we have the following result:

$$f(k + 1) - f(k) = \begin{cases} 1 \text{ if the binary representation of } 2k + 1 \\ \quad \text{contains three digits 1,} \\ 0 \text{ otherwise.} \end{cases} \tag{1}$$

Another way to derive formula (1) is the following: Consider going from computing $f(k)$ to computing $f(k + 1)$. The sets of integers used to compute these are nearly the same. The difference is just that we replace $k + 1$ with $2k + 1$ and $2k + 2$. Since $k + 1$ and $2k + 2$ have the same number of ones in their binary representations, we get (1).

It proves that the function f increases by at most 1 from k to $k + 1$.

Since $g(2^n) = \binom{n}{3}$ and $f(2^n) = \binom{n+1}{3} - \binom{n}{3} = \binom{n}{2}$, it follows that f is an unbounded function. If we combine the above property with the observation that $f(4) = 1$, we find that the range of f is the set of all nonnegative integers.

Also, we can obtain the formula $f(2^n) = \binom{n}{2}$ in a different way: The elements of $\{2^n+1, \ldots, 2^{n+1}\}$ whose binary representation has exactly three 1's are exactly the $(n + 1)$-digit binary numbers whose leading digit is 1 and that have ones in 2 of the remaining n places. Hence there are $f(n) = \binom{n}{2}$ of them.

(b) Let us suppose that the equation $f(k) = m$ has a unique solution. It follows that

$$f(k + 1) - f(k) = f(k) - f(k + 1) = 1.$$

By (1), it follows that the binary representations of $2k + 1$ and $2k - 1$ contain three digits 1. Then the binary representation of k contains two digits 1. From $2k - 1 = 2(k - 1) + 1$ one obtains that the binary representation of $k - 1$ also contains two digits 1. Hence, the last digit of $k - 1$ is 1, and the last-but-one digit is 0. Thus, $k - 1 = 2^n + 1$ and $k = 2^n + 2$, where $n \geq 2$.

For such a number we have

$$f(2^n + 2) = g(2^{n+1} + 4) - g(2^n + 2) = 1 + g(2^{n+1}) - g(2^n) = 1 + \binom{n}{2}.$$

Thus, we have proved that the equation $f(k) = m$ has a unique solution if and only if m is a number of the form $m = 1 + \binom{n}{2}$, $n \geq 2$.

Problem 1.7.20. *For each positive integer n, let $S(n)$ be the sum of digits in the decimal representation of n. Any positive integer obtained by removing several (at least one) digits from the right-hand end of the decimal representation of n is called a stump of n. Let $T(n)$ be the sum of all stumps of n. Prove that $n = S(n) + 9T(n)$.*

(2001 Asian Pacific Mathematical Olympiad)

Solution. Let d_i be the digit associated with 10^i in the base-10 representation of n, so that $n = \overline{d_m d_{m-1} \ldots d_0}$ for some integer $m \geq 0$ (where $d_m \neq 0$). The stumps of n are $\sum_{j=k}^{m} d_j 10^{j-k}$ for $k = 1, 2, \ldots, m$, and their sum is

$$T(n) = \sum_{k=1}^{m} \sum_{j=k}^{m} d_j 10^{j-k} = \sum_{j=1}^{m} d_j \sum_{k=1}^{j} 10^{j-k}$$

$$= \sum_{j=1}^{m} d_j \sum_{k=0}^{j-1} 10^k = \sum_{j=1}^{m} d_j \frac{10^j - 1}{10 - 1}.$$

Hence,

$$9T(n) = \sum_{j=1}^{m} d_j (10^j - 1) = \sum_{j=1}^{m} 10^j d_j - \sum_{j=1}^{m} d_j$$

$$= \sum_{j=0}^{m} 10^j d_j - \sum_{j=0}^{m} d_j = n - S(n),$$

as desired.

Problem 1.7.21. *Let p be a prime number and m a positive integer. Show that there exists a positive integer n such that there exist m consecutive zeros in the decimal representation of p^n.*

(2001 Japanese Mathematical Olympiad)

Solution. It is well known that if $\gcd(s, t) = 1$, then $s^k \equiv 1 \pmod{t}$ for some $k > 0$: indeed, of all the positive powers of s, some two $s^{k_1} < s^{k_2}$ must be congruent modulo t, and then $s^{k_2 - k_1} \equiv 1 \pmod{t}$.

First suppose that $p \neq 2, 5$. Then $\gcd(p, 10^{m+1}) = 1$, so there exists a $k > 1$ such that $p^k \equiv 1 \pmod{10^{m+1}}$. Then $p^k = a \cdot 10^{m+1} + 1$, so there are m consecutive zeros in the decimal representation of p^k.

Now suppose that $p = 2$. We claim that for any a, some power of 2 has the following final a digits: $a - \lceil \log 2^a \rceil$ zeros, followed by the $\lceil \log 2^a \rceil$ digits of 2^a. Because $\gcd(2, 5^a) = 1$, there exists k such that $2^k \equiv 1 \pmod{5^a}$. Let $b = k + a$. Then $2^b \equiv 2^a \pmod{5^a}$, and $2^b \equiv 0 \equiv 2^a \pmod{2^a}$. Hence, $2^b \equiv 2^a \pmod{10^a}$. Because $2^a < 10^a$, it follows that 2^b has the required property.

Now simply choose a such that $a - \lceil \log 2^a \rceil \geq m$ (for instance, we could choose $a = \lceil \frac{m+1}{1 - \log 2} \rceil$). Then 2^b contains at least m consecutive zeros, as desired.

Finally, the case $p = 5$ is done analogously to the case $p = 2$.

Remark. In fact, the property holds for every integer $p \geq 2$. If p is a power of 2, it is trivial. Otherwise, one can prove using Kronecker's[1] theorem (stating that for

---

[1] Leopold Kronecker (1823–1891), German mathematician with many contributions in the theory of equations. He made major contributions in elliptic functions and the theory of algebraic numbers.

$\alpha \in \mathbb{R} \setminus \mathbb{Q}$ the set of $\{n_\alpha\}$ with $n \in \mathbb{N}$ is dense in $[0, 1]$) that the numbers p^n can start with any combination of digits we may need, in particular with $1\underbrace{00\ldots0}_{m \text{ times}}$.

Problem 1.7.22. *Knowing that 2^{29} is a 9-digit number whose digits are distinct, without computing the actual number determine which of the ten digits is missing. Justify your answer.*

Solution. It is not difficult to see that when divided by 9, the remainder is 5. The ten-digit number containing all digits 0, 1, 2, 3, 4, 5, 6, 7, 8, 9 is a multiple of 9, because the sum of its digits has this property. So, in our nine-digit number, 4 is missing.

Problem 1.7.23. *It is well known that the divisibility tests for division by 3 and 9 do not depend on the order of the decimal digits. Prove that 3 and 9 are the only positive integers with this property. More exactly, if an integer $d > 1$ has the property that $d \mid n$ implies $d \mid n_1$, where n_1 is obtained from n through an arbitrary permutation of its digits, then $d = 3$ or $d = 9$.*

Solution. Let d be a k-digit number. Then among the $(k+2)$-digit numbers starting with 10 there is at least one that is divisible by d. Denote it by $\overline{10a_1a_2\cdots a_k}$. The assumption implies that both numbers $\overline{a_1a_2\cdots a_k10}$ and $\overline{a_1a_2\cdots a_k01}$ are divisible by d, and then so is their difference. This difference equals 9, and the proof is finished, since d may be only some divisor of 9.

Remark. The following problem, given in an old Russian Mathematical Olympiad, is much more restrictive and difficult:

Suppose that $d > 1$ has the property that $d \mid n$ implies $d \mid n_1$, where n_1 is obtained from n by reversing the order of its digits. Then $d \mid 99$. Try to solve this problem.

2

Powers of Integers

2.1 Perfect Squares

Problem 2.1.14. *Let x, y, z be positive integers such that*

$$\frac{1}{x} - \frac{1}{y} = \frac{1}{z}.$$

Let h be the greatest common divisor of x, y, z. Prove that $hxyz$ and $h(y-x)$ are perfect squares.

(1998 United Kingdom Mathematical Olympiad)

Solution. Let $x = ha$, $y = hb$, $z = hc$. Then a, b, c are positive integers such that $\gcd(a, b, c) = 1$. Let $\gcd(a, b) = g$. So $a = ga'$, $b = gb'$ and a' and b' are positive integers such that

$$\gcd(a', b') = \gcd(a' - b', b') = \gcd(a', a' - b') = 1.$$

We have

$$\frac{1}{a} - \frac{1}{b} = \frac{1}{c} \iff c(b - a) = ab \iff c(b' - a') = a'b'g.$$

Since $\gcd(a, b, c) = 1$, we have $\gcd(g, c) = 1$ and hence $g \mid b' - a'$. Since $\gcd(b' - a', a') = \gcd(b' - a', b') = 1$, we also have $b' - a' \mid g$. Hence $b' - a' = g$ and $c = a'b'$. Thus $hxyz = h^4g^2c^2$ and $h(y - x) = h^2g^2$ are perfect squares.

Problem 2.1.15. *Let b be an integer greater than 5. For each positive integer n, consider the number*

$$x_n = \underbrace{11\ldots1}_{n-1}\underbrace{22\ldots2}_{n}5,$$

written in base b. Provethat the following condition holds if and only if $b = 10$:

T. Andreescu and D. Andrica, *Number Theory*, DOI: 10.1007/b11856_13,
© Birkhäuser Boston, a part of Springer Science + Business Media, LLC 2009

There exists a positive integer M such that for every integer n greater than M, the number x_n is a perfect square.

(44th International Mathematical Olympiad Shortlist)

First solution. Assume that $b \geq 6$ has the required property. Consider the sequence $y_n = (b - 1)x_n$. From the definition of x_n we easily find that

$$y_n = b^{2n} + b^{n+1} + 3b - 5.$$

Then $y_n y_{n+1} = (b - 1)^2 x_n x_{n+1}$ is a perfect square for all $n > M$. Also, straightforward calculation implies

$$\left(b^{2n+1} + \frac{b^{n+2} + b^{n+1}}{2} - b^3\right)^2 < y_n y_{n+1} < \left(b^{2n+1} + \frac{b^{n+2} + b^{n+1}}{2} + b^3\right)^2.$$

Hence for every $n > M$ there is an integer a_n such that $|a_n| < b^3$ and

$$y_n y_{n+1} = (b^{2n} + b^{n+1} + 3b - 5)(b^{2n+2} + b^{n+2} + 3b - 5)$$
$$= \left(b^{2n+1} + \frac{b^{n+1}(b + 1)}{2} + a_n\right)^2. \tag{1}$$

Now considering this equation modulo b^n we obtain $(3b - 5)^2 \equiv a_n^2$, so that assuming that $n > 3$, we get $a_n = \pm(3b - 5)$.

If $a_n = 3b - 5$, then substituting in (1) yields

$$\tfrac{1}{4}b^{2n}(b^4 - 14b^3 + 45b^2 - 52b + 20) = 0,$$

with $b = 10$ the only solution greater than 5. Also, if $a_n = -3b + 5$, we similarly obtain

$$\tfrac{1}{4}(b^4 - 14b^3 - 3b^2 + 28b + 20) - 2b^{n+1}(3b^2 - 2b - 5) = 0$$

for each n, which is impossible.

For $b = 10$ we have $x_n = \left(\frac{10^n + 5}{3}\right)^2$ for all n (see Problem 2.1.8). This proves the statement.

Second solution. In problems of this type, computing $z_n = \sqrt{x_n}$ asymptotically usually works.

From $\lim_{n \to \infty} \frac{b^{2n}}{(b-1)x_n} = 1$ we infer that $\lim_{n \to \infty} \frac{b^n}{z_n} = \sqrt{b - 1}$. Furthermore, from

$$(bz_n + z_{n+1})(bz_n - z_{n+1}) = b^2 x_n - x_{n+1} = b^{n+2} + 3b^2 - 2b - 5$$

we obtain

$$\lim_{n \to \infty} (bz_n - z_{n+1}) = \frac{b\sqrt{b - 1}}{2}.$$

Since the z_n's are integers for all $n \geq M$, we conclude that

$$bz_n - z_{n+1} = \frac{b\sqrt{b-1}}{2}$$

for all n sufficiently large. Hence $b-1$ is a perfect square, and moreover, b divides $2z_{n+1}$ for all large n. Hence b divides $2x_{n+1} \equiv 10 \pmod{b}$ for all large n. It follows that $b \mid 10$; hence the only possibility is $b = 10$.

Problem 2.1.16. *Do there exist three natural numbers greater than 1, such that the square of each, minus one, is divisible by each of the others?*

(1996 Russian Mathematical Olympiad)

Solution. Such integers do not exist. Suppose $a \geq b \geq c$ satisfy the desired condition. Since $a^2 - 1$ is divisible by b, the numbers a and b are relatively prime. Hence the number $c^2 - 1$, which is divisible by a and b, must be a multiple of ab, so in particular $c^2 - 1 \geq ab$. But $a \geq c$ and $b \geq c$, so $ab \geq c^2$, a contradiction.

Problem 2.1.17. *(a) Find the first positive integer whose square ends in three 4's.*
(b) Find all positive integers whose squares end in three 4's.
(c) Show that no perfect square ends with four 4's.

(1995 United Kingdom Mathematical Olympiad)

Solution. It is easy to check that $38^2 = 1444$ is the first positive integer whose square ends in three 4's. Now let n be any such positive integer. Then $n^2 - 38^2 = (n - 38)(n + 38)$ is divisible by $1000 = 2^3 \cdot 5^3$. Hence at least one of $n - 38$, $n + 38$ is divisible by 4, and thus both are, since their difference is $76 = 4 \cdot 19$. Since $5 \nmid 76$, then 5 divides only one of the two factors. Consequently $n - 38$ or $n + 38$ is a multiple of $4 \cdot 5^3 = 500$, so we have $n = 500k \pm 38$. It is easy to check that the square of all numbers of this form (where k is a positive integer) end in three 4's.

Note that (c) follows from Problem 2.1.12.

Problem 2.1.18. *Let $\overline{abc}$ be a prime. Prove that $b^2 - 4ac$ cannot be a perfect square.*

(Mathematical Reflections)

First solution. Assume that $b^2 - 4ac$ is a perfect square and then let $b^2 - 4ac = k^2$, $k \in \mathbb{N}$. We have

$$4a \cdot \overline{abc} = 4a \cdot (100a + 10b + c) = 400a^2 + 40ab + 4ac$$
$$= (20a + b)^2 - (b^2 - 4ac) = (20a + b + k)(20a + b - k). \quad (*)$$

Since $a, b, k \in \mathbb{N}$, then $(20a + b + k) \in \mathbb{Z}$ and $(20a + b - k) \in \mathbb{Z}$. Since $\overline{abc}$ is a prime, then according to $(*)$,

$$\overline{abc} \mid (20a + b + k) \quad \text{or} \quad \overline{abc} \mid (20a + b - k).$$

It follows that $\overline{abc} \leq 20a + b + k$ or $\overline{abc} \leq 20a + b - k$. This leads to a contradiction, since $20a + b + k < \overline{abc}$ and $20a + b - k < \overline{abc}$. Hence, $\overline{abc}$ cannot be a perfect square. This completes our proof.

Second solution. It is clear that $a, b, c \in \mathbb{N}$, $a \neq 0$, and $\gcd(a, b, c) = 1$. If $x_1 = u$ and $x_2 = v$ are the solutions of the equation $ax^2 + bx + c = 0$, then we obtain the factorization $ax^2 + bx + c = a(x - u)(x - v)$. On the other hand, if the discriminant $D = b^2 - 4ac = h^2$, $h' \in \mathbb{N}$, is a perfect square, the solutions of the equation $ax^2 + bx + c$ are rational. The factorization is such that

$$a\left(x - \frac{-b+h}{2a}\right)\left(x - \frac{-b-h}{2a}\right) = p,$$

where p is prime. We have $x = 10$ and $\overline{abc} = a \cdot 10^2 + b \cdot 10 + c = p$; thus

$$(2ax + b - h)(2ax + b + h) = 4ap.$$

Since b and h have the same parity, we get

$$\left(ax + \frac{b-h}{2}\right)\left(ax + \frac{b+h}{2}\right) = ap.$$

One of the factors on the left-hand side should be divisible by p, but clearly $\left(ax + \frac{b-h}{2}\right), \left(ax + \frac{b+h}{2}\right) \leq 100$, a contradiction. Thus $b^2 - 4ac$ cannot be a perfect square.

Problem 2.1.19. *For each positive integer n, denote by $s(n)$ the greatest integer such that for all positive integers $k \leq s(n)$, n^2 can be expressed as a sum of squares of k positive integers.*
   (a) Prove that $s(n) \leq n^2 - 14$ for all $n \geq 4$.
   (b) Find a number n such that $s(n) = n^2 - 14$.
   (c) Prove that there exist infinitely many positive integers n such that

$$s(n) = n^2 - 14.$$

(33rd International Mathematical Olympiad)

Solution. (a) Representing n^2 as a sum of $n^2 - 13$ squares is equivalent to representing 13 as a sum of numbers of the form $x^2 - 1$, $x \in \mathbb{N}$, such as $0, 3, 8, 15, \ldots$. But it is easy to check that this is impossible, and hence $s(n) \leq n^2 - 14$.
   (b) Let us prove that $s(13) = 13^2 - 14 = 155$. Observe that

$$13^2 = 8^2 + 8^2 + 4^2 + 4^2 + 3^2$$
$$= 8^2 + 8^2 + 4^2 + 4^2 + 2^2 + 2^2 + 1^2$$
$$= 8^2 + 8^2 + 4^2 + 3^2 + 3^2 + 2^2 + 1^2 + 1^2 + 1^2.$$

Given any representation of n^2 as a sum of m squares one of which is even, we can construct a representation as a sum of $m + 3$ squares by dividing the square into four equal squares. Thus the first equality enables us to construct representations with $5, 8, 11, \ldots, 155$ squares, the second to construct ones with $7, 10, 13, \ldots, 154$ squares, and the third with $9, 12, \ldots, 153$ squares. It remains only to represent 13^2 as a sum of $k = 2, 3, 4, 6$ squares. This can be done as follows:

$$13^2 = 12^2 + 5^2 = 12^2 + 4^2 + 3^2$$
$$= 11^2 + 4^2 + 4^2 + 4^2$$
$$= 12^2 + 3^2 + 2^2 + 2^2 + 2^2 + 2^2.$$

(c) We shall prove that whenever $s(n) = n^2 - 14$ for some $n \geq 13$, it also holds that $s(2n) = (2n)^2 - 14$. This will imply that $s(n) = n^2 - 14$ for any $n = 2^t \cdot 13$.

If $n^2 = x_1^2 + \cdots + x_r^2$, then we have $(2n)^2 = (2x_1)^2 + \cdots + (2x_r)^2$. Replacing $(2x_i)^2$ with $x_i^2 + x_i^2 + x_i^2 + x_i^2$ as long as it is possible, we can obtain representations of $(2n)^2$ consisting of $r, r+3, \ldots, 4r$ squares. This gives representations of $(2n)^2$ into k squares for any $k \leq 4n^2 - 62$. Further, we observe that each number $m \geq 14$ can be written as a sum of $k \geq m$ numbers of the form $x^2 - 1$, $x \in \mathbb{N}$, which is easy to verify. Therefore if $2n^2 \leq k \leq 4n^2 - 14$, it follows that $4n^2 - k$ is a sum of k numbers of the form $x^2 - 1$ (since $k \geq 4n^2 - k \geq 14$), and consequently $4n^2$ is a sum of k squares.

Remark. One can find exactly the value of $s(n)$ for each n:

$$s(n) = \begin{cases} 1, & \text{if } n \text{ has a prime divisor congruent to 3 mod 4,} \\ 2, & \text{if } n \text{ is of the form } 5 \cdot 2^k, \ k \text{ a positive integer,} \\ n^2 - 14, & \text{otherwise.} \end{cases}$$

Problem 2.1.20. *Let A be the set of positive integers representable in the form $a^2 + 2b^2$ for integers a, b with $b \neq 0$. Show that if $p^2 \in A$ for a prime p, then $p \in A$.*
(1997 Romanian International Mathematical Olympiad Team Selection Test)

Solution. The case $p = 2$ is easy, so assume $p > 2$. Note that if $p^2 = a^2 + 2b^2$, then $2b^2 = (p - a)(p + a)$. In particular, a is odd, and since a cannot be divisible by p, $\gcd(p - a, p + a) = \gcd(p - a, 2p) = 2$. By changing the sign of a, we may assume that $p - a$ is not divisible by 4, and so

$$|p + a| = m^2, \quad |p - a| = 2n^2.$$

Since $|a| < |p|$, both $p + a$ and $p - a$ are actually positive, so we have $2p = m^2 + 2n^2$, so $p = n^2 + 2(m/2)^2$.

Problem 2.1.21. *Is it possible to find* 100 *positive integers not exceeding* 25000 *such that all pairwise sums of them are different?*

(42nd International Mathematical Olympiad Shortlist)

Solution. Yes. The desired result is an immediate consequence of the following fact applied to $p = 101$.

Lemma. *For any odd prime number* p, *there exist* p *positive integers less than* $2p^2$ *with all sums distinct.*

Proof. We claim that the numbers $a_n = 2np + (n^2)_p$, $n = 0, 1, \ldots, p - 1$, have the desired property, where $(x)_p$ denotes the remainder of x upon division by p.

Suppose that $a_k + a_l = a_m + a_n$. By the construction of a_i, we have

$$2p(k + l) \leq a_k + a_l \leq 2p(k + l + 1).$$

Hence we must have $k + l = m + n$, and therefore also

$$(k^2)_p + (l^2)_p = (m^2)_p + (n^2)_p.$$

Thus

$$k + l \equiv m + n \quad \text{and} \quad k^2 + l^2 \equiv m^2 + n^2 \pmod{p}.$$

But then it holds that

$$(k - l)^2 = 2(k^2 + l^2) - (k + l)^2 \equiv (m - n)^2 \pmod{p},$$

so $k - l \equiv \pm(m - n)$, which leads to $\{k, l\} = \{m, n\}$. This proves the lemma.

Problem 2.1.22. *Do there exist* 10 *distinct integers, the sum of any* 9 *of which is a perfect square?*

(1999 Russian Mathematical Olympiad)

Solution. Yes, there do exist 10 such integers. Write $S = a_1 + a_2 + \cdots + a_{10}$, and consider the linear system of equations

$$S - a_1 = 9 \cdot 1^2$$
$$S - a_2 = 9 \cdot 2^2$$
$$\cdots$$
$$S - a_{10} = 9 \cdot 10^2.$$

Adding all these gives

$$9S = 9 \cdot (1^2 + 2^2 + \cdots + 10^2),$$

so that
$$a_k = S - 9k^2 = 1^2 + 2^2 + \cdots + 10^2 - 9k^2.$$

Then all the a_k's are distinct integers, and any nine of them add up to a perfect square.

Problem 2.1.23. *Let n be a positive integer such that n is a divisor of the sum*

$$1 + \sum_{i=1}^{n-1} i^{n-1}.$$

Prove that n is square-free.

(1995 Indian Mathematical Olympiad)

Solution. If $n = mp^2$ for some prime p, then

$$1 + \sum_{i=1}^{n-1} i^{n-1} = 1 + \sum_{j=0}^{p-1} \sum_{k=0}^{mp-1} (kp + j)^{n-1}$$

$$\equiv 1 + (mp)\left(\sum_{j=0}^{p-1} j^{n-1}\right) \equiv 1 \pmod{p},$$

and the sum is not even a multiple of p. Hence if the sum is a multiple of n, n must have no repeated prime divisors, or equivalently no square divisors greater than 1.

Remark. The famous Giuga's conjecture states that if $n > 1$ satisfies $n \mid 1 + \sum_{i=1}^{n-1} i^{n-1}$, then n is a prime.

The reader can prove instead that for any such n we have that for any prime divisor p of n, $p - 1 \mid \frac{n}{p} - 1$ and $p \mid \frac{n}{p} - 1$.

Problem 2.1.24. *Let n, p be integers such that n > 1 and p is a prime. If n | (p − 1) and p | (n³ − 1), show that 4p − 3 is a perfect square.*

(2002 Czech–Polish–Slovak Mathematical Competition)

Solution. From $n \mid p - 1$ it follows $p - 1 \geq n$ and $p > n$. Because

$$p \mid n^3 - 1 = (n - 1)(n^2 + n + 1)$$

we get $p \mid n^2 + n + 1$, i.e., $pk = n^2 + n + 1$ for some positive integer k.

On the other hand, $n \mid p - 1$ implies $p \equiv 1 \pmod{n}$ and $pk \equiv k \pmod{n}$. We obtain $n^2 + n + 1 \equiv k \pmod{n}$; hence $k \equiv 1 \pmod{n}$.

It follows that $p = an + 1, k = bn + 1$ for some integers $a > 0, b \geq 0$. We can write

$$(an + 1)(bn + 1) = n^2 + n + 1,$$

so

$$abn^2 + (a + b)n + 1 = n^2 + n + 1,$$

i.e.,

$$abn + (a + b) = n + 1.$$

If $b \geq 1$, then $abn + (a+b) \geq n+2 > n+1$. So $b = 0, k = 1, p = n^2+n+1$. Therefore

$$4p - 3 = 4n^2 + 4n + 4 - 3 = 4n^2 + 4n + 1 = (2n + 1)^2.$$

Problem 2.1.25. *Show that for any positive integer $n > 10000$, there exists a positive integer m that is a sum of two squares and such that $0 < m - n < 3\sqrt[4]{n}$.*

(Russian Mathematical Olympiad)

Solution. Suppose $k^2 \leq n + 1 < (k + 1)^2$ and write $n + 1 = k^2 + r$. Then $r \leq 2k \leq 2\sqrt{n + 1}$. Suppose $l^2 < r \leq (l + 1)^2$ and write $r = (l + 1)^2 - s$. Then $0 \leq s \leq 2l < 2\sqrt{r} \leq 2^{3/2}(n + 1)^{1/4}$. Let $m = k^2 + (l + 1)^2$. Then

$$1 \leq m - n = s + 1 < 2^{3/2}(n + 1)^{1/4} + 1.$$

Thus it remains only to show that for $n > 10000$ we have $2^{3/2}(n+1)^{1/4} + 1 < 3n^{1/4}$. For this we note that

$$\frac{2^{3/2}(n + 1)^{1/4} + 1}{n^{1/4}} = 2^{3/2}(1 + 1/n)^{1/4} + 1/n^{1/4}$$

$$\leq 2^{3/2}(1 + 1/10000)^{1/4} + 1/10000^{1/4} \approx 2.928 < 3.$$

Problem 2.1.26. *Show that a positive integer m is a perfect square if and only if for each positive integer n, at least one of the differences*

$$(m + 1)^2 - m, \quad (m + 2)^2 - m, \quad \ldots, \quad (m + n)^2 - m$$

is divisible by n.

(2002 Czech and Slovak Mathematical Olympiad)

Solution. First, assume that m is a perfect square. If $m = a^2$, then

$$(m + c)^2 - m = (m + c)^2 - a^2 = (m + c + a)(m + c - a).$$

Clearly, there exists some c, with $1 \leq c \leq n$, for which $m + c + a$ is divisible by n. Thus, one of the given differences is divisible by n if m is a perfect square.

Now we assume that m is not a perfect square and show that there exists n for which none of the given differences is divisible by n. Clearly, there exists a prime p and positive integer k such that p^{2k-1} is the highest power of p that divides m.

We may let $m = bp^{2k-1}$, with b and p being relatively prime. Furthermore, pick $n = p^{2k}$. For the sake of contradiction, assume that there exists a positive integer c for which $(m+c)^2 - m$ is divisible by n. By expanding $(m+c)^2 - m$, we note that

$$p^{2k} \mid \left(2bcp^{2k-1} + c^2 - bp^{2k-1}\right).$$

If p^{2k} divides the quantity, then so does p^{2k-1}. Thus, $p^{2k-1} \mid c^2$, and so $p^k \mid c$. Let $c = rp^k$. Then, we have

$$p^{2k} \mid \left(2brp^{3k-1} + r^2p^{2k} - bp^{2k-1}\right).$$

However, this implies that $p \mid b$, which contradicts the original assumption that b and p are relatively prime. Therefore, if m is not a perfect square, n may be chosen so that none of the given differences are divisible by n. This completes the proof.

2.2 Perfect Cubes

Problem 2.2.5. *Find all positive perfect cubes that are not divisible by 10 such that the number obtained by erasing the last three digits is also a perfect cube.*

Solution. We have $(10m + n)^3 = 1000a^3 + b$, where $1 \le n \le 9$ and $b < 1000$. The equality gives

$$(10m + n)^3 - (10a)^3 = b < 1000,$$

so

$$(10m + n - 10a)[(10m + n)^2 + (10m + n) \cdot 10a + 100a^2] < 1000.$$

Since $(10m+n)^2 + (10m+n) \cdot 10a + 100a^2 > 100$, we obtain $10m + n - 10a < 10$; hence $m = a$.

If $m \ge 2$, then $n(300m^2 + 30mn + n^2) > 1000$, false.

Then $m = 1$ and $n(300 + 30n + n^2) < 1000$; hence $n \le 2$. For $n = 2$, we obtain $12^3 = 1728$, and for $n = 1$, we get $11^3 = 1331$.

Problem 2.2.6. *Find all positive integers n less than 1999 such that n^2 is equal to the cube of the sum of n's digits.*

(1999 Iberoamerican Mathematical Olympiad)

Solution. In order for n^2 to be a cube, n must be a cube itself. Because $n < 1000$ we must have $n = 1^3, 2^3, \ldots$, or 9^3. Quick checks show that $n = 1$ and $n = 27$ work, while $n = 8, 64$, and 125 don't. As for $n \ge 6^3 = 216$, we have $n^2 \ge 216^2 > 27^2$. However, the sum of n's digits is at most $9 + 9 + 9 = 27$, implying that no $n \ge 6^3$ has the desired property. Thus $n = 1, 27$ are the only answers.

Problem 2.2.7. *Prove that for any nonnegative integer n, the number*

$$A = 2^n + 3^n + 5^n + 6^n$$

is not a perfect cube.

Solution. We will use modular arithmetic. A perfect cube has the form $7k, 7k+1$, or $7k - 1$, since

$$(7x + 1)^3 \equiv (7x + 2)^3 = (7x + 4)^3 \equiv 1 \quad (\text{mod } 7)$$

and

$$(7x + 3)^3 \equiv (7x + 5)^3 \equiv (7x + 6)^3 \equiv -1 \quad (\text{mod } 7).$$

Now observe that

$$2^6 = 4^3 \equiv 1 \quad (\text{mod } 7),$$
$$3^6 = 9^3 \equiv 2^3 \equiv 1 \quad (\text{mod } 7),$$
$$5^6 = (-2)^6 = 2^6 \equiv 1 \quad (\text{mod } 7),$$
$$6^6 \equiv (-1)^6 \equiv 1 \quad (\text{mod } 7).$$

It follows that $2^{6k} \equiv 3^{6k} \equiv 5^{7k} \equiv 6^{6k} \equiv 1 \pmod{7}$.

Let $a_n = 2^n + 3^n + 5^n + 6^n$ for $n \geq 0$. Set $n = 6k + r$, with $r \in \{0, 1, 2, 3, 4, 5, 6\}$. Since $2^n \equiv 2^r \pmod{7}$, $3^n \equiv 3^r \pmod{7}$, $5^n \equiv 5^r \pmod{7}$, and $6^n \equiv 6^r \pmod{7}$, we have $a_n \equiv a_r \pmod{7}$.

It is easy to observe that $a_0 \equiv a_2 \equiv a_6 \equiv 4 \pmod{7}$, $a_1 \equiv a_4 \equiv 2 \pmod{7}$, and $a_3 \equiv 5 \pmod{7}$. Therefore, a_n is not a perfect cube.

Problem 2.2.8. *Prove that every integer is a sum of five cubes.*

Solution. For any integer n we have the identity

$$6n = (n + 1)^3 + (n - 1)^3 + (-n)^3 + (-n)^3. \qquad (1)$$

For an arbitrary integer m we choose the integer v such that $v^3 \equiv m \pmod{6}$. It follows that $m - v^3 = 6n$ for some integer n and we apply identity (1).

The actual representations are given by (1) and

$$6n + 1 = 6n + 1^3,$$
$$6n + 2 = 6(n - 1) + 2^3,$$
$$6n + 3 = 6(n - 4) + 3^3,$$
$$6n + 4 = 6(n + 1) + (-2)^3,$$
$$6n + 5 = 6(n + 1) + (-1)^3.$$

Remark. A direct solution is given by the representation

$$m = m^3 + \left[\binom{m+1}{3} + 1\right]^3 + \left[\binom{m+1}{3} - 1\right]^3$$
$$+ \left(-\binom{m+1}{3}\right)^3 + \left(-\binom{m+1}{3}\right)^3$$

Problem 2.2.9. *Show that every rational number can be written as a sum of three cubes.*

Solution. Let x be a rational number. We would be done if we could find a relation of the form $a^3(x) + b^3(x) + c^3(x) = x$, where a, b, c are rational functions. To make the arithmetic easier, it will actually be convenient to look for a relation $a^3(x) + b^3(x) + c^3(x) = nx$ for some integer n. Rewrite this as $a^3(x) + b^3(x) = nx - c^3(x)$. Writing $a(x) = f(x)/h(x)$, $b(x) = g(x)/h(x)$ for polynomials f, g, h and clearing denominators gives

$$f^3(x) + g^3(x) = (nx - c^3(x))h^3(x).$$

To build such an equation let $\varepsilon = \cos\frac{2\pi}{3} + i\sin\frac{2\pi}{3}$. Then we can write

$$f^3(x) + g^3(x) = (f(x) + g(x))(f(x) + \varepsilon g(x))(f(x) + \varepsilon^2 g(x)).$$

It would be convenient if two of the factors on the right were cubes. Then we could combine them into h and we could choose c so that the third factor is $nx - c^3(x)$. Since we want f and g to be real, we try

$$f(x) + \varepsilon g(x) = (u + \varepsilon v)^3$$
$$f(x) + \varepsilon^2 g(x) = (u + \varepsilon^2 v)^3.$$

Solving this system (using $\varepsilon^2 = -1 - \varepsilon$) gives $f(x) = u^3 - 3uv^2 + v^3$, $g = 3u^2 v - 3uv^2$, and hence we are left with solving

$$nx - c^3(x) = f(x) + g(x) = u^3 + 3u^2 v - 6uv^2 + v^3.$$

Notice that the right-hand side is $(u + v)^3 - 9uv^2$. Thus we can take $u = x$, $v = 1$, $n = -9$, and $c(x) = u + v = x + 1$. Solving back through the calculation gives $f(x) = x^3 - 3x + 1$, $g(x) = 3x^2 - 3x$, and $h(x) = x^2 - x + 1$. Hence we get

$$\left(\frac{x^3 - 3x + 1}{x^2 - x + 1}\right)^2 + \left(\frac{3x^2 - 3x}{x^2 - x + 1}\right)^3 + (x + 1)^3 = -9x,$$

and the desired conclusion follows by applying this for x equal to the desired rational number divided by -9.

Remark. There are rational numbers that are not the sum of two cubes. We suggest to the reader to find a such example.

2.3   *k*th Powers of Integers, *k* at least 4

Problem 2.3.6. *Let p be a prime number and a, n positive integers. Prove that if*

$$2^p + 3^p = a^n,$$

then n = 1.

(1996 Irish Mathematical Olympiad)

Solution. If $p = 2$, we have $2^2 + 3^2 = 13$ and $n = 1$. If $p > 2$, then p is odd, so 5 divides $2^p + 3^p$ and so 5 divides a. Now if $n > 1$, then 25 divides a^n and 5 divides

$$\frac{2^p + 3^p}{2 + 3} = 2^{p-1} - 2^{p-2} \cdot 3 + \cdots + 3^{p-1} \equiv p2^{p-1} \quad (\bmod 5),$$

a contradiction if $p \neq 5$. Finally, if $p = 5$, then $2^5 + 3^5 = 375$ is not a perfect power, so $n = 1$ again.

Problem 2.3.7. *Let x, y, p, n, k be natural numbers such that*

$$x^n + y^n = p^k.$$

Prove that if n > 1 is odd, and p is an odd prime, then n is a power of p.

(1996 Russian Mathematical Olympiad)

Solution. Let $m = \gcd(x, y)$. Then $x = mx_1$, $y = my_1$, and by virtue of the given equation, $m^n(x_1^n + y_1^n) = p^k$, and so $m = p^\alpha$ for some nonnegative integer α. It follows that

$$x_1^n + y_1^n = p^{k-n\alpha}. \tag{1}$$

Since n is odd,

$$\frac{x_1^n + y_1^n}{x_1 + y_1} = x_1^{n-1} - x_1^{n-2}y_1 + x_1^{n-3}y_1^2 - \cdots - x_1y_1^{n-2} + y_1^{n-1}. \tag{2}$$

Let A denote the right side of equation (2). By the condition $p > 2$, it follows that at least one of x_1, y_1 is greater than 1, so since $n > 1$, $A > 1$.

From (1) it follows that $A(x_1 + y_1) = p^{k-n\alpha}$, so since $x_1 + y_1 > 1$ and $A > 1$, both of these numbers are divisible by p; moreover, $x_1 + y_1 = p^\beta$ for some natural number β. Thus

$$\begin{aligned} A &= x_1^{n-1} - x_1^{n-2}(p^\beta - x_1) + \cdots - x_1(p^\beta - x_1)^{n-2} + (p^\beta - x_1)^{n-1} \\ &= nx_1^{n-1} + Bp. \end{aligned}$$

Since A is divisible by p and x_1 is relatively prime to p, it follows that n is divisible by p.

Let $n = pq$. Then $x^{pq} + y^{pq} = p^k$ or $(x^p)^q + (y^p)^q = p^k$. If $q > 1$, then by the same argument, p divides q. If $q = 1$, then $n = p$. Repeating this argument, we deduce that $n = p^l$ for some natural number l.

Problem 2.3.8. *Prove that a product of three consecutive integers cannot be a power of an integer.*

Solution. Let n be an integer and assume by contradiction that

$$n(n+1)(n+2) = x^z$$

for some integers x and z, where $z \geq 2$. We note that $n(n+2) = (n+1)^2 - 1$ and that $n+1$ and $(n+1)^2 - 1$ are relatively prime. It follows that

$$n + 1 = a^z,$$
$$(n+1)^2 - 1 = b^z,$$

for some integers a and b. It follows that $a^{2z} - b^z = 1$, i.e.,

$$(a^2 - b)((a^2)^{z-1} + (a^2)^{z-2}b + \cdots + b^{z-1}) = 1.$$

We get $a^2 - b = 1$; hence $a^2 = b + 1$. The equation $(b+1)^z - b^z = 1$ has the unique solution $z = 1$, a contradiction.

Remark. A famous theorem of Erdős and Selfridge, answering a conjecture of more than 150 years, states that the product of consecutive integers is never a power.

Problem 2.3.9. *Show that there exists an infinite set A of positive integers such that for any finite nonempty subset $B \subset A$, $\sum_{x \in B} x$ is not a perfect power.*

(Kvant)

Solution. The set

$$A = \{2^n 3^{n+1} : n \geq 1\}$$

has the desired property. Indeed, if $B = \{2^{n_1} 3^{n_1+1}, \ldots, 2^{n_k} 3^{n_k+1}\}$ is a finite subset of A, where $n_1 < \cdots < n_k$, then

$$\sum_{x \in B} x = 2^{n_1} 3^{n_1+1}(1 + 2^{n_2-n_1} 3^{n_2-n_1} + \cdots + 2^{n_k-n_1} 3^{n_k-n_1}) = 2^{n_1} 3^{n_1+1} N,$$

where $\gcd(N, 2) = \gcd(N, 3) = 1$. Taking into account that n_1 and $n_1 + 1$ are relatively prime, it follows that $\sum_{x \in B} x$ is not a perfect power.

Problem 2.3.10. *Prove that there is no infinite arithmetic progression consisting only of perfect powers.*

Solution. Assume that we have such an arithmetic progression, $an + b$, $n = 1, 2, \ldots$. It is well known that

$$\sum_{n \geq 1} \frac{1}{an + b} = \infty. \tag{1}$$

But on the other hand, we have

$$\sum_{n \geq 1} \frac{1}{an + b} \leq \sum_{m,s \geq 2} \frac{1}{m^s} < +\infty,$$

contradicting (1).

Remark. There are alternative solutions to Problems 2.3.9 and 2.3.10 using the following observation, which is a nice result in its own right.

Lemma. *There are arbitrarily long stretches of consecutive positive integers that contain no perfect powers.*

Proof. This follows immediately from our observation that $\sum_{m,s \geq 2} \frac{1}{m^s}$ converges, or more elementarily by the Chinese remainder theorem by finding an n with $n \equiv 2 \pmod 4$, $n + 1 \equiv 3 \pmod 9$, etc., or with a little analysis by showing that if $p(N)$ is the number of perfect powers less than N then $\lim_{N \to \infty} p(N)/N = 0$. $\qquad\square$

From this Problem 2.3.10 is obvious: there will be infinitely many of these stretches longer than the difference of the arithmetic progression and hence the progression cannot cross them. For Problem 2.3.9, we build A inductively. Then the nth element a_n of A is chosen to be the first element of a stretch of $1 + \sum_{k=1}^{n-1} a_k$ consecutive elements with no perfect power.

3

Floor Function and Fractional Part

3.1 General Problems

Problem 3.1.10. *Let n be a positive integer. Find with proof a closed formula for the sum*

$$\left\lfloor \frac{n+1}{2} \right\rfloor + \left\lfloor \frac{n+2}{2^2} \right\rfloor + \cdots + \left\lfloor \frac{n+2^k}{2^{k+1}} \right\rfloor + \cdots .$$

(10th International Mathematical Olympiad)

Solution. The sum is n. We rewrite the sum as

$$\left\lfloor \frac{n}{2} + \frac{1}{2} \right\rfloor + \left\lfloor \frac{n}{2^2} + \frac{1}{2} \right\rfloor + \cdots + \left\lfloor \frac{n}{2^{k+1}} + \frac{1}{2} \right\rfloor + \cdots ,$$

and use a special case of Hermite's identity:

$$\left\lfloor x + \frac{1}{2} \right\rfloor = \lfloor 2x \rfloor - \lfloor x \rfloor.$$

This allows us to write the sum as

$$\lfloor n \rfloor - \left\lfloor \frac{n}{2} \right\rfloor + \left\lfloor \frac{n}{2} \right\rfloor - \left\lfloor \frac{n}{2^2} \right\rfloor + \cdots + \left\lfloor \frac{n}{2^k} \right\rfloor - \left\lfloor \frac{n}{2^{k+1}} \right\rfloor + \cdots .$$

The sum telescopes, and $\lfloor n/2^{k+1} \rfloor = 0$ for large enough k's.

Problem 3.1.11. *Compute the sum*

$$\sum_{0 \le i < j \le n} \left\lfloor \frac{x+i}{j} \right\rfloor,$$

where x is a real number.

T. Andreescu and D. Andrica, *Number Theory*, DOI: 10.1007/b11856_14,
© Birkhäuser Boston, a part of Springer Science + Business Media, LLC 2009

First solution. Denote the sum in question by S_n. Then

$$S_n - S_{n-1} = \left\lfloor \frac{x}{n} \right\rfloor + \left\lfloor \frac{x+1}{n} \right\rfloor + \cdots + \left\lfloor \frac{x+n-1}{n} \right\rfloor$$

$$= \left\lfloor \frac{x}{n} \right\rfloor + \left\lfloor \frac{x}{n} + \frac{1}{n} \right\rfloor + \cdots + \left\lfloor \frac{x}{n} + \frac{n-1}{n} \right\rfloor,$$

and according to Hermite's identity,

$$S_n - S_{n-1} = \left\lfloor n\frac{x}{n} \right\rfloor = \lfloor x \rfloor.$$

Because $S_1 = \lfloor x \rfloor$, it follows that $S_n = n\lfloor x \rfloor$ for all n.

Second solution. By Hermite's identity applied to x/j we have

$$\sum_{i=0}^{j-1} \left\lfloor \frac{x+i}{j} \right\rfloor = \left\lfloor j\frac{x}{j} \right\rfloor = \lfloor x \rfloor.$$

Summing this over j gives

$$\sum_{j=1}^{n} \sum_{i=0}^{j-1} \left\lfloor \frac{x+i}{j} \right\rfloor = n\lfloor x \rfloor.$$

Problem 3.1.12. *Evaluate the difference between the numbers*

$$\sum_{k=0}^{2000} \left\lfloor \frac{3^k + 2000}{3^{k+1}} \right\rfloor \quad and \quad \sum_{k=0}^{2000} \left\lfloor \frac{3^k - 2000}{3^{k+1}} \right\rfloor.$$

Solution. We can write each term of the difference in question as

$$\left\lfloor \tfrac{1}{3} + v_k \right\rfloor - \left\lfloor \tfrac{1}{3} - v_k \right\rfloor,$$

where $v_k = 2000/3^{k+1}$. Since $-\lfloor u \rfloor = \lfloor -u \rfloor + 1$ for each nonintegral value of u, and since $\tfrac{1}{3} - v_k$ is never an integer, we have to examine the sum

$$\sum_{k=0}^{2000} \left(\left\lfloor v_k + \tfrac{1}{3} \right\rfloor + \left\lfloor v_k - \tfrac{1}{3} \right\rfloor + 1 \right).$$

Taking $n = 3$ and $x = v - \tfrac{1}{3}$ in (1) yields

$$\left\lfloor v + \tfrac{1}{3} \right\rfloor + \left\lfloor v - \tfrac{1}{3} \right\rfloor + 1 = \lfloor 3v \rfloor - \lfloor v \rfloor.$$

Hence the desired difference becomes

$$\sum_{k=0}^{2000} \left(\left\lfloor \frac{2000}{3^k} \right\rfloor - \left\lfloor \frac{2000}{3^{k+1}} \right\rfloor \right),$$

which telescopes to

$$\lfloor 2000 \rfloor - \left\lfloor \frac{2000}{3} \right\rfloor + \left\lfloor \frac{2000}{3} \right\rfloor - \left\lfloor \frac{2000}{3^2} \right\rfloor + \cdots = 2000.$$

Problem 3.1.13. *(a) Prove that there are infinitely many rational positive numbers x such that*

$$\{x^2\} + \{x\} = 0.99.$$

(b) Prove that there are no rational numbers x > 0 such that

$$\{x^2\} + \{x\} = 1.$$

(2004 Romanian Mathematical Olympiad)

Solution. (a) Since $0.99 = \frac{99}{100}$, it is natural to look for a rational x of the form $\frac{n}{10}$, for some positive integer n. It is not difficult to see that $x = \frac{13}{10}$ satisfies the given equality and then that $x = 10k + \frac{13}{10}$ also satisfies the equality for any positive integer k.

(b) Suppose that $x = p/q$, with p, q positive integers, $\gcd(p, q) = 1$, satisfies $\{x^2\} + \{x\} = 1$. We can see that $\frac{p^2 + pq - q^2}{q^2} = x^2 + x - 1 \in \mathbb{Z}$; thus $q \mid p^2$, and since $\gcd(p, q) = 1$, one has $q = 1$. Thus $x \in \mathbb{Z}$, and this is obviously impossible.

Problem 3.1.14. *Show that the fractional part of the number $\sqrt{4n^2 + n}$ is not greater than 0.25.*

(2003 Romanian Mathematical Olympiad)

Solution. From the inequalities $4n^2 < 4n^2 + n < 4n^2 + n + \frac{1}{16}$ one obtains $2n < \sqrt{4n^2 + n} < 2n + \frac{1}{4}$. So $\lfloor \sqrt{4n^2 + n} \rfloor = 2n$ and $\{\sqrt{4n^2 + n}\} < 1/4$.

Problem 3.1.15. *Prove that for every natural number n,*

$$\sum_{k=1}^{n^2} \{\sqrt{k}\} \leq \frac{n^2 - 1}{2}.$$

(1999 Russian Mathematical Olympiad)

Solution. We prove the claim by induction on n. For $n = 1$, we have $0 \le 0$. Now supposing that the claim is true for n, we prove that it is true for $n + 1$.

Each of the numbers $\sqrt{n^2 + 1}, \sqrt{n^2 + 2}, \ldots, \sqrt{n^2 + 2n}$ is between n and $n + 1$. Thus

$$\{\sqrt{n^2 + i}\} = \sqrt{n^2 + i} - n < \sqrt{n^2 + i + \frac{i^2}{4n^2}} - n = \frac{i}{2n}, \quad i = 1, 2, \ldots, 2n.$$

Therefore we have

$$\sum_{k=1}^{(n+1)^2} \{\sqrt{k}\} = \sum_{k=1}^{n^2} \{\sqrt{k}\} + \sum_{k=n^2+1}^{(n+1)^2} \{\sqrt{k}\} < \frac{n^2 - 1}{2} + \frac{1}{2n} \sum_{i=1}^{2n} i + 0$$

$$= \frac{n^2 - 1}{2} + \frac{2n + 1}{2} = \frac{(n + 1)^2 - 1}{2},$$

completing the inductive step and the proof.

Problem 3.1.16. *The rational numbers* $\alpha_1, \ldots, \alpha_n$ *satisfy*

$$\sum_{i=1}^{n} \{k\alpha_i\} < \frac{n}{2}$$

for any positive integer k.
  (a) Prove that at least one of $\alpha_1, \ldots, \alpha_n$ *is an integer.*
  (b) Do there exist $\alpha_1, \ldots, \alpha_n$ *that satisfy, for every positive integer* k,

$$\sum_{i=1}^{n} \{k\alpha_i\} \le \frac{n}{2},$$

such that no α_i *is an integer?*

(2002 Belarusian Mathematical Olympiad)

Solution. (a) Assume the contrary. The problem would not change if we replace α_i with $\{\alpha_i\}$. So we may assume $0 < \alpha_i < 1$ for all $1 \le i \le n$. Because α_i is rational, let $\alpha_i = \frac{p_i}{q_i}$, and $D = \prod_{i=1}^{n} q_i$. Because $(D - 1)\alpha_i + \alpha_i = D\alpha_i$ is an integer, and α_i is not an integer, $\{(D - 1)\alpha_i\} + \{\alpha_i\} = 1$. Then

$$n > \sum_{i=1}^{n} \{(D - 1)\alpha_i\} + \sum_{i=1}^{n} \{\alpha_i\} = \sum_{i=1}^{n} 1 = n,$$

contradiction. Therefore, one of the α_i has to be an integer.

  (b) Yes. Let $\alpha_i = \frac{1}{2}$ for all i. Then $\sum_{i=1}^{n} \{k\alpha_i\} = 0$ when k is even and $\sum_{i=1}^{n} \{k\alpha_i\} = \frac{n}{2}$ when k is odd.

3.2 Floor Function and Integer Points

Problem 3.2.3. *Prove that*

$$\sum_{k=1}^{n}\left\lfloor\frac{n^2}{k^2}\right\rfloor = \sum_{k=1}^{n^2}\left\lfloor\frac{n}{\sqrt{k}}\right\rfloor$$

for all integers $n \geq 1$.

Solution. Consider the function $f : [1, n] \to [1, n^2]$,

$$f(x) = \frac{n^2}{x^2}.$$

Note that f is decreasing and bijective, and

$$f^{-1}(x) = \frac{n}{\sqrt{x}}.$$

Using the formula in Theorem 3.2.3, we obtain

$$\sum_{k=1}^{n}\left\lfloor\frac{n^2}{k^2}\right\rfloor - \sum_{k=1}^{n^2}\left\lfloor\frac{n}{\sqrt{k}}\right\rfloor = n\lceil 1 - 1\rceil - n^2\lceil 1 - 1\rceil = 0,$$

hence

$$\sum_{k=1}^{n}\left\lfloor\frac{n^2}{k^2}\right\rfloor = \sum_{k=1}^{n^2}\left\lfloor\frac{n}{\sqrt{k}}\right\rfloor, \quad n \geq 1,$$

as desired.

Problem 3.2.4. *Let θ be a positive irrational number. Then, for any positive integer m,*

$$\sum_{k=1}^{m}\lfloor k\theta\rfloor + \sum_{k=1}^{\lfloor m\theta\rfloor}\left\lfloor\frac{k}{\theta}\right\rfloor = m\lfloor m\theta\rfloor.$$

Solution. Consider the function $f : [0, m] \to [0, m\theta]$, $f(x) = \theta x$. Because θ is irrational, we have that $n(G_f) = 1$ cancels the $\lceil a - 1\rceil\lceil c - 1\rceil = (-1)^2 = 1$ term, and the conclusion follows from Theorem 3.2.1.

Problem 3.2.5. *Let p and q be relatively prime positive integers and let m be a real number such that $1 \leq m < p$.*
    (1) If $s = \lfloor\frac{mq}{p}\rfloor$, then

$$\sum_{k=1}^{\lfloor m\rfloor}\left\lfloor\frac{kq}{p}\right\rfloor + \sum_{k=1}^{s}\left\lfloor\frac{kp}{q}\right\rfloor = \lfloor m\rfloor s.$$

(2) (Landau) If p and q are odd, then

$$\sum_{k=1}^{\frac{p-1}{2}} \left\lfloor \frac{kq}{p} \right\rfloor + \sum_{k=1}^{\frac{q-1}{2}} \left\lfloor \frac{kp}{q} \right\rfloor = \frac{(p-1)(q-1)}{4}.$$

Solution. (1) Let $f : [0, m] \to [0, \frac{mq}{p}]$, $f(x) = \frac{q}{p}x$. Because $\gcd(p, q) = 1$ and $m < p$, we have $n(G_f) = 1$, and the desired equality follows from Theorem 3.2.1.

(2) In the previous identity we take $m = \frac{p}{2}$. It follows that $s = \frac{q-1}{2}$, and the conclusion follows.

3.3   A Useful Result

Problem 3.3.3. *Let p be an odd prime and let q be an integer that is not divisible by p. Show that*

$$\sum_{k=1}^{p-1} \left\lfloor (-1)^k k^2 \frac{q}{p} \right\rfloor = \frac{(p-1)(q-1)}{2}.$$

(2005 "Alexandru Myller" Romanian Regional Contest)

Solution. For $f : \mathbb{Z}_+^* \to \mathbb{R}$, $f(s) = (-1)^s s^2$, conditions (i) and (ii) in Theorem 3.3.1 are both satisfied. We obtain

$$\sum_{k=1}^{p-1} \left\lfloor (-1)^k k^2 \frac{q}{p} \right\rfloor = \frac{q}{p}(-1^2 + 2^2 - \cdots + (p-1)^2) - \frac{p-1}{2}$$

$$= \frac{q}{p} \cdot \frac{p(p-1)}{2} - \frac{p-1}{2};$$

hence

$$\sum_{k=1}^{p-1} \left\lfloor (-1)^k k^2 \frac{q}{p} \right\rfloor = \frac{(p-1)(q-1)}{2}.$$

Remarks. (1) By taking $q = 1$ we get

$$\sum_{k=1}^{p-1} \left\lfloor (-1)^k \frac{k^2}{p} \right\rfloor = 0.$$

Using now the identity $\lfloor -x \rfloor = 1 - \lfloor x \rfloor$, $x \in \mathbb{R}$, the last display takes the form

$$\sum_{k=1}^{p-1} (-1)^k \left\lfloor \frac{k^2}{p} \right\rfloor = \frac{1-p}{2}.$$

(2) Similarly, applying Theorem 3.3.1 to $f : \mathbb{Z}_+^* \to \mathbb{R}$, $f(s) = (-1)^s s^4$ yields

$$\sum_{k=1}^{p-1} \left\lfloor (-1)^k k^4 \frac{q}{p} \right\rfloor = \frac{q(p-1)(p^2-p-1)}{2} - \frac{p-1}{2}.$$

Taking $q = 1$ gives

$$\sum_{k=1}^{p-1} \left\lfloor (-1)^k \frac{k^4}{p} \right\rfloor = \frac{(p-2)(p-1)(p+1)}{2}.$$

Problem 3.3.4. *Let p be an odd prime. Show that*

$$\sum_{k=1}^{p-1} \frac{k^p - k}{p} \equiv \frac{p+1}{2} \quad (\text{mod } p).$$

(2006 "Alexandru Myller" Romanian Regional Contest)

Solution. For $f(s) = \frac{s^p}{p}$, conditions (i) and (ii) in Theorem 3.3.1 are also satisfied, and for $q = 1$ we have

$$\sum_{k=1}^{p-1} \left\lfloor \frac{k^p}{p^2} \right\rfloor = \frac{1}{p} \sum_{k=1}^{p-1} \frac{k^p}{p} - \frac{p-1}{2}$$

$$= \frac{1}{p} \sum_{k=1}^{p-1} \frac{k^p}{p} - \frac{1}{p^2} \sum_{k=1}^{p-1} k + \frac{1}{p^2} \frac{p(p-1)}{2} - \frac{p-1}{2}$$

$$= \frac{1}{p} \sum_{k=1}^{p-1} \frac{k^p - k}{p} - \frac{1}{p} \cdot \frac{(p-1)^2}{2}.$$

It follows that

$$\sum_{k=1}^{p-1} \frac{k^p - k}{p} - \frac{(p-1)^2}{2} = p \sum_{k=1}^{p} \left\lfloor \frac{k^p}{p^2} \right\rfloor,$$

i.e.,

$$\sum_{k=1}^{p-1} \frac{k^p - k}{p} \equiv \frac{(p-1)^2}{2} \quad (\text{mod } p).$$

The conclusion follows since

$$\frac{(p-1)^2}{2} \equiv \frac{p^2+1}{2} \equiv \frac{p+1}{2} \quad (\text{mod } p).$$

Remarks. (1) For each $k = 1, 2, \ldots, p - 1$ denote by r_k the remainder when k^p is divided by p^2. We have

$$k^p = \left\lfloor \frac{k^p}{p^2} \right\rfloor p^2 + r_k, \quad k = 1, 2, \ldots, p - 1,$$

hence

$$\sum_{k=1}^{p-1} k^p = p^2 \sum_{k=1}^{p-1} \left\lfloor \frac{k^p}{p^2} \right\rfloor + \sum_{k=1}^{p-1} r_k = -\frac{p^2(p-1)}{2} + \sum_{k=1}^{p-1} r_k + \sum_{k=1}^{p-1} k^p.$$

It follows that

$$r_1 + r_2 + \cdots + r_{p-1} = \frac{p^2(p-1)}{2}.$$

(2) The formula in our problem shows that the sum of the quotients obtained when $k^p - k$ is divided by p (Fermat's little theorem) is congruent to $\frac{p+1}{2}$ modulo p.

4

Digits of Numbers

4.1 The Last Digits of a Number

Problem 4.1.4. *In how may zeros can the number $1^n + 2^n + 3^n + 4^n$ end for $n \in \mathbb{N}$?*

(1998 St. Petersburg City Mathematical Olympiad)

Solution. There can be no zeros (e.g., $n = 4$), one zero ($n = 1$), or two zeros ($n = 2$). In fact, for $n \geq 3$, 2^n and 4^n are divisible by 8, while $1^n + 3^n$ is congruent to 2 or 4 mod 8. Thus the sum cannot end in three or more zeros.

Problem 4.1.5. *Find the last five digits of the number 5^{1981}.*

Solution. First, we prove that $5^{1981} = 5^5 \pmod{10^5}$. We have

$$
\begin{aligned}
5^{1981} - 5^5 &= (5^{1976} - 1)5^5 = 5^5[(5^8)^{247} - 1] \\
&= \mathcal{M}[5^5(5^8 - 1)] = \mathcal{M}[5^5(5^4 - 1)(5^4 + 1)] \\
&= \mathcal{M}[5^5(5 - 1)(5 + 1)(5^2 + 1)(5^4 + 1)] \\
&= \mathcal{M}5^2 2^5 = \mathcal{M}100{,}000.
\end{aligned}
$$

Therefore $5^{1981} = \mathcal{M}100{,}000 + 5^5 = \mathcal{M}100{,}000 + 3125$, so 03125 are the last five digits of the number 5^{1981}. Of course, the relation $a = \mathcal{M}b$ means that a is a multiple of b.

Problem 4.1.6. *Consider all pairs (a, b) of natural numbers such that the product $a^a b^b$, written in base 10, ends with exactly 98 zeros. Find the pair (a, b) for which the product ab is smallest.*

(1998 Austrian–Polish Mathematics Competition)

Solution. Let a_2 be the maximum integer such that $2^{a_2} \mid a$. Define a_5, b_2, and b_5 similarly. Our task translates into the following: find a, b such that

T. Andreescu and D. Andrica, *Number Theory*, DOI: 10.1007/b11856_15,
© Birkhäuser Boston, a part of Springer Science + Business Media, LLC 2009

$\min\{a_5a + b_5b, a_2a + b_2b\} = 98$ and ab is minimal. Since $5 \mid a_5a + b_5b$, we have $a_5a + b_5b > 98$ and $\min\{a_5a + b_5b, a_2a + b_2b\} = a_2a + b_2b = 98$. Note that if $5 \mid \gcd(a, b)$, then $a_2a + b_2b \neq 98$, contradiction. Without loss of generality, suppose that $a_5 \geq 1$ and $b_5 = 0$. Let $a = 2^{a_2}5^{a_5}x$ and $2^{b_2}y$, $\gcd(2, x) = \gcd(5, x) = \gcd(2, y) = 1$. Then $a_5a = a_5(2^{a_2}5^{a_5}x) > 98$ and $a_2a = a_2(2^{a_2}5^{a_5}x) \leq 98$. So $a_5 > a_2$. We consider the following cases.

(a) $a_2 = 0$. Then $b_2(2^{b_2}y) = 98$. So $b_2 = 1$, $y = 49$, $b = 98$. Since $a_5(5^{a_5}x) \geq 98$ and x is odd, $a = 5^{a_5}x \geq 125$ for $a_5 \geq 3$; $x \geq 3$ and $a \geq 75$ for $a_5 = 2$; $x \geq 21$ and $a \geq 105$ for $a_5 = 1$. Hence for $a_2 = 0$, $b = 98$, $a \geq 75$.

(b) $a_2 \geq 1$. Then $a_5 \geq 2$. We have $2^{a_2}5^{a_5}x \leq 98$ and $5^{a_5}x \leq 49$. Thus $a_5 = 2$, $x = 1$, $a_2 = 1$, $a = 50$. Then $b_2b = 48$. Let $b = 2^{b_2}y$. Then $b_2(2^{b_2}y) = 48$, which is impossible.

From the above, we have $(a, b) = (75, 98)$ or $(98, 75)$.

4.2 The Sum of the Digits of a Number

Problem 4.2.7. *Show that there exist infinitely many natural numbers n such that $S(3^n) \geq S(3^{n+1})$.*

(1997 Russian Mathematical Olympiad)

Solution. If $S(3^n) < S(3^{n+1})$ for large n, we have (since powers of 3 are divisible by 9, as are their digit sums) $S(3^n) \leq S(3^{n+1}) - 9$. Thus $S(3^n) \geq 9(n - c)$ for some c, which is eventually a contradiction, since for large n, $3^n < 10^{n-c}$.

Problem 4.2.8. *Do there exist three natural numbers a, b, c such that $S(a+b) < 5$, $S(b+c) < 5$, $S(c+a) < 5$, but $S(a+b+c) > 50$?*

(1998 Russian Mathematical Olympiad)

Solution. The answer is yes. It is easier to focus on the numbers $a + b$, $b + c$, $c + a$ instead. Each of these has digit sum at most 4. Hence their sum $2(a+b+c)$ has digit sum at most 12. However, half this $a + b + c$ has digit sum at least 51. The only way this can happen is if $a + b + c$ has digits either ten 5's and a 1 or nine 5's and a 6. Trying the former, we take $a + b + c = 105555555555$ and $2(a + b + c) = 211111111110$ (many other choices also work). Each of $a + b$, $b + c$, and $c + a$ must have digit sum 4, and they must add to $2(a + b + c)$, so there can be no carries. One such choice is

$$a + b = 100001110000, \quad b + c = 11110000000, \quad c + a = 100000001110.$$

From these we get

$$a = 105555555555 - 11110000000 = 94445555555,$$
$$b = 105555555555 - 100000001110 = 5555554445,$$
$$c = 105555555555 - 100001110000 = 5554445555.$$

Problem 4.2.9. *Prove that there exist distinct positive integers $\{n_i\}_{1\leq i\leq 50}$ such that*

$$n_1 + S(n_1) = n_2 + S(n_2) = \cdots = n_{50} + S(n_{50}).$$

(1999 Polish Mathematical Olympiad)

Solution. We show by induction on k that there exist positive integers $n_1, \ldots, n_k$ with the desired property. For $k = 1$ the statement is obvious. For $k > 1$, let $m_1 < \cdots < m_{k-1}$ satisfy the induction hypothesis for $k - 1$. Note that we can make all the m_i arbitrarily large by adding some large power of 10 to all of them, which preserves the described property. Then, choose m with $1 \leq m \leq 9$ and $m \equiv m_1 + 1 \pmod 9$. Observing that $S(x) \equiv x \pmod 9$, we have $m_1 - m + S(m_1) - S(m) + 11 = 9l$ for some integer l. By choosing the m_i large enough we can ensure that $10^l > m_{k-1}$. Now let $n_i = 10^{l+1} + m_i$ for $i < k$ and $n_k = m + 10^{l+1} - 10$. It is obvious that $n_i + S(n_i) = n_j + S(n_j)$ for $i, j < k$, and

$$
\begin{aligned}
n_1 + S(n_1) &= (10^{l+1} + m_1) + (1 + S(m + 1)) \\
&= (m_1 + S(m_1) + 1) + 10^{l+1} \\
&= (9l + S(m) + m - 10) + 10^{l+1} \\
&= (m + 10^{l+1} - 10) + (9l + S(m)) \\
&= n_k + S(n_k),
\end{aligned}
$$

as needed.

Problem 4.2.10. *The sum of the decimal digits of the natural number n is 100, and that of $44n$ is 800. What is the sum of the digits of $3n$?*

(1999 Russian Mathematical Olympiad)

Solution. The sum of the digits of $3n$ is 300.

Suppose that d is a digit between 0 and 9, inclusive. If $d \leq 2$ then $S(44d) = 8d$, and if $d = 3$ then $S(8d) = 6 < 8d$. If $d \geq 4$, then $44d \leq 44(9)$ has at most three digits so that $S(44d) \leq 27 < 8d$.

Now write $n = \sum n_i \cdot 10^i$, so that the n_i are the digits of n in base 10. Then

$$
\sum 8n_i = S(44n) \leq \sum S(44n_i \cdot 10^i)
$$
$$
= \sum S(44n_i) \leq \sum 8n_i,
$$

so equality must occur in the second inequality, that is, each of the n_i must equal 0, 1, or 2. Then each digit of $3n$ is simply three times the corresponding digit of n, and $S(3n) = 3S(n) = 300$, as claimed.

Alternative solution. Using properties (3) and (5) involving the sum of digits, we have

$$S(3n) \leq 3S(n) = 300$$

and

$$800 = S(11 \cdot 3n + 11n) \leq S(11 \cdot 3n) + S(11n)$$
$$\leq S(11)S(3n) + S(11)S(n) = 2S(3n) + 200,$$

whence $S(3n) \geq 300$. Thus, $S(3n) = 300$.

Problem 4.2.11. *Consider all numbers of the form $3n^2 + n + 1$, where n is a positive integer.*

(a) How small can the sum of the digits (in base 10) of such a number be?

(b) Can such a number have the sum of its digits (in base 10) equal to 1999?

(1999 United Kingdom Mathematical Olympiad)

Solution. (a) Let $f(n) = 3n^2 + n + 1$. When $n = 8$, the sum of the digits of $f(8) = 201$ is 3. Suppose that there were some m such that $f(m)$ had a smaller sum of digits. Then the last digit of $f(m)$ must be either 0, 1, or 2. Because $f(n) \equiv 1 \pmod 2$ for all n, $f(m)$ must have units digit 1.

Because $f(n)$ can never equal 1, this means we must have $3m^2 + m + 1 = 10^k + 1$ for some positive integer k, and $m(3m + 1) = 10^k$. Because m and $3m + 1$ are relatively prime, and $m < 3m+1$, we must have either $(m, 3m+1) = (1, 10^k)$, which is impossible, or $(m, 3m + 1) = (2^k, 5^k)$. For $k = 1$, $5^k \neq 3 \cdot 2^k + 1$; for $k > 1$, we have

$$5^k = 5^{k-2} \cdot 25 > 2^{k-2} \cdot (12 + 1) \geq 3 \cdot 2^k + 1.$$

Therefore, $f(m)$ cannot equal $10^k + 1$, and 3 is indeed the minimum value for the sum of digits.

(b) Consider $n = 10^{222} - 1$. Then

$$f(n) = 3 \cdot 10^{444} - 6 \cdot 10^{222} + 3 + 10^{222}.$$

Thus, its decimal expansion is

$$2\underbrace{9 \ldots 9}_{221}5\underbrace{0 \ldots 0}_{221}3,$$

and the sum of digits in $f(10^{222} - 1)$ is 1999.

Problem 4.2.12. *Consider the set A of all positive integers n with the following properties: the decimal expansion contains no 0, and the sum of the (decimal) digits of n divides n.*

(a) Prove that there exist infinitely many elements in A with the following property: the digits that appear in the decimal expansion of A appear the same number of times.

(b) Show that for each positive integer k, there exists an element in A with exactly k digits.

(2001 Austrian–Polish Mathematics Competition)

Solution. (a) We can take $n_k = \underbrace{11\ldots1}_{3^k \text{ times}}$ and prove by induction that $3^{k+2} \mid 10^{3^k} - 1$. Alternatively, one can observe that

$$10^{3^k} - 1 = (10 - 1)(10^2 + 10 + 1)(10^{2\cdot3} + 10^3 + 1) \cdots (10^{2\cdot3^{k-1}} + 10^{3^{k-1}} + 1)$$

and that $9 \mid 10 - 1$ and $3 \mid 10^{2\cdot3^i} + 10^{3^i} + 1$ for $0 \le i \le k - 1$.

(b) We will need the following lemmas.

Lemma 1. *For every $d > 0$ there exists a d-digit number that contains only ones and twos in its decimal expansion and is a multiple of 2^d.*

Proof. Exactly in the same way as in the proof of Theorem 1.7.1 one can prove that any two d-digit numbers that have only ones and twos give different residues mod 2^d. Since there are 2^d such numbers, one of them is a multiple of 2^d. $\square$

Lemma 2. *For each $k > 2$ there exists $d \le k$ such that the following inequality holds: $k + d \le 2^d \le 9k - 8d$.*

Proof. For $3 \le k \le 5$, $d = 3$ satisfies the inequalities. For $5 \le k \le 10$, $d = 4$ satisfies the inequalities. We will show that $d = \lfloor \log_2 4k \rfloor$ satisfies the inequality for all $k > 10$. If $k > 3$, then $\log_2 4k \le 2^k$, so $d < k$. Additionally, $k + d \le 2k \le 2^d$. If $k > 10$, then $16k^2 \le 2^k$, so $4k \le 2^{k/2} \le 2^{5k/8}$, $d \le \log_2 4k \le \frac{5}{8}n$, and $8k - 8d \ge 4k \ge 2^d$. $\square$

Now return to the original problem. For $k = 1$, $n = 1$ has the desired property. For $k = 2$, $n = 12$ has the desired property. Now, for each $k > 2$ we have some number d satisfying the condition Lemma 2. Consider a k-digit integer n such that the last d digits of n have the property described in the first lemma. We can choose each of the other digits of n to be any number between zero and nine. We know that the sum of the last d digits of n is between d and $2d$, and we can choose the sum of the other $k - d$ digits to be any number between $k - d$ and $9(k - d)$. Since $k - d + 2d \le 2^d \le 9(k - d) + d$, we can choose the other digits such that the sum of the digits of n is 2^d. This completes the proof because n is a multiple of 2^d.

Remarks. (1) Suppose $3^m \le k < 3^{m+1}$ and choose an integer r with $k + 1 - 3^m$ decimal digits and $S(r) = 3$, for example, $r = 10^{k-3^m} + 2$. Then the desired number is $n = r \cdot 111\ldots1$, with 3^m ones. Since $S(r) = 3$, $3 \mid r$, and we saw in part (a) that $3^m \mid 111\ldots1$. Hence $3^{m+1} \mid n$. Also since $S(r) = 3 < 10$, no carries occur in multiplying out to compute n. Hence n has k decimal digits and $S(n) = 3^m S(r) = 3^{m+1}$.

(2) A number divisible by the sum of its digits is called a *Niven*[1] *number*. It has been proved recently that the number of Niven numbers smaller than x is $\left(\frac{14}{27}\log 10 + o(1)\right)\frac{x}{\log x}$. The courageous reader may try to prove that there are arbitrarily long sequences of consecutive numbers that are not Niven numbers (which is easily implied by the above result; yet there is an elementary proof). For more details one can read the article "Large and small gaps between consecutive Niven numbers," *Journal of Integer Sequences*, 6 (2003), by J.-M. Koninck and N. Doyon.

4.3   Other Problems Involving Digits

Problem 4.3.3. *A wobbly number is a positive integer whose digits in base 10 are alternately nonzero and zero, the units digit being nonzero. Determine all positive integers that do not divide any wobbly number.*

(35th International Mathematical Olympiad Shortlist)

Solution. If n is a multiple of 10, then the last digit of any multiple of n is 0. Hence it is not wobbly. If n is a multiple of 25, then the last two digits of any multiple of n are 25, 50, 75 or 00. Hence it is not wobbly. We now prove that these are the only numbers not dividing any wobbly number.

We first consider odd numbers m not divisible by 5. Then $\gcd(m, 10) = 1$, and we have $\gcd((10^k - 1)m, 10) = 1$, for any $k \geq 1$. It follows that there exists a positive integer l such that $10^l \equiv 1 \pmod{(10^k - 1)m}$, and we have $10^{kl} \equiv 1 \pmod{(10^k - 1)m}$. Now

$$10^{kl} - 1 = (10^k - 1)(10^{k(l-1)} + 10^{k(l-2)} + \cdots + 10^k + 1).$$

Hence $x_k = 10^{k(l-1)} + 10^{k(l-2)} + \cdots + 10^k + 1$ is a multiple of m for any $k \geq 1$. In particular, x_2 is a wobbly multiple of m. If m is divisible by 5, then $5x_2$ is a wobbly multiple of m.

Next, we consider powers of 2. We prove by induction on t that 2^{2t+1} has a wobbly multiple w_t with precisely t nonzero digits. For $t = 1$, take $w_1 = 8$. Suppose w_t exists for some $t \geq 1$. Then $w_t = 2^{2t+1}d$ for some d. Let $w_{t+1} = 10^{2t}c + w_t$, where $c \in \{1, 2, 3, \ldots, 9\}$ is to be chosen later. Clearly, w_{t+1} is wobbly, and has precisely $t + 1$ nonzero digits. Since $w_{t+1} = 2^{2t}(5^{2t}c + 2d)$, it is divisible by 2^{2t+3} if and only if $5^{2t}c + 2d \equiv 0 \pmod 8$ or $c \equiv 6d \pmod 8$. We

---

[1] Ivan Niven (1915–1999), Canadian mathematician with contributions in the areas of Diophantine approximation, the study of irrationality and transcendence of numbers, and combinatorics.

can always choose c to be one of 8, 6, 4, and 2 in order to satisfy this congruence. Thus the inductive argument is completed. It now follows that every power of 2 has a wobbly multiple.

Finally, consider numbers of the form $2^t m$, where $t \geq 1$ and $\gcd(m, 10) = 1$. Such a number has $w_t x_{2t}$ as a wobbly multiple.

Problem 4.3.4. *A positive integer is called monotonic if its digits in base 10, read from left right, are in nondecreasing order. Prove that for each $n \in \mathbb{N}$, there exists an n-digit monotonic number that is a perfect square.*

$$\text{(2000 Belarusian Mathematical Olympiad)}$$

Solution. Any 1-digit perfect square (namely, 1, 4, or 9) is monotonic, proving the claim for $n = 1$. We now assume $n > 1$.

If n is odd, write $n = 2k - 1$ for an integer $k \geq 2$, and let

$$x_k = (10^k + 2)/6 = 1 \underbrace{66 \ldots 6}_{k-2} 7.$$

Then

$$x_k^2 = \frac{10^{2k} + 4 \cdot 10^k + 4}{36} = \frac{10^{2k}}{36} + \frac{10^k}{9} + \frac{1}{9}. \tag{1}$$

Observe that

$$\frac{10^{2k}}{36} = 10^{2k-2}\left(\frac{72}{36} + \frac{28}{36}\right)$$

$$2 \cdot 10^{2k-2} + 10^{2k-2} \cdot \frac{7}{9} = 2 \underbrace{77 \ldots 7}_{2k-2} + \frac{7}{9}.$$

Thus, the right-hand side of (1) equals

$$\left(2 \underbrace{77 \ldots 7}_{2k-2} + \frac{7}{9}\right) + \left(\underbrace{11 \ldots 1}_{k} + \frac{1}{9}\right) + \frac{1}{9} = 2 \underbrace{77 \ldots 7}_{k-2} \underbrace{88 \ldots 8}_{k-1} 9,$$

an n-digit monotonic perfect square.

If n is even, write $n = 2k$ for an integer $k \geq 1$, and let

$$y_k = \frac{10^k + 2}{3} = \underbrace{33 \ldots 3}_{k-1} 4.$$

Then

$$y_k^2 = \frac{1}{9}(10^{2k} + 4 \cdot 10^k + 4) = \frac{10^{2k}}{9} + 4 \cdot \frac{10^k}{9} + \frac{4}{9}$$

$$= \left(\underbrace{11 \ldots 1}_{2k} + \frac{1}{9}\right) + \left(\underbrace{44 \ldots 4}_{k} + \frac{4}{9}\right) + \frac{4}{9} = \underbrace{11 \ldots 1}_{k} \underbrace{55 \ldots 5}_{k-1} 6,$$

an n-digit monotonic perfect square. This completes the proof.

5

Basic Principles in Number Theory

5.1   Two Simple Principles

Problem 5.1.7. *Let $n_1 < n_2 < \cdots < n_{2000} < 10^{100}$ be positive integers. Prove that one can find two disjoint subsets A and B of $\{n_1, n_2, \ldots, n_{2000}\}$ such that*

$$|A| = |B|, \quad \sum_{x \in A} x = \sum_{x \in B} x, \quad and \quad \sum_{x \in A} x^2 = \sum_{x \in B} x^2.$$

(2001 Polish Mathematical Olympiad)

Solution. Given any subset $S \subseteq \{n_1, n_2, \ldots, n_{2000}\}$ of size 1000, we have

$$0 < \sum_{x \in S} x < 1000 \cdot 10^{100},$$

$$0 < \sum_{x \in S} x^2 < 1000 \cdot 10^{200}.$$

Thus, as S varies, there are fewer than $(1000 \cdot 10^{100})(1000 \cdot 10^{200}) = 10^{306}$ values of $\left(\sum_{x \in S} x, \sum_{x \in S} x^2 \right)$.

Because $\sum_{k=0}^{2000} \binom{2000}{k} = 2^{2000}$ and $\binom{2000}{1000}$ is the biggest term in the sum, $\binom{2000}{1000} > \frac{2^{2000}}{2001}$. There are

$$\binom{2000}{1000} > \frac{2^{2000}}{2001} > \frac{10^{600}}{2001} > 10^{306}$$

distinct subsets of size 1000. By the pigeonhole principle, there exist distinct subsets C and D of size 1000 such that $\sum_{x \in C} x^2 = \sum_{x \in D} x^2$ and $\sum_{x \in C} x = \sum_{x \in D} x$. Removing the common elements from C and D yields sets A and B with the required properties.

T. Andreescu and D. Andrica, *Number Theory*, DOI: 10.1007/b11856_16,
© Birkhäuser Boston, a part of Springer Science + Business Media, LLC 2009

Problem 5.1.8. *Find the greatest positive integer n for which there exist n nonnegative integers $x_1, x_2, \ldots, x_n$, not all zero, such that for any sequence $\varepsilon_1, \varepsilon_2, \ldots, \varepsilon_n$ of elements $\{-1, 0, 1\}$, not all zero, n^3 does not divide $\varepsilon_1 x_1 + \varepsilon_2 x_2 + \cdots + \varepsilon_n x_n$.*

(1996 Romanian Mathematical Olympiad)

Solution. The statement holds for $n = 9$ by choosing $1, 2, 2^2, \ldots, 2^8$, since in that case

$$|\varepsilon_1 + \cdots + \varepsilon_9 2^8| \leq 1 + 2 + \cdots + 2^8 < 9^3.$$

However, if $n \geq 10$, then $2^{10} > 10^3$, so by the pigeonhole principle, there are two subsets A and B of $\{x_1, \ldots, x_{10}\}$ whose sums are congruent modulo 10^3. Let $\varepsilon_i = 1$ if x_i occurs in A but not in B, -1 if x_i occurs in B but not in A, and 0 otherwise; then $\sum \varepsilon_i x_i$ is divisible by n^3.

Problem 5.1.9. *Given a positive integer n, prove that there exists $\varepsilon > 0$ such that for any n positive real numbers $a_1, a_2, \ldots, a_n$, there exists a real number $t > 0$ such that*

$$\varepsilon < \{ta_1\}, \{ta_2\}, \ldots, \{ta_n\} < \frac{1}{2}.$$

(1998 St. Petersburg City Mathematical Olympiad)

Solution. More generally, we prove by induction on n that for any real number $0 < r < 1$, there exists $0 < \varepsilon < r$ such that for $a_1, \ldots, a_n$ any positive real numbers, there exists $t > 0$ with

$$\{ta_1\}, \ldots, \{ta_n\} \in (\varepsilon, r).$$

The case $n = 1$ needs no further comment.

Assume without loss of generality that a_n is the largest of the a_i. By hypothesis, for any $r' > 0$ (which we will specify later) there exists $\varepsilon' > 0$ such that for any $a_1, \ldots, a_{n-1} > 0$, there exists $t' > 0$ such that

$$\{t'a_1\}, \ldots, \{t'a_{n-1}\} \in (\varepsilon', r').$$

Let N be an integer also to be specified later. A standard argument using the pigeonhole principle shows that one of $t'a_n, 2t'a_n, \ldots, Nt'a_n$ has fractional part in $(-1/N, 1/N)$. Let $st'a_n$ be one such term, and take $t = st' + c$ for $c = (r - 1/N)/a_n$. Then

$$ta_n - \lfloor st's_n \rfloor \in (r - 2/N, r).$$

So we choose N such that $0 < r - 2/N$, thus making $\{ta_n\} \in (r - 2/N, r)$. Note that this choice of N makes $c > 0$ and $t > 0$, as well.

As for the other ta_i, for each i we have $k_i + \varepsilon' < t'a_i < k_i + r'$ for some integer k_i, so $sk_i + s\varepsilon' < st'a_i < sk_i + sr'$ and

$$sk_i + \varepsilon' < (st' + c)a_i < sk_i + sr' + \frac{a_i(r - 1/N)}{a_n} \le sk_i + Nr' + r - 1/N.$$

So we choose r' such that $Nr' - 1/N < 0$, thus making $\{ta_i\} \in (\varepsilon', r)$. Therefore, letting $\varepsilon = \min\{r - 2/N, \varepsilon'\}$, we have

$$0 < \varepsilon < \{ta_1\}, \{ta_2\}, \ldots, \{ta_n\} < r$$

for any choices of a_i. This completes the inductive step, and the claim is true for all natural numbers n.

Problem 5.1.10. *We have 2^n prime numbers written on the blackboard in a line. We know that there are fewer than n different prime numbers on the blackboard. Prove that there is a subsequence of consecutive numbers in that line whose product is a perfect square.*

Solution. Suppose that $p_1, p_2, \ldots, p_m$ $(m < n)$ are primes that we met in the sequence $a_1, a_2, \ldots, a_{2^n}$ written on the blackboard. It is enough to prove that there is a subsequence in which each prime occurs an even number of times. Denote by c_{ij} the exponent of the prime p_i $(1 \le i \le m)$ in the product $a_1 \cdots a_2 \cdots a_j$ of the first j numbers from our sequence. Let d_{ij} be the residue modulo 2 of c_{ij}, so we can write $c_{ij} = 2t_{ij} + d_{ij}$, $d_{ij} \in \{0, 1\}$. Every system $(d_{1j}, d_{2j}, \ldots, d_{mj})$ is formed from m zeros and ones. The number of possible such systems is 2^m, which is less than 2^n. Hence by the pigeonhole principle there exist two identical systems

$$(d_{1k}, d_{2k}, \ldots, d_{mk}) = (d_{1l}, d_{2l}, \ldots, d_{ml}), \quad 1 \le k < l \le 2^n.$$

We have $d_{ik} = d_{il}$ for $1 \le i \le m$, and therefore

$$c_{il} - c_{ik} = 2(t_{il} - t_{ik}) + (d_{il} - d_{ik}) = 2(t_{il} - t_{ik}),$$

and $c_{il} - c_{ik}$ is divisible by 2 for $1 \le i \le m$.

Thus the exponent of the p_i in the product $a_{k+1}a_{k+2} \cdots a_l = \frac{a_1 a_2 \cdots a_l}{a_1 a_2 \cdots a_k}$ is equal to $c_{il} - c_{ik}$, so every number p_i has an even exponent in the product $a_{k+1}a_{k+2} \cdots a_l$. Hence $a_{k+1}a_{k+2} \cdots a_l$ is a perfect square.

Problem 5.1.11. *Let $x_1 = x_2 = x_3 = 1$ and $x_{n+3} = x_n + x_{n+1}x_{n+2}$ for all positive integers n. Prove that for any positive integer m there is an integer $k > 0$ such that m divides x_k.*

Solution. Observe that setting $x_0 = 0$, the condition is satisfied for $n = 0$.

We prove that there is an integer $k \le m^3$ such that x_k divides m. Let r_t be the remainder of x_t when divided by m for $t = 0, 1, \ldots, m^3 + 2$. Consider the triples (r_0, r_1, r_2), (r_1, r_2, r_3), ..., $(r_{m^3}, r_{m^3+1}, r_{m^3+2})$. Since r_t can take m values, it follows by the pigeonhole principle that at least two triples are equal. Let p be the smallest number such that the triple (r_p, r_{p+1}, r_{p+2}) is equal to another triple (r_q, r_{q+1}, r_{q+2}), $p < q \le m^3$. We claim that $p = 0$.

Assume by way of contradiction that $p \ge 1$. Using the hypothesis, we have

$$r_p \equiv r_{p-1} + r_p r_{p+1} \pmod{m} \quad \text{and} \quad r_{q+2} \equiv r_{q-1} + r_q r_{q+1} \pmod{m}.$$

Since $r_p = r_q, r_{p+1} = r_{q+1}$ and $r_{p+2} = r_{q+2}$, it follows that $r_{p-1} = r_{q-1}$, so $(r_{p-1}, r_p, r_{p+1}) = (r_{q-1}, r_q, r_{q+1})$, which is a contradiction to the minimality of p. Hence $p = 0$, so $r_q = r_0 = 0$, and therefore $x_q \equiv 0 \mod m$.

Problem 5.1.12. Prove that among seven arbitrary perfect squares there are two whose difference is divisible by 20.

(Mathematical Reflections)

First solution. It is easy to check that perfect squares can give one of the following residues: $1, 2, 4, 8, 16 \pmod{20}$.

By the pigeonhole principle we conclude that among seven perfect squares we must have at least two that have the same residue modulo 20. Hence their difference is divisible by 20 and our proof is complete.

Second solution. Note that for all integers x we have $x^2 \equiv 1, 2, 4, 8, 16 \pmod{m}$ and we have six distinct possible residues. If we have seven arbitrary perfect squares $x_1^2, x_2^2, x_3^2, x_4^2, x_5^2, x_6^2, x_7^2$, by the pigeonhole principle, there are two squares x_i^2 and x_j^2 with the same residue and they satisfy the requirement.

Third solution. Observe that by the pigeonhole principle, there are at least four perfect squares that all have the same parity. Now note that for any integer n, we have $n^2 \equiv -1, 0, 1 \pmod{5}$. Again by the pigeonhole principle, out of these four perfect squares, we have at least two perfect squares, say a^2 and b^2, such that $a^2 \equiv b^2 \pmod{5}$. This implies that $5 \mid a^2 - b^2$. Also, $2 \mid a - b$ and $2 \mid a + b$, since both a and b have the same parity. Hence, $4 \mid a^2 - b^2$, but $\gcd(5, 4) = 1$; thus we have $20 \mid a^2 - b^2$, and we are done.

5.2 Mathematical Induction

Problem 5.2.7. Let p be an odd prime. The sequence $(a_n)_{n \ge 0}$ is defined as follows: $a_0 = 0, a_1 = 1, \ldots, a_{p-2} = p - 2$, and for all $n \ge p - 1$, a_n is the least positive integer that does not form an arithmetic sequence of length p with any of the preceding terms. Prove that for all n, a_n is the number obtained by writing n in base $p - 1$ and reading the result in base p.

(1995 USA Mathematical Olympiad)

Solution. Our proof uses the following result.

Lemma. *Let $B = \{b_0, b_1, b_2, \dots\}$, where b_n is the number obtained by writing n in base $p - 1$ and reading the result in base p. Then*

(a) for every $a \notin B$, there exists $d > 0$ such that $a - kd \in B$ for $k = 1, 2, \dots, p - 1$; and

(b) B contains no p-term arithmetic progression.

Proof. Note that $b \in B$ if and only if the representation of b in base p does not use the digit $p - 1$.

(a) Since $a \notin B$, when a is written in base p at least one digit is $p - 1$. Let d be the positive integer whose representation in base p is obtained from that of a by replacing each $p - 1$ by 1 and each digit other than $p - 1$ by 0. Then none of the numbers $a - d, a - 2d, \dots, a - (p - 1)d$ has $p - 1$ as a digit when written in base p, and the result follows.

(b) Let $a, a + d, a + 2d, \dots, a + (p - 1)d$ be an arbitrary p-term arithmetic progression of nonnegative integers. Let δ be the rightmost nonzero digit when d is written in base p, and let α be the corresponding digit in the representation of a in base p. Then $\alpha, \alpha + \delta, \dots, \alpha + (p - 1)\delta$ is a complete set of residues modulo p. It follows that at least one of the numbers $a, a + d, \dots, a + (p - 1)d$ has $p - 1$ as a digit when written in base p. Hence at least one term of the given arithmetic progression does not belong to B. $\qquad \square$

Let $(a_n)_{n \geq 0}$ be the sequence defined in the problem. To prove that $a_n = b_n$ for all $n \geq 0$, we use mathematical induction. Clearly $a_0 = b_0 = 0$. Assume that $a_k = b_k$ for $0 \leq k \leq n - 1$, where $n \geq 1$. Then a_n is the smallest integer greater than b_{n-1} such that $\{b_0, b_1, \dots, b_{n-1}, a_n\}$ contains no p-term arithmetic progression. By part (i) of the proposition, $a_n \in B$, so $a_n \geq b_n$. By part (ii) of the proposition, the choice of $a_n = b_n$ does not yield a p-term arithmetic progression with any of the preceding terms. It follows by induction that $a_n = b_n$ for all $n \geq 0$.

Problem 5.2.8. *Suppose that x, y, and z are natural numbers such that $xy = z^2 + 1$. Prove that there exist integers $a, b, c,$ and d such that $x = a^2 + b^2$, $y = c^2 + d^2$, and $z = ac + bd$.*

(Euler's problem)

Solution. We prove the claim by strong induction on z. For $z = 1$, we have $(x, y) = (1, 2)$ or $(2, 1)$; in the former (resp. latter) case, we can set $(a, b, c, d) = (1, 0, 1, 1)$ (resp. $(0, 1, 1, 1)$).

Suppose that the claim is true whenever $z < z_0$, and that we wish to prove it for $(x, y, z) = (x_0, y_0, z_0)$, where $x_0 y_0 = z_0^2 + 1$. Without loss of generality, assume that $x_0 \leq y_0$. Consider the triple $(x_1, y_1, z_1) = (x_0, x_0 + y_0 - 2z_0, z_0 - x_0)$, so that $(x_0, y_0, z_0) = (x_1, x_1 + y_1 + 2z_1, x_1 + z_1)$.

First, using the fact that $x_0 y_0 = z_0^2 + 1$, it is easy to check that $(x, y, z) = (x_1, y_1, z_1)$ satisfies $xy = z^2 + 1$.

Second, we claim that $x_1, y_1, z_1 > 0$. This is obvious for x_1. Next, note that $y_1 = x_0 + y_0 - 2z_0 \geq 2\sqrt{x_0 y_0} - 2z_0 > 2z_0 - 2z_0 = 0$. Finally, because $x_0 \leq y_0$ and $x_0 y_0 = z_0^2 + 1$, we have $x_0 \leq \sqrt{z_0^2 + 1}$, or $x_0 \leq z_0$. However, $x_0 \neq z_0$, because this would imply that $z_0 y_0 = z_0^2 + 1$, but $z_0 \nmid (z_0^2 + 1)$ when $z_0 > 1$. Thus, $z_0 - x_0 > 0$, or $z_1 > 0$.

Therefore, (x_1, y_1, z_1) is a triple of positive integers (x, y, z) satisfying $xy = z^2 + 1$ and with $z < z_0$. By the inductive hypothesis, we can write $x_1 = a^2 + b^2$, $y_1 = c^2 + d^2$, and $z_1 = ac + bd$. Then

$$
\begin{aligned}
(ac + bd)^2 = z_1^2 &= x_1 y_1 - 1 \\
&= (a^2 + b^2)(c^2 + d^2) - 1 \\
&= (a^2 c^2 + b^2 d^2 + 2abcd) + (a^2 d^2 + b^2 c^2 - 2abcd) - 1 \\
&= (ac + bd)^2 (ad - bc)^2 - 1,
\end{aligned}
$$

so that $|ad - bc| = 1$.

Now note that $x_0 = x_1 = a^2 + b^2$ and $y_0 = x_1 + y_1 + 2z_1 = a^2 + b^2 + c^2 + d^2 + 2(ac + bd) = (a + c)^2 + (b + d)^2$. In other words, $x_0 = a'^2 + b'^2$ and $y_0 = c'^2 + d'^2$ for $(a', b', c', d') = (a, b, a + c, b + d)$. Then $|a'd' - b'c'| = |ad - bc| = 1$, implying (by logic analogous to the reasoning in the previous paragraph) that $z_0 = a'c' + b'd'$, as desired. This completes the inductive step, and the proof.

Problem 5.2.9. *Find all pairs of sets A, B, which satisfy the following conditions:*
(i) $A \cup B = \mathbb{Z}$;
(ii) *if $x \in A$, then $x - 1 \in B$;*
(iii) *if $x \in B$ and $y \in B$, then $x + y \in A$.*

(2002 Romanian International Mathematical Olympiad Team Selection Test)

Solution. We shall prove that either $A = B = \mathbb{Z}$ or A is the set of even numbers and B the set of odd numbers.

First, assume that $0 \in B$. Then we have $x \in B$, $x + 0 \in A$, and so $B \subset A$. Then $\mathbb{Z} = A \cup B \subset A$, and so $A = \mathbb{Z}$. From (ii) we also find that $B = \mathbb{Z}$. Now suppose that $0 \notin B$; thus $0 \in A$ and $-1 \in B$. Then, using (ii) we obtain $-2 \in A$, $-3 \in B$, $-4 \in A$, and by induction $-2n \in A$ and $-2n - 1 \in B$, for all $n \in \mathbb{N}$. Of course, $2 \in A$ (otherwise $2 \in B$ and $1 = 2 + (-1) \in A$ and $0 = 1 - 1 \in B$, false), and so $1 = 2 - 1 \in B$. Let $n > 1$ be minimal with $2n \in B$. Then $2n - 1 \in A$ and $2(n - 1) \in B$, contradiction. This shows that $2\mathbb{N} \subset A \setminus B$ and all odd integers are in $B \setminus A$. One can also observe that $-1 \notin A$ (otherwise $-2 \in B$ implies $-1 \in B$, i.e., $-1 \notin A$), and so $A = 2\mathbb{Z}$, $B = 2\mathbb{Z} + 1$.

Problem 5.2.10. *Find all positive integers n such that*

$$
n = \prod_{k=0}^{m} (a_k + 1),
$$

where $\overline{a_m a_{m-1} \cdots a_0}$ is the decimal representation of n.

(2001 Japanese Mathematical Olympiad)

Solution. We claim that the only such n is 18. If $n = \overline{a_m \cdots a_1 a_0}$, then let

$$P(n) = \prod_{j=0}^{m} (a_j + 1).$$

Note that if $s \geq 1$ and t is a single-digit number, then $P(10s + t) = (t + 1)P(s)$. Using this we will prove the following two statements.

Lemma 1. *If $P(s) \leq s$, then $P(10s + t) < 10s + t$.*

Proof. Indeed, if $P(s) \leq s$, then

$$10s + t \geq 10s \geq 10P(s) \geq (t + 1)P(s) = P(10s + t).$$

Equality must fail either in the first inequality (if $t \neq 0$) or in the third inequality (if $t \neq 9$). $\qquad\square$

Lemma 2. $P(n) \leq n + 1$ *for all n.*

Proof. We prove this by induction on the number of digits of n. First, we know that for all one-digit n, $P(n) = n + 1$. Now suppose that $P(n) \leq n + 1$ for all m-digit numbers n. Any $(m + 1)$-digit number n is of the form $10s + t$, where s is an m-digit number. Then

$$t(P(s) - 1) \leq 9((s + 1) - 1),$$
$$tP(s) - 10s - t \leq -s,$$
$$P(s)(t + 1) - 10s - t \leq P(s) - s,$$
$$P(10s + t) - (10s + t) \leq P(s) - s \leq 1,$$

completing the inductive step. Thus, $P(n) \leq n + 1$ for all n. $\qquad\square$

If $P(n) = n$, then n has more than one digit and we may write $n = 10s + t$. From the first statement, we have $P(s) \geq s + 1$. From the second one, we have $P(s) \leq s + 1$. Thus, $P(s) = s + 1$. Hence,

$$(t + 1)P(s) = P(10s + t) = 10s + t,$$
$$(t + 1)(s + 1) = 10s + t,$$
$$1 = (9 - t)s.$$

This is possible if $t = 8$ and $s = 1$, so the only possible n such that $P(n) = n$ is 18. Indeed, $P(18) = (1 + 1)(8 + 1) = 18$.

Problem 5.2.11. *The sequence* $(u_n)_{n \geq 0}$ *is defined as follows:* $u_0 = 2$, $u_1 = \frac{5}{2}$ *and*

$$u_{n+1} = u_n(u_{n-1}^2 - 2) - u_1 \ for \ n = 1, 2, \ldots .$$

Prove that $\lfloor u_n \rfloor = 2^{\frac{2^n - (-1)^n}{3}}$, *for all* $n > 0$ ($\lfloor x \rfloor$ *denotes the integer part of* x).

(18th International Mathematical Olympiad)

Solution. To start, we compute a few members of the sequence. Write

$$u_1 = \tfrac{5}{2} = 2 + \tfrac{1}{2}.$$

Then:

$$u_2 = u_1(u_0^2 - 2) - \left(2 + \tfrac{1}{2}\right) = \left(2 + \tfrac{1}{2}\right)(2^2 - 2) - \left(2 + \tfrac{1}{2}\right) = 2 + \tfrac{1}{2},$$

$$u_3 = u_2(u_1^2 - 2) - \left(2 + \tfrac{1}{2}\right) = \left(2 + \tfrac{1}{2}\right)\left[\left(2 + \tfrac{1}{2}\right)^2 - 2\right] - \left(2 + \tfrac{1}{2}\right)$$

$$= \left(2 + \tfrac{1}{2}\right)\left(2^2 + \tfrac{1}{2^2}\right) - \left(2 + \tfrac{1}{2}\right) = \left(2 + \tfrac{1}{2}\right)\left(2^2 - 1 + \tfrac{1}{2^2}\right) = 2^3 + \tfrac{1}{2^3},$$

$$u_4 = \left(2^3 + \tfrac{1}{2^3}\right)\left[\left(2 + \tfrac{1}{2}\right)^2 - 2\right] - \left(2 + \tfrac{1}{2}\right)$$

$$= \left(2^3 + \tfrac{1}{2^3}\right)\left(2^2 + \tfrac{1}{2^2}\right) - \left(2 + \tfrac{1}{2}\right)$$

$$= 2^5 + \tfrac{1}{2} + 2 + \tfrac{1}{2^5} - \left(2 + \tfrac{1}{2}\right) = 2^5 + \tfrac{1}{2^5},$$

$$u_5 = \left(2^5 + \tfrac{1}{2^5}\right)\left[\left(2^3 + \tfrac{1}{2^3}\right)^2 - 2\right] - \left(2 + \tfrac{1}{2}\right)$$

$$= \left(2^5 + \tfrac{1}{2^5}\right)\left(2^6 + \tfrac{1}{2^6}\right) - \left(2 + \tfrac{1}{2}\right) = 2^{11} + \tfrac{1}{2^{11}}.$$

Taking into account the required result, we claim that $u_n = 2^{a_n} + 2^{-a_n}$, where $a_n = \frac{2^n - (-1)^n}{3}$, for all $n \geq 1$. We observe that a_n is a positive integer, because $2^n \equiv (-1)^n \pmod 3$.

Observe that the claimed formula is true for $n = 1, 2, 3, 4, 5$. Using induction and the inductive formula that defined u_n, we have

$$u_{n+1} = (2^{a_n} + 2^{-a_n})[(2^{a_{n-1}} + 2^{-a_{n-1}}) - 2] - \left(2 + \tfrac{1}{2}\right)$$

$$= (2^{a_n} + 2^{-a_n})(2^{2a_{n-1}} + 2^{-2a_{n-1}}) - \left(2 + \tfrac{1}{2}\right)$$

$$= 2^{a_n + 2a_{n-1}} + 2^{-a_n - 2a_{n-1}} + 2^{2a_{n-1} - a_n} + 2^{a_n - 2a_{n-1}} - 2 - 2^{-1}.$$

We have only to consider the equalities

$$a_n + 2a_{n-1} = a_{n+1},$$
$$2a_{n-1} - a_n = (-1)^n,$$

which are easy to check. Hence, we obtain the general formula

$$u_n = 2^{\frac{2^n - (-1)^n}{3}} + \frac{1}{2^{\frac{2^n - (-1)^n}{3}}}, \text{ for all } n \geq 1.$$

The required result,

$$\lfloor u_n \rfloor = 2^{\frac{2^n - (-1)^n}{3}},$$

is now obvious.

Second solution. We have $u_0 \geq 2$, $u_1 \geq \frac{5}{2}$. We prove by induction that

$$u_n \geq \frac{5}{2}, \text{ for all } n \geq 1.$$

$$u_{n+1} = u_n(u_{n-1}^2 - 2) - \frac{5}{2} \geq \frac{5}{2}\left(\frac{25}{4} - 2\right) - \frac{5}{2} = \frac{5}{2}\left(\frac{25}{4} - 3\right) > \frac{5}{2}.$$

The equation

$$x + \frac{1}{x} = u_n$$

has a unique real solution x_n, with $x_n > 1$. Indeed, write the equation in the form

$$x^2 - u_n x + 1 = 0,$$

and we observe that $\Delta = u_n^2 - 4 \geq \frac{25}{4} - 4 > 0$. The equation has two positive real solutions, only one being greater than 1.

Therefore, there exists a unique real sequence $(x_n)_{n \geq 1}$ such that $x_n > 1$ and

$$x_n + \frac{1}{x_n} = u_n.$$

Put this formula in the definition for u_{n+1} and obtain

$$x_{n+1} + \frac{1}{x_{n+1}} = x_n x_{n-1}^2 + \frac{1}{x_n x_{n-1}^2} + \left(\frac{x_n}{x_{n-1}^2} + \frac{x_{n-1}^2}{x_n}\right) - \frac{5}{2}.$$

We claim that the sequence $(x_n)_{n \geq 1}$ is uniquely defined by the conditions

$$x_{n+1} = x_n x_{n-1}^2, \tag{1}$$

$$\frac{x_{n+1}}{x_{n-1}^2} = 2^{(-1)^{n-1}}. \tag{2}$$

Actually, from condition (1) and $x_1 = 2$, $x_2 = 2$ we deduce

$$x_3 = 2^{1+2} = 2^3, \quad x_4 = 2^{1+2} \cdot 2^{1 \cdot 2} = 2^5,$$

and generally, $x_n = 2^{\frac{2^n - (-1)^n}{3}}$. After that, the solution follows as in the first part.

5.3   Infinite Descent

Problem 5.3.2. *Find all primes p for which there exist positive integers x, y, and n such that $p^n = x^3 + y^3$.*

(2000 Hungarian Mathematical Olympiad)

Solution. Observe that $2^1 = 1^3 + 1^3$ and $3^2 = 2^3 + 1^3$. We will prove that the only answers are $p = 2$ and $p = 3$. Assume by contradiction that there exists $p \geq 5$ such that $p^n = x^3 + y^3$ with x, y, n positive integers and n of the smallest possible value. Hence at least one of x and y is greater than 1. We have $x^3 + y^3 = (x + y)(x^2 - xy + y^2)$ with $x + y \geq 3$ and $x^2 - xy + y^2 = (x - y)^2 + xy \geq 2$. It follows that both $x + y$ and $x^2 - xy + y^2$ are divisible by p. Therefore $(x + y)^2 - (x^2 - xy + y^2) = 3xy$ is also divisible by p. However, 3 is not divisible by p, so at least one of x and y must be divisible by p. Since $x + y$ is divisible by p, both x and y are divisible by p. Then $x^3 + y^3 \geq 2p^3$ and necessarily $n > 3$. We obtain

$$p^{n-3} = \frac{p^n}{p^3} = \frac{x^3}{p^3} + \frac{y^3}{p^3} = \left(\frac{x}{p}\right)^3 + \left(\frac{y}{p}\right)^3,$$

and this contradicts the minimality of n (see the remark after FMID Variant 1, Part I, Section 5.3).

5.4   Inclusion–Exclusion

Problem 5.4.2. *The numbers from 1 to 1,000,000 can be colored black or white. A permissible move consists in selecting a number from 1 to 1,000,000 and changing the color of that number and each number not relatively prime to it. Initially all of the numbers are black. Is it possible to make a sequence of moves after which all of the numbers are colored white?*

(1999 Russian Mathematical Olympiad)

First solution. It is possible. We begin by proving the following lemma:

Lemma. *Given a set S of positive integers, there is a subset $T \subseteq S$ such that every element of S divides an odd number of elements in T.*

Proof. We prove the claim by induction on $|S|$, the number of elements in S. If $|S| = 1$ then let $T = S$.

If $|S| > 1$, then let a be the smallest element of S. Consider the set $S' = S \setminus \{a\}$, the set of the largest $|S| - 1$ elements in S. By induction there is a subset $T' \subseteq S'$ such that every element in S' divides an odd number of elements in T'.

If a also divides an odd number of elements in T', then the set $T = T'$ suffices. Otherwise, consider the set $T = T' \cup \{a\}$. Thus a divides an odd number of elements in T. Every other element in S is bigger than a and can't divide it, but divides an odd number of elements in $T' = T \setminus \{a\}$. Hence T suffices, completing the induction and the proof of the lemma. □

Now, write each number $n > 1$ in its prime factorization

$$n = p_1^{a_1} p_2^{a_2} \cdots p_k^{a_k},$$

where the p_i are distinct primes and the a_i ate positive integers. Notice that the color of n will always be the same as the color of $P(n) = p_1 p_2 \cdots p_k$.

Apply the lemma to the set S consisting of all $P(i)$ for $i = 2, 3, \ldots, 1000000$ to find a subset $T \subset S$ such that every element of S divides an odd number of elements in T. For each $q \in S$, let $t(q)$ equal the number of elements in T that q divides, and let $u(q)$ equal the number of primes dividing q.

Select all the numbers in T, and consider how the color of a number $n > 1$ changes. By the inclusion–exclusion principle, the number of elements in T not relatively prime to n equals

$$\sum_{q \mid P(n), q > 1} (-1)^{u(q)+1} t(q).$$

If $x \in T$ and either $\gcd(x, P(n))$ or equivalently $\gcd(x, n)$ is divisible by exactly m primes, then it is counted for all $q > 1$ that divide $\gcd(x, P(n))$. Thus it is counted $\binom{m}{1} - \binom{m}{2} + \binom{m}{3} - \cdots = 1$ time in the sum. (For example, if $n = 6$, then the number of elements in T divisible by 2 or 3 equals $t(2) + t(3) - t(6)$.)

By the definition of T, each of the values $t(q)$ is odd. Because there are $2^k - 1$ divisors $q > 1$ of $P(n)$, the above quantity is the sum of $2^k - 1$ odd numbers and is odd itself. Therefore after selecting T, every number $n > 1$ will switch color an odd number of times and will turn white.

Finally, select 1 to turn 1 white to complete the process.

Note. In fact, a slight modification of the above proof shows that T is unique if you restrict it to square-free numbers. With some work, this stronger result implies that there is in essence exactly one way to make all the numbers white up to trivial manipulations.

Second solution. Yes, it is possible. We prove a more general statement, where we replace 1000000 in the problem by some arbitrary positive integer m. We also focus on the numbers divisible by just a few primes instead of all the primes.

Lemma. *For a finite set of distinct primes* $S = \{p_1, p_2, \ldots, p_n\}$, *let* $Q_m(S)$ *be the set of numbers between 2 and m divisible only by primes in* S. *The elements of* $Q_m(S)$ *can be colored black or white. A permissible move consists in selecting a number in* $Q_m(S)$ *and changing the color of that number and each number not relatively prime to it. Then it is possible to reverse the coloring of* $Q_m(S)$ *by selecting several numbers in a subset* $R_m(S) \subseteq Q_m(S)$.

Proof. We prove the lemma by induction on n. If $n = 1$, then selecting p_1 suffices. Now suppose $n > 1$, and assume without loss of generality that the numbers are all black to start with.

Let $T = \{p_1, p_2, \ldots, p_{n-1}\}$, and define t to be the largest integer such that $t p_n \le m$. We can assume $t \ge 1$ because otherwise we could ignore p_n and just use the smaller set T, and we'd be done by our induction hypothesis.

Now select the numbers in $R_m(T)$, $R_t(T)$, and $p_n R_t(T) = \{p_n x \mid x \in R_t(T)\}$, and consider the effect of this action on a number y:

- y is not a multiple of p_n. Selecting the numbers in $R_m(T)$ makes y white. If selecting $x \in R_t(T)$ changes y's color, selecting $x p_n$ will change it back, so that y will become white.

- y is a power of p_n. Selecting the numbers in $R_m(T)$ and $R_t(T)$ has no effect on y, but each of the $|R_t(T)|$ numbers in $x R_t(T)$ changes y's color.

- $p_n \mid y$ but y is not a power of p_n. Selecting the numbers in $R_m(T)$ makes y white. Because $y \ne p_n^i$, it is divisible by some prime in T, so selecting the numbers in $R_t(T)$ makes y black again. Finally, each of the $|R_t(T)|$ numbers in $x R_t(T)$ changes y's color.

Therefore, all the multiples of p_n are the same color (black if $|R_t(T)|$ is even, white if $|R_t(T)|$ is odd), while all the other numbers in $Q_m(S)$ are white. If the multiples of p_n are still black, we can select p_n to make them white, and we are done. □

We now return to the original problem. Set $m = 1000000$, and let S be the set of all primes under 1000000. From the lemma, we can select numbers between 2 and 1000000 so that all the numbers $2, 3, \ldots, 1000000$ are white. Finally, complete the process by selecting 1.

Third solution. Define $P(n)$ as in the first solution and note that n and $P(n)$ are always the same color. Fix $n = p_1 p_2 \cdots p_k$ square-free and consider the effect of selecting in succession every divisor $q > 1$ of n. If s is divisible by exactly m of the p_i, then 2^{k-m} divisors of n are relatively prime to s, and thus the color of s is changed by $2^k - 2^{k-m}$ choices of q. This is even unless $m = k$. Thus the net effect of choosing all these values of q is to change the color of all multiples of n and only these.

Now we argue by induction on n that for any n we can make every number $1, 2, \ldots, n$ white. For $n = 1$, we simply choose 1. For the inductive step, suppose we have found a way to make $1, 2, \ldots, n-1$ white and we wish to make n white. If it is already white, then we are done. If it is black, then n is square-free (otherwise n and $P(n) < n$ are the same color; hence white), and hence applying the construction above, we can change the color of every multiple of n. This leaves $1, 2, \ldots, n-1$ white and flips n to make it white.

6

Arithmetic Functions

6.1   Multiplicative Functions

Problem 6.1.6. *Let f be a function from the positive integers to the integers satisfying $f(m + n) \equiv f(n)$ (mod m) for all $m, n \geq 1$ (e.g., a polynomial with integer coefficients). Let $g(n)$ be the number of values (including repetitions) of $f(1), f(2), \ldots, f(n)$ divisible by n, and let $h(n)$ be the number of these values relatively prime to n. Show that g and h are multiplicative functions related by*

$$h(n) = n \sum_{d \mid n} \mu(d) \frac{g(d)}{d} = n \prod_{j=1}^{k} \left(1 - \frac{g(p_j)}{p_j}\right),$$

where $n = p_1^{\alpha_1} \cdots p_k^{\alpha_k}$ is the prime factorization of n.

(American Mathematical Monthly)

Solution. Let m and n be positive integers such that $\gcd(m, n) = 1$ and let $1 \leq a \leq m$, $1 \leq b \leq n$. From the Chinese remainder theorem and the properties of f it follows that $m \mid f(a)$ and $n \mid f(b)$ if and only if $mn \mid f(x)$, where $x = x(a, b)$ is the unique integer such that $x \equiv a$ (mod m), $x \equiv b$ (mod n), and $1 \leq x \leq \min\{m, n\}$. Thus g is multiplicative. For $d \mid n$, the number of values of $f(1), \ldots, f(n)$ divisible by d is just $\frac{n}{d} g(d)$. By a straightforward inclusion–exclusion count,

$$h(n) = n - \sum_{i=1}^{k} \frac{n}{p_i} g(p_i) + \sum_{1 \leq i < j \leq k} \frac{n}{p_i p_j} (p_i p_j) - \cdots,$$

and we get

$$h(n) = n \prod_{j=1}^{k} \left(1 - \frac{g(p_j)}{p_j}\right).$$

T. Andreescu and D. Andrica, *Number Theory*, DOI: 10.1007/b11856_17,
© Birkhäuser Boston, a part of Springer Science + Business Media, LLC 2009

Problem 6.1.7. *Define* $\lambda(1) = 1$, *and if* $n = p_1^{\alpha_1} \cdots p_k^{\alpha_k}$, *define*

$$\lambda(n) = (-1)^{\alpha_1 + \cdots + \alpha_k}.$$

(1) Show that λ *is completely multiplicative.*

(2) Prove that

$$\sum_{d|n} \lambda(d) = \begin{cases} 1 & \text{if } n \text{ is a square,} \\ 0 & \text{otherwise.} \end{cases}$$

(3) Find the convolutive inverse of λ.

Solution. (1) Assume $m = p_1^{\alpha_1} \cdots p_k^{\alpha_k}$ and $n = p_1^{\beta_1} \cdots p_k^{\beta_k}$, where $\alpha_1, \ldots, \alpha_k$, $\beta_1, \ldots, \beta_k \geq 0$. Then $mn = p_1^{\alpha_1 + \beta_1} \cdots p_k^{\alpha_k + \beta_k}$ and

$$\lambda(mn) = (-1)^{\alpha_1 + \beta_1 + \cdots + \alpha_k + \beta_k} = (-1)^{\alpha_1 + \cdots + \alpha_k} (-1)^{\beta_1 + \cdots + \beta_k} = \lambda(m)\lambda(n).$$

(2) Because λ is multiplicative, according to Theorem 6.1.2, it follows that its summation function Λ also has this property. Therefore, it is sufficient to calculate Λ for a power of a prime. we have

$$\Lambda(p^{\alpha}) = \Lambda(1) + \Lambda(p) + \cdots + \Lambda(p^{\alpha}) = \begin{cases} 1 & \text{if } \alpha \text{ even,} \\ 0 & \text{if } \alpha \text{ odd.} \end{cases}$$

If $n = p_1^{\alpha_1} \cdots p_k^{\alpha_k}$, then $\Lambda(n) = \Lambda(p_1^{\alpha_1}) \cdots \Lambda(p_k^{\alpha_k}) = 1$ if all $\alpha_1, \ldots, \alpha_k$ are even and 0 otherwise. Hence

$$\Lambda(n) = \begin{cases} 1 & \text{if } n \text{ is a square,} \\ 0 & \text{otherwise.} \end{cases}$$

(3) Let g be the convolutive inverse of λ. From Problem 6.1.4(2) it follows that g is multiplicative; hence it is determined by its values on powers of primes. From $g * \lambda = \varepsilon$ we get $(g * \lambda)(p) = g(1)\lambda(p) + g(p)\lambda(1) = -1 + g(p) = 0$, i.e., $g(p) = 1$ for any prime p. Also, $(g * \lambda)(p^2) = 0$ implies $1 - 1 + g(p^2) = 0$, i.e., $g(p^2) = 0$. A simple inductive argument shows that $g(p^{\alpha}) = 0$ for any positive integer $\alpha \geq 2$. It follows that

$$g(n) = \begin{cases} 1 & \text{if } n = 1, \\ 0 & \text{if } p^2 \mid n \text{ for some prime } p > 1, \\ 1 & \text{if } n = p_1 \cdots p_k, \text{ where } p_1, \ldots, p_k \text{ are distinct primes,} \end{cases}$$

i.e., $g = \mu^2$, where μ is the Möbius function. Hence $g(n) = 1$ if n is square-free, and 0 otherwise.

Problem 6.1.8. *Let an integer* $n > 1$ *be factored into primes:* $n = p_1^{\alpha_1} \cdots p_m^{\alpha_m}$ *(p_i distinct) and let its own positive integral exponents be factored similarly. The process is to be repeated until it terminates with a unique "constellation" of prime* numbers. *For example, the constellation for* 192 *is* $192 = 2^{2 \cdot 3} \cdot 3$ *and for* 10000 *is* $10000 = 2^{2^2} \cdot 5^2$. *Call an arithmetic function g generally multiplicative if* $g(ab) = g(a)g(b)$ *whenever the constellations for a and b have no prime in common.*

(1) Prove that every multiplicative function is generally multiplicative. Is the converse true?

(2) Let h be an additive function (i.e., $h(ab) = h(a) + h(b)$ whenever $\gcd(a, b) = 1$). Call a function k generally additive if $k(ab) = k(a) + k(b)$ whenever the constellations for a and b have no prime in common. Prove that every additive function is generally additive. Is the converse true?

(American Mathematical Monthly)

Solution. (1) Let f be multiplicative. If the constellations for a and b have no prime in common, then the same is true of their factorizations, so $f(ab) = f(a)f(b)$. Hence f is generally multiplicative.

The converse is not true. Indeed, define $g(a)$ to the product of all primes in the constellation of a, taken once only, regardless of how many times they appear in the constellation. Then g is clearly generally multiplicative, but $g(9) = 6$, $g(2) = 2$, and $g(18) = 6$, so $g(9 \cdot 2) \neq g(9)g(2)$.

(2) The statement "additive implies generally additive" can be proved in the same way. If $k(a)$ is the sum of all primes in the constellation of a each taken once only, then k is generally additive, but $k(9) = 5$, $k(2) = 2$, and $k(18) = 5$.

6.2 Number of Divisors

Problem 6.2.5. *Does there exist a positive integer such that the product of its proper divisors ends with exactly* 2001 *zeros?*

(2001 Russian Mathematical Olympiad)

Solution. Yes. Given an integer n with $\tau(n)$ divisors, the product of its divisors is

$$\sqrt{\left(\prod_{d|n} d\right)\left(\prod_{d|n} (n/d)\right)} = \sqrt{\prod_{d|n} d(n/d)} = \sqrt{n^{\tau(n)}}.$$

Thus, the product of all proper positive divisors of n equals

$$n^{\frac{1}{2}\tau(n)-1}.$$

Since this number ends in exactly 2001 zeros, $\frac{1}{2}\tau(n) - 1$ divides 2001. Suppose $\frac{1}{2}\tau(n) - 1 = 2001$. Then $10 \mid n$ but $100 \nmid n$ and $\tau(n) = 4004 = 2 \cdot 2 \cdot 7 \cdot 11 \cdot 13$. One way to arrange this is to take $n = 2^1 \cdot 5^1 \cdot 7^{\cdot} 11^{10} \cdot 13^{12}$.

Problem 6.2.6. *Prove that the number of divisors of the form* $4k + 1$ *of each positive integer is not less than the number of its divisors of the form* $4k + 3$.

Solution. To solve the problem, consider the function

$$f(n) = \begin{cases} 0, & \text{if } n \text{ is even,} \\ 1, & \text{if } n \equiv 1 \pmod{4}, \\ -1, & \text{if } n \equiv 3 \pmod{4}. \end{cases}$$

It follows directly from this definition that $f(n)$ is multiplicative. Now we apply (1). The even divisors of n do not influence its left-hand side. Each divisor of the form $4k + 1$ contributes a 1, and each divisor of the form $4k + 3$ contributes $a - 1$. Consequently, it suffices to prove that the summation function of f, $\sum_{d|n} f(d)$ is nonnegative for each positive integer n.

Take any prime divisor p_i of n. If $p_i \equiv 1 \pmod{4}$, then the same congruence holds for all powers of p_i, so the ith factor in the right-hand side of (1) is positive. If p_i is congruent to 3 modulo 4, then so are its odd powers, while the even powers are congruent to 1 modulo 4. In this case the ith factor in the right-hand side has the form $1 - 1 + 1 - 1 + \cdots$, and it equals 1 or 0 according as α_i is even or odd. Summing up, we conclude that the sum in question is nonnegative.

Problem 6.2.7. *Let* $d_1, d_2, \ldots, d_l$ *be all positive divisors of a positive integer. For each* $i = 1, 2, \ldots, l$ *denote by* a_i *the number of positive divisors of* d_i. *Then*

$$a_1^3 + a_2^3 + \cdots + a_l^3 = (a_1 + a_2 + \cdots + a_l)^2.$$

Solution. The basic ingredient in the proof is the well-known identity

$$\sum_{k=1}^{n} k^3 = \left(\frac{n(n+1)}{2}\right)^2 = \left(\sum_{k=1}^{n} k\right)^2.$$

We have

$$a_1 + a_2 + \cdots + a_l = \sum_{d|n} \tau(d) = \prod_{i=1}^{k} \left(1 + \tau(p_i) + \cdots + \tau(p_i^{\alpha_i})\right),$$

$$a_1^3 + a_2^3 + \cdots + a_l^3 = \sum_{d|n} \tau(d)^3 = \prod_{i=1}^{k} \left(1 + \tau(p_i)^3 + \cdots + \tau(p_i^{\alpha_i})^3\right),$$

where $n = p_1^{\alpha_1} \cdots p_k^{\alpha_k}$ is the prime factorization of n.

Since

$$1 + \tau(p_i) + \cdots + \tau(p_i^{\alpha_i}) = 1 + 2 + \cdots + (\alpha_i + 1)$$

and

$$1 + \tau(p_i)^3 + \cdots + \tau(p_i^{\alpha_i})^3 = 1^3 + 2^3 + \cdots + (\alpha+i+1)^3 = [1 + 2 + \cdots + (\alpha+i+1)]^2,$$

the conclusion follows.

For example, if $n = 12$ we have $d_1 = 1$, $d_2 = 2$, $d_3 = 3$, $d_4 = 4$, $d_5 = 6$, $d_6 = 12$; $a_1 = 1$, $a_2 = 2$, $a_3 = 2$, $a_4 = 3$, $a_5 = 4$, $a_6 = 6$, and

$$1^3 + 2^3 + 2^3 + 3^3 + 4^3 + 6^3 = 324 = (1 + 2 + 2 + 3 + 4 + 6)^2.$$

Remark. The above identity shows that solving the equation

$$(x_1 + x_2 + \cdots + x_n)^2 = x_1^3 + x_2^3 + \cdots + x_n^3$$

in positive integers is a very difficult job. If we assume that $x_i \neq x_j$ for $i \neq j$, there are only a few solutions. Try to prove this last assertion.

6.3 Sum of Divisors

Problem 6.3.5. *For any $n \geq 2$,*

$$\sigma(n) < n\sqrt{2\tau(n)}.$$

(1999 Belarusian Mathematical Olympiad)

Solution. Let $d_1, d_2, \ldots, d_{\tau(n)}$ be the divisors of n. They can be rewritten in the form

$$\frac{n}{d_1}, \frac{n}{d_2}, \ldots, \frac{n}{d_{\tau(n)}}.$$

By the power mean inequality,

$$\sigma(n) \leq \sqrt{\tau(n) \sum_{i=1}^{\tau(n)} d_i^2}.$$

Now,

$$\frac{1}{n^2}\left(\sum_{i=1}^{\tau(n)} d_i^2\right) = \sum_{i=1}^{\tau(n)} \frac{1}{d_i^2} \leq \sum_{i=1}^{\tau(n)} \frac{1}{j^2} < \sum_{j=1}^{\infty} \frac{1}{j^2} = \frac{\pi^2}{6}.$$

Hence

$$\sigma(n) \leq \sqrt{\tau(n) \sum_{i=1}^{\tau(n)} d_i} < \sqrt{\tau(n)\frac{n^2\pi^2}{6}} < n\sqrt{2\tau(n)}.$$

Problem 6.3.6. *Find all the four-digit numbers whose prime factorization has the property that the sum of the prime factors is equal to the sum of the exponents.*

Solution. (1) If the number has at least four prime divisors, then $n \geq 2^{14} \cdot 3 \cdot 5 \cdot 7 > 9999$, a contradiction.

(2) If n has three prime divisors, these must be 2, 3, and 5. The numbers are

$$2^8 \cdot 3 \cdot 5 = 3840, \ 2^7 \cdot 3^2 \cdot 5 = 5760, \ 2^6 \cdot 3^3 \cdot 5 = 8640, \text{ and } 2^7 \cdot 3 \cdot 5^2 = 9600.$$

(3) If n has 2 prime divisors, at least one of them must be 2, since if neither is 2, they are at least 3 and 5 and hence $n \geq 3^7 \cdot 5 = 10935$ has more than four digits. The numbers

$$2^4 \cdot 5^3 = 2000, \ 2^3 \cdot 5^4 = 5000, \ 2^8 \cdot 7 = 1792, \ 2^7 \cdot 7^2 = 6272$$

satisfy the solutions.

(4) If n has only one prime factor, then $5^5 = 3125$.

Therefore there are nine solutions.

Problem 6.3.7. *Let m, n, k be positive integers with $n > 1$. Show that $\sigma(n)^k \neq n^m$.*

(2001 St. Petersburg City Mathematical Olympiad)

Solution. Let $n = p_1^{e_1} p_2^{e_2} \cdots p_k^{e_k}$. Because $\sigma(n) > n$, if $\sigma(n)^k = n^m$, then $\sigma(n) = p_1^{f_1} p_2^{f_2} \cdots p_k^{f_k}$ where $f_i > e_i$. This implies $f_i \geq e_i + 1$, for all i, and

$$\sigma(n) \geq p_1^{1+e_1} p_2^{1+e_2} \cdots p_k^{1+e_k} > \frac{p_1^{1+e_1} - 1}{p_1 - 1} \frac{p_2^{1+e_2} - 1}{p_2 - 1} \cdots \frac{p_k^{1+e_k} - 1}{p_k - 1}$$
$$= (1 + p_1 + \cdots + p_1^{e_1})(1 + p_2 + \cdots + p_2^{e_2}) \cdots (1 + p_k + \cdots + p_k^{e_k})$$
$$= \sigma(n).$$

This is a contradiction.

6.4 Euler's Totient Function

Problem 6.4.5. *For a positive integer n, let $\psi(n)$ be the number of prime factors of n. Show that if $\varphi(n)$ divides $n - 1$ and $\psi(n) \leq 3$, then n is prime.*

(1998 Korean Mathematical Olympiad)

Solution. Note that for prime p, if $p^2 \mid n$ then $p \mid \varphi(n)$ but $p \nmid n - 1$, contradiction. So we need only show that $n \neq pq, n \neq pqr$ for primes $p < q < r$.

First assume $n = pq$, so $(p - 1)(q - 1) \mid pq - 1$. Note that $q \geq 3$ implies that the left side is even, so the right is too and p, q are odd. But if $p = 3, q = 5$ then

$$\frac{pq - 1}{(p - 1)(q - 1)} < 2;$$

the left side is decreasing in each variable and is always greater than 1, so it cannot be an integer, contradiction.

Now let $n = pqr$. As before, p, q, r are odd; if $p = 3$, $q = 7$, and $r = 11$ then

$$\frac{pqr - 1}{(p - 1)(q - 1)(r - 1)} < 2$$

and again the left side is decreasing and greater than 1; this eliminates all cases except where $p = 3$, $q = 5$. Then for $r = 7$ we have

$$\frac{pqr - 1}{(p - 1)(q - 1)(r - 1)} < 3,$$

so the only integer value ever attainable is 2. Note that $(15r - 1)/8(r - 1) = 2$ gives $r = 15$, which is not a prime, and we have eliminated all cases.

Remarks. (1) The problem is a direct consequence of Problem 1.1.16.

(2) A long standing conjecture due to Lehmer asserts that if $\varphi(n) \mid n - 1$, then n is a prime. This has been proved so far for $\psi(n) \le 14$. The proofs are very long and computational and no further progress has been made on this conjecture.

Problem 6.4.6. *Show that the equation* $\varphi(n) = \tau(n)$ *has only the solutions* $n = 1, 3, 8, 10, 18, 24, 30$.

Solution. We check directly that the listed integers satisfy the equation and there are no others ≤ 30 with this property. We will prove that for $n \ge 31$, $\varphi(n) > \tau(n)$. For this we consider the multiplicative function $f(n) = \varphi(n)/\tau(n)$. If n is a prime, we have $f(n) = \frac{n-1}{2}$; hence f increases on the set of primes.

For a prime p, define $S_p = \{p^\alpha \mid \alpha \ge 1\}$. Because

$$f(p^\alpha) = \frac{p^{\alpha-1}(p - 1)}{\alpha + 1} \quad \text{and} \quad \frac{p}{\alpha + 2} \ge \frac{2}{\alpha + 2} > \frac{1}{\alpha + 1},$$

we obtain $f(p^{\alpha+1}) > f(p^\alpha)$, that is, f increases on S_p. Using the fact that $\min_{p,\alpha} f(p^\alpha) = f(2) = \frac{1}{2}$, it follows that in order to solve the given equation we need to consider the integers p^α with $f(p^\alpha) \le 2$. These are 2, 3, 4, 5, 8, 9, 16, with $f(2) = \frac{1}{2}$, $f(3) = \frac{2}{3}$, $f(3) = f(8) = 1$ and $f(5) = f(9) = f(16) = 2$. The only way to write 1 as a product of these values of f is to use only ones or a single $\frac{1}{2}$, a single 2, and possibly some 1's. These gives the possibilities $n = 1, 3, 8, 3 \cdot 8 = 24$, and $n = 2 \cdot 5 = 10$, $2 \cdot 9 = 18$, $2 \cdot 3 \cdot 5 = 30$, respectively. Thus these are exactly the values given in the statement.

Problem 6.4.7. *Let* $n > 6$ *be an integer and let* $a_1, a_2, \ldots, a_k$ *be all positive integers less than* n *and relatively prime to* n. *If*

$$a_2 - a_1 = a_3 - a_2 = \cdots = a_k - a_{k-1} > 0,$$

prove that n *must be either a prime number or a power of 2.*

(32nd International Mathematical Olympiad)

Solution. It is given that the reduced system of residues mod n chosen from the set $\{1, 2, \ldots, n - 1\}$ is an arithmetic progression. We write it as an increasing sequence $1 = a_1 < a_2 < \cdots < a_k = n - 1$.

Suppose the reduced system of residues for n is an arithmetic progression with difference 1. Since 1 and $n - 1$ are relatively prime to n, this system must be $1 < 2 < \cdots < n - 1$. Hence n has no factors in $\{2, \ldots, n - 1\}$ and n is prime. If the reduced system of residues for n is an arithmetic progression with difference 2, then it must be $1 < 3 < \cdots < n - 1$. Hence n has no odd factors and n is a power of 2. The problem asks us to prove that only these cases can appear.

Let a_2 be the second member of the progression. Because $a_2 > 1$ is the least positive number relatively prime to n, it is a prime number, say p and $p > 3$. Then, the difference of the progression is $a_2 - a_1 = p - 1$, and $a_k = n - 1 = 1 + (k - 1)(p - 1)$. We obtain a "key" formula:

$$n - 2 = (k - 1)(p - 1).$$

Remembering the choice of p, n is divisible by 3, and then $n - 2 \equiv 1 \pmod{3}$. Thus, by the key formula we cannot have $p \equiv 1 \pmod{3}$. Since $p > 3$, we have $p \equiv 2 \pmod{3}$. Then $a_3 = 1 + 2(p - 1) \equiv 0 \pmod{3}$, and this contradicts the supposition that a_3 and n are relatively prime numbers.

6.5 Exponent of a Prime and Legendre's Formula

Problem 6.5.7. (a) If p is a prime, prove that for any positive integer n,

$$-\left\lfloor \frac{\ln n}{\ln p} \right\rfloor + n \sum_{k=1}^{\left\lfloor \frac{\ln n}{\ln p} \right\rfloor} \frac{1}{p^k} < e_p(n) < \frac{n}{p - 1}.$$

(b) Prove that

$$\lim_{n \to \infty} \frac{e_p(n)}{n} = \frac{1}{p - 1}.$$

Solution. (a) From Legendre's formula,

$$e_p(n) = \sum_{k \geq 1} \left\lfloor \frac{n}{p^k} \right\rfloor \leq \sum_{k \geq 1} \frac{n}{p^k} < n \sum_{j=1}^{\infty} \frac{1}{p^j} = \frac{n}{p - 1}.$$

For the left bound note that $\left\lfloor \frac{\ln n}{\ln p} \right\rfloor$ is the least nonnegative integer s such that $n < p^{s+1}$. That is, $\left\lfloor \frac{n}{p^k} \right\rfloor = 0$ for $k \geq s + 1$. It follows that

$$e_p(n) = \sum_{k=1}^{s} \left\lfloor \frac{n}{p^k} \right\rfloor > \sum_{k=1}^{s} \left(\frac{n}{p^k} - 1 \right) = n \sum_{k=1}^{s} \frac{1}{p^k} - s,$$

and we are done.

(b) From the inequalities

$$-\frac{1}{n}\left\lfloor\frac{\ln n}{\ln p}\right\rfloor+\sum_{k=1}^{\left\lfloor\frac{\ln n}{\ln p}\right\rfloor}\frac{1}{p^k}<\frac{e_p(n)}{n}<\frac{1}{p-1}$$

and the fact that

$$\lim_{n\to\infty}\frac{1}{n}\left\lfloor\frac{\ln n}{\ln p}\right\rfloor=0\quad\text{and}\quad\lim_{n\to\infty}\sum_{k=1}^{\left\lfloor\frac{\ln n}{\ln p}\right\rfloor}\frac{1}{p^k}=\frac{1}{p-1},$$

the desired formula follows.

Remark. An easier to understand lower bound on $e_p(n)$ is

$$e_p(n)>\frac{n}{p-1}-\left\lceil\frac{\log n}{\log p}\right\rceil,$$

which follows easily from the fact that n has at most $\left\lceil\frac{\log n}{\log p}\right\rceil$ digits in base p. This lower bound suffices to prove (b).

Problem 6.5.8. *Show that for all nonnegative integers m, n the number*

$$\frac{(2m)!(2n)!}{m!n!(m+n)!}$$

is also an integer.

(14th International Mathematical Olympiad)

Solution. It is sufficient to prove that for any prime number p,

$$e_p(2m)+e_p(2n)\geq e_p(m)+e_p(n)+e_p(m+n).$$

Again, it is sufficient to prove that for all $i, j\geq 1$, the following inequality holds:

$$\left\lfloor\frac{2m}{p^i}\right\rfloor+\left\lfloor\frac{2n}{p^i}\right\rfloor\geq\left\lfloor\frac{m}{p^i}\right\rfloor+\left\lfloor\frac{n}{p^i}\right\rfloor+\left\lfloor\frac{m+n}{p^i}\right\rfloor.$$

This follows from a more general result.

Lemma. *For any real numbers a, b,*

$$\lfloor 2a\rfloor+\lfloor 2b\rfloor\geq\lfloor a\rfloor+\lfloor b\rfloor+\lfloor a+b\rfloor.$$

Proof. Let $a=\lfloor a\rfloor+x, b=\lfloor b\rfloor+y$, where $0\leq x, y<1$. If $x+y<1$ we have $\lfloor a+b\rfloor=\lfloor a\rfloor+\lfloor b\rfloor$, and the required inequality becomes

$$\lfloor 2a\rfloor+\lfloor 2b\rfloor\geq 2(\lfloor a\rfloor+\lfloor b\rfloor).$$

In this form, it is obvious.

Let $1 \le x + y < 2$. Then $2x \ge 1$ or $2y \ge 1$. Let $2x \ge 1$. Then

$$\lfloor 2a \rfloor = 2\lfloor a \rfloor + 1 \quad \text{and} \quad \lfloor a + b \rfloor = \lfloor a \rfloor + \lfloor b \rfloor + 1.$$

Thus

$$\lfloor 2a \rfloor + \lfloor 2b \rfloor = 2\lfloor a \rfloor + 1 + \lfloor 2b \rfloor \ge 2\lfloor a \rfloor + 1 + 2\lfloor b \rfloor = \lfloor a \rfloor + \lfloor b \rfloor + \lfloor a + b \rfloor.$$

The other cases follow in a similar way.

Problem 6.5.9. *Prove that* $\frac{(3a+3b)!(2a)!(3b)!(2b)!}{(2a+3b)!(a+2b)!(a+b)!a!(b!)^2}$ *is an integer for all pairs of positive integers* a, b.

(American Mathematical Monthly)

Solution. First, let us clarify something. When we write

$$\left\lfloor \frac{n}{p} \right\rfloor + \left\lfloor \frac{n}{p^2} \right\rfloor + \left\lfloor \frac{n}{p^3} \right\rfloor + \cdots,$$

we write in fact $\sum_{k \ge 1} \lfloor \frac{n}{p^k} \rfloor$, and this sum has clearly a finite number of nonzero terms. Now let us take a prime p and apply Legendre's formula as well as the first observations. We find that

$$v_p((3a + 3b)!(2a)!(3b)!(2b)!) = \sum_{k \ge 1} \left(\left\lfloor \frac{3a + 3b}{p^k} \right\rfloor + \left\lfloor \frac{2a}{p^k} \right\rfloor + \left\lfloor \frac{3b}{p^k} \right\rfloor + \left\lfloor \frac{2b}{p^k} \right\rfloor \right)$$

and also

$$v_p\big((2a + 3b)!(a + 2b)!(a + b)!a!(b!)^2\big)$$
$$= \sum_{k \ge 1} \left(\left\lfloor \frac{2a + 3b}{p^k} \right\rfloor + \left\lfloor \frac{a + 2b}{p^k} \right\rfloor + \left\lfloor \frac{a + b}{p^k} \right\rfloor + \left\lfloor \frac{a}{p^k} \right\rfloor + 2\left\lfloor \frac{b}{p^k} \right\rfloor \right).$$

Of course, it is enough to prove that for each $k \ge 1$, the term corresponding to k in the first sum is greater than or equal to the term corresponding to k in the second sum. With the substitution $x = \frac{a}{p^k}$, $y = \frac{b}{p^k}$, we have to prove that for any nonnegative real numbers x, y we have

$$\lfloor 3x + 3y \rfloor + \lfloor 2x \rfloor + \lfloor 3y \rfloor + \lfloor 2y \rfloor \ge \lfloor 2x + 3y \rfloor + \lfloor x + 2y \rfloor + \lfloor x + y \rfloor + \lfloor x \rfloor + 2\lfloor y \rfloor.$$

This isn't easy, but with another useful idea the inequality will become easy. The idea is that

$$\lfloor 3x + 3y \rfloor = 3\lfloor x \rfloor + 3\lfloor y \rfloor + \lfloor 3\{x\} + 3\{y\} \rfloor,$$

and similar relations for the other terms of the inequality. After this operation, we see that it suffices to prove the inequality only for $0 \le x, y < 1$. Because we can easily compute all terms, after splitting in some cases, it suffices to see when $\lfloor 2\{x\} \rfloor$, $\lfloor 3\{y\} \rfloor$, $\lfloor 2\{y\} \rfloor$ are 0, 1, or 2.

Remark. A graphical solution to this problem is the following. The problem reduces (as in the text) to showing that

$$\lfloor 3x + 3y \rfloor + \lfloor 2x \rfloor + \lfloor 3y \rfloor + \lfloor 2y \rfloor \geq \lfloor 2x + 3y \rfloor + \lfloor x + 2y \rfloor + \lfloor x + y \rfloor + \lfloor x \rfloor + 2 \lfloor y \rfloor.$$

Now observe that it suffices to show this for $0 \leq x < 1$ and $0 \leq y < 1$. For this draw two copies of the unit square. On the first plot the regions on which the left-hand side is constant and the corresponding value, and for the second do the same thing for the right-hand side. Then one just compares and sees that the left-hand side is always greater.

Problem 6.5.10. *Prove that there exists a constant c such that for any positive integers a, b, n for which $a! \cdot b! | n!$ we have $a + b < n + c \ln n$.*

(Paul Erdős)

Solution. This time, the second formula for $e_p(n)$ is useful. Of course, there is no reasonable estimation of this constant, so we should see what happens if $a! \cdot b! \mid n!$. Then $e_2(a) + e_2(b) \leq e_2(n)$, which can be translated as $a - S_2(a) + b - S_2(b) \leq n - S_2(n) < n$. So, we have found almost exactly what we needed: $a + b < n + S_2(a) + S_2(b)$. Now we need another observation: the sum of digits of a number A when written in binary is at most the number of digits of A in base 2, which is $1 + \lfloor \log_2 A \rfloor$ (this follows from the fact that $2^{k-1} \leq A < 2^k$, where k is the number of digits of A in base 2). So, we have the estimations $a + b < n + S_2(a) + S_2(b) \leq n + 2 + \log_2 ab \leq n + 2 + 2 \log_2 n$ (since we have of course $a, b \leq n$). And now the conclusion is immediate.

Problem 6.5.11. *Prove that for any integer $k \geq 2$, the equation*

$$\frac{1}{10^n} = \frac{1}{n_1!} + \frac{1}{n_2!} + \cdots + \frac{1}{n_k!}$$

does not have integer solutions such that $1 \leq n_1 < n_2 < \cdots < n_k$.

(Tuymaada Olympiad)

Solution. Suppose we have found a solution of the equation and let us consider

$$P = n_1! n_2! \cdots n_k!.$$

We have

$$10^n \big((n_1 + 1) \cdots (n_k - 1) n_k + \cdots + (n_{k-1} + 1) \cdots (n_k - 1) n_k + 1 \big) = n_k!,$$

which shows that n_k divides 10^n. Let us write $n_k = 2^x \cdot 5^y$. First of all, suppose that x, y are positive. Thus,

$$(n_1 + 1) \cdots (n_k - 1) n_k + \cdots + (n_{k-1} + 1) \cdots (n_k - 1) n_k + 1$$

is relatively prime to 10, and it follows that $e_2(n_k) = e_5(n_k)$. This implies of course that $\lfloor \frac{n_k}{2^j} \rfloor = \lfloor \frac{n_k}{5^j} \rfloor$ for all j (because we clearly have $\lfloor \frac{n_k}{2^j} \rfloor > \lfloor \frac{n_k}{5^j} \rfloor$). Take $j = 1$, then from

$$\frac{n_k}{2} \geq \left\lfloor \frac{n_k}{2} \right\rfloor \geq \left\lfloor \frac{n_k}{5} \right\rfloor \geq \frac{n_k}{5} - 1$$

we get $n_k \leq 3$. Checking by hand shows that the inequality does not hold for $n_k = 2$ or $n_k = 3$, so we get only $k = 1$, which is not possible, since $k \geq 2$.

Next, suppose that $y = 0$. Then

$$(n_1 + 1) \cdots (n_k - 1)n_k + \cdots + (n_{k-1} + 1) \cdots (n_k - 1)n_k + 1$$

is odd and thus $e_2(n_k) = n \leq e_5(n_k)$. Again this implies $e_2(n_k) = e_5(n_k)$, and we have seen that this gives no solution. So, actually $x = 0$. A crucial observation is that if $n_k > n_{k-1} + 1$, then

$$(n_1 + 1) \cdots (n_k - 1)n_k + \cdots + (n_{k-1} + 1) \cdots (n_k - 1)n_k + 1$$

is again odd, and thus we find again that $e_2(n_k) = n \leq e_5(n_k)$, impossible. So, $n_k = n_{k-1} + 1$. But then, taking into account that n_k is a power of 5, we deduce that

$$(n_1 + 1) \cdots (n_k - 1)n_k + \cdots + (n_{k-1} + 1) \cdots (n_k - 1)n_k + 1$$

is congruent to 2 modulo 4 and thus $e_2(n_k) = n + 1 \leq e_5(n_k) + 1$. It follows that $\lfloor \frac{n_k}{2} \rfloor \leq 1 + \lfloor \frac{n_k}{5} \rfloor$ and thus $n_k \leq 6$. Since n_k is a power of 5, we find that $n_k = 5$, $n_{k-1} = 1$, and a quick search of all possibilities shows that there are no solutions.

7

More on Divisibility

7.1 Congruences Modulo a Prime: Fermat's Little Theorem

Problem 7.1.11. *Let $3^n - 2^n$ be a power of a prime for some positive integer n. Prove that n is a prime.*

Solution. Let $3^n - 2^n = p^\alpha$ for some prime p and some $\alpha \geq 1$, and let q be a prime divisor of n. Assume that $q \neq n$; then $n = kq$, where $k > 1$. Since $p^\alpha = 3^{kq} - 2^{kq} = (3^k)^q - (2^k)^q$, we observe that p^α is divisible by $3^k - 2^k$. Hence $3^k - 2^k = p^\beta$ for some $\beta \geq 1$. Now we have

$$p^\alpha = (2^k + p^\beta)^q - 2^{kq}$$
$$= q2^{k(q-1)}p^\beta + \frac{q(q-1)}{2}2^{k(q-2)}p^{2\beta} + \cdots + p^{q\beta}.$$

Since $\alpha > \beta$ (because $p^\beta = 3^k - 2^k$ is less than $p^\alpha = 3^{kq} - 2^{kq}$), it follows that p^α is divisible by a power of p at least as great as $p^{\beta+1}$. Then the above equality implies that p divides $q2^{k(q-1)}$. On the other hand, p is obviously odd, and hence it divides q. Being a prime, q must then be equal to p. Therefore $n = kq = kp$, and $p^\alpha = (3^p)^k - (2^p)^k$ is divisible by $3^p - 2^p$, implying $3^p - 2^p = p^\gamma$ for some $\gamma \geq 1$. In particular, we infer that $3^p \equiv 2^p \pmod{p}$. Now, observing that $p \neq 2, 3$, we reach a contradiction to Fermat's little theorem, by which

$$3^p \equiv 3 \pmod{p}, \quad 2^p \equiv 2 \pmod{p}.$$

Problem 7.1.12. *Let $f(x_1, \ldots, x_n)$ be a polynomial with integer coefficients of total degree less than n. Show that the number of ordered n-tuples $(x_1, \ldots, x_n)$ with $0 \leq x_i \leq 12$ such that $f(x_1, \ldots, x_n) \equiv 0 \pmod{13}$ is divisible by 13.*

(1998 Turkish Mathematical Olympiad)

T. Andreescu and D. Andrica, *Number Theory*, DOI: 10.1007/b11856_18,
© Birkhäuser Boston, a part of Springer Science + Business Media, LLC 2009

Solution. (All congruences in this problem are modulo 13.) We claim that

$$\sum_{x=0}^{12} x^k \equiv 0 \text{ for } 0 \le k < 12.$$

The case $k = 0$ is obvious, so suppose $k > 0$.

Note that the twelve powers $1, 2, 4, \ldots, 2^{11}$ represent all twelve nonzero residues mod 13. Thus $2^k \equiv 1 \pmod{13}$ if and only if $12 \mid k$. Since the numbers $2, 4, \ldots, 24$ are congruent (in some order) to $1, 2, \ldots, 12 \pmod{13}$, we have

$$\sum_{x=0}^{12} x^k \equiv \sum_{x=0}^{12} (2x)^k = 2^k \sum_{x=0}^{12} x^k;$$

since $g^k \not\equiv 1$, we must have $\sum_{x=0}^{12} x^k \equiv 0$. This proves our claim.

Now let $S = \{(x_1, \ldots, x_n) \mid 0 \le x_i \le 12\}$. It suffices to show that the number of n-tuples $(x_1, \ldots, x_n) \in S$ with $f(x_1, \ldots, x_n) \not\equiv 0$ is divisible by 13, since $|S| = 13^n$ is divisible by 13. Consider the sum

$$\sum_{(x_1, \ldots, x_n) \in S} (f(x_1, \ldots, x_n))^{12}.$$

This sum counts mod 13 the number of n-tuples $(x_1, \ldots, x_n) \in S$ such that $f(x_1, \ldots, x_n) \not\equiv 0$, since by Fermat's little theorem,

$$(f(x_1, \ldots, x_n))^{12} \equiv \begin{cases} 1, \text{ if } f(x_1, \ldots, x_n) \not\equiv 0, \\ 0, \text{ if } f(x_1, \ldots, x_n) \equiv 0. \end{cases}$$

On the other hand, we can expand $(f(x_1, \ldots, x_n))^{12}$ in the form

$$(f(x_1, \ldots, x_n))^{12} = \sum_{j=1}^{N} c_j \prod_{i=1}^{n} x_i^{e_{ji}}$$

for some integers N, c_j, e_{ji}. Since f is a polynomial of total degree less than n, we have $e_{j1} + e_{j2} + \cdots + e_{jn} < 12n$ for every j, so for each j there exists an i such that $e_{ji} < 12$. Thus by our claim,

$$\sum_{(x_1, \ldots, x_n) \in S} c_j \prod_{i=1}^{n} x_i^{e_{ji}} = c_j \prod_{i=1}^{n} \sum_{x=0}^{12} x_i^{e_{ji}} \equiv 0,$$

since one of the sums in the product is 0. Therefore

$$\sum_{(x_1, \ldots, x_n) \in S} (f(x_1, \ldots, x_n))^{12} = \sum_{(x_1, \ldots, x_n) \in S} \sum_{j=1}^{N} c_j \prod_{i=1}^{n} x_i^{e_{ji}} \equiv 0,$$

so the number of $(x_1, \ldots, x_n)$ such that $f(x_1, \ldots, x_n) \not\equiv 0 \pmod{13}$ is divisible by 13, and we are done.

Problem 7.1.13. *Find all pairs (m, n) of positive integers, with $m, n \geq 2$, such that $a^n - 1$ is divisible by m for each $a \in \{1, 2, \ldots, n\}$.*

(2001 Romanian International Mathematical Olympiad Team Selection Test)

Solution. The solution is the set of all $(p, p - 1)$, for odd primes p. The fact that all of these pairs are indeed solutions follows immediately from Fermat's little theorem. Now we show that no other solutions exist.

Suppose that (m, n) is a solution. Let p be a prime dividing m. We first observe that $p > n$. Otherwise, we could take $a = p$, and then $p^n - 1$ would not be divisible by p, let alone m. Then because $n \geq 2$, we have $p \geq 3$, and hence p is odd.

Now we prove that $p < n + 2$. Suppose to the contrary that $p \geq n + 2$. If n is odd, then $n + 1$ is even and less than p. Otherwise, if n is even, then $n + 2$ is even and hence less than p as well, because p is odd. In either case, there exists an even d such that $n < d < p$ with $\frac{d}{2} \leq n$. Setting $a = 2$ and $a = \frac{d}{2}$ in the given condition, we find that

$$d^n = 2^n \left(\frac{d}{2}\right)^n \equiv 1 \cdot 1 \equiv 1 \pmod{m},$$

so that $d^n - 1 \equiv 0 \pmod{m}$ as well. Because $n < d < p < m$, we see that $1, 2, \ldots, n, d$ are $n + 1$ distinct roots of the polynomial congruence $x^n - 1 \equiv 0 \pmod{p}$. By Lagrange's theorem, however, this congruence can have at most n roots, a contradiction.

Thus, we have sandwiched p between n and $n + 2$, and the only possibility is that $p = n + 1$. Therefore, all solutions are of the form $(p^k, p - 1)$ with p an odd prime. It remains to prove that $k = 1$. Using $a = n = p - 1$, it suffices to prove that

$$p^k \nmid ((p - 1)^{p-1} - 1).$$

Expanding the term $(p - 1)^{p-1}$ modulo p^2, and recalling that p is odd, we have

$$
\begin{aligned}
(p - 1)^{p-1} &= \sum_{i=0}^{p-1} \binom{p-1}{i}(-1)^{p-1-i}p^i \\
&\equiv \binom{p-1}{0}(-1)^{p-1} + \binom{p-1}{1}(-1)^{p-2}p \\
&\equiv 1 - p(p - 1) \\
&\equiv p + 1 \not\equiv 1 \pmod{p^2}.
\end{aligned}
$$

It follows immediately that k cannot be greater than 1, completing the proof.

Problem 7.1.14. *Let p be a prime and b_0 an integer, $0 < b_0 < p$. Prove that there exists a unique sequence of base-p digits $b_0, b_1, b_2, \ldots, b_n, \ldots$ with the following property: If the base-p representation of a number x ends in the group of digits $b_n b_{n-1} \ldots b_1 b_0$, then so does the representation of x^p.*

Solution. We are looking for a sequence $b_0, b_1, b_2, \ldots, b_n, \ldots$ of base p digits such that the numbers $x_n = b_0 + b_1 p + \cdots + b_n p^n$ and x_n^p are congruent modulo p^{n+1} for each $n = 0, 1, 2, \ldots$ Of course, the choice of the first term b_0 is predetermined, and given in the problem statement; let us note that the numbers $x_0 = b_0$ and x_0^p are congruent modulo p by Fermat's little theorem. Suppose that the base p digits $b_1, b_2, \ldots, b_n$ are already chosen in such a way that $x_n^p \equiv x_n$ (mod p^{n+1}). We shall prove that there is a unique digit b_{n+1} such that

$$(x_n + b_{n+1} p^{n+1})^p \equiv x_n + b_{n+1} p^{n+1} \pmod{p^{n+2}};$$

this proves the existence and the uniqueness at the same time. Since

$$(x_n + b_{n+1} p^{n+1})^p = x_n^p + \binom{p}{1} x_n^{p-1} b_{n+1} p^{n+1} + C p^{n+2}$$

for some integer constant C, and since $\binom{p}{1}$ is divisible by p, we get

$$(x_n + b_{n+1} p^{n+1})^p \equiv x_n^p \pmod{p^{n+2}}.$$

Hence b_{n+1} should satisfy the congruence

$$x_n^p - x_n - b_{n+1} p^{n+1} \equiv 0 \pmod{p^{n+2}}. \tag{1}$$

By the induction hypothesis, the number $x_n^p - x_n$ is divisible by p^{n+1}. This implies that its $(n+2)$nd base p digit (from right to left) is indeed the only choice for b_{n+1} such that (1) holds. The inductive proof is complete.

Problem 7.1.15. *Determine all integers $n > 1$ such that $\frac{2^n+1}{n^2}$ is an integer.*

(31st International Mathematical Olympiad)

Solution. We will prove that the problem has only the solution $n = 3$. First, observe that n is an odd number. Then, we prove that $3 \mid n$.

Let p be the least prime divisor of n. Since $n^2 \mid 2^n + 1, 2^n + 1 \equiv 0$ (mod p) and $2^{2n} \equiv 1$ (mod p). By Fermat's little theorem, $2^{p-1} \equiv 1$ (mod p). Then $2^d \equiv 1$ (mod p), where $d = \gcd(p - 1, 2n)$. By the definition of p, d has no prime divisor greater than 2, which shows that $d = 2$. It follows that $p = 3$.

Let $n = 3^k m$, where $k \geq 1$ and $(3, m) = 1$. Using the identity

$$x^{3^k} + 1 = (x + 1)(x^2 - x + 1)(x^{2 \cdot 3} - x^3 + 1) \cdots (x^{2 \cdot 3^{k-1}} - x^{3^{k-1}} + 1)$$

we obtain the decomposition

$$2^{3^k m}+1 = (2^m+1)(2^{2m}-2^m+1)(2^{2\cdot3m}-2^{3m}+1)\cdots(2^{2\cdot3^{k-1}m}-2^{3^{k-1}m}+1). \quad (1)$$

Let us remark that $2^3 \equiv -1 \pmod 9$; hence $2^{3^k} \equiv -1 \pmod 9$ for all $k \geq 1$. Since $2^{2s} - 2^s + 1 \equiv 3 \pmod 9$ for s of the form 3^j, we obtain in (1) that

$$3^k \mid (2^{2m} - 2^m + 1)(2^{2\cdot3m} - 2^{3m} + 1)\cdots(2^{2\cdot3^{k-1}m} - 2^{3^{k-1}m} + 1)$$

but 3^{k+1} does not divides the product. Therefore, $3^k \mid 2^m + 1$. Since 3 does not divide m and

$$2^m +1 = (3-1)^m +1 = 3^m - \binom{m}{1}3^{m-1}+\cdots-\binom{m}{m-1}3 \equiv -3m \pmod 9,$$

we obtain $k = 1$.

Now we have $n = 3m$ and $9m^2 \mid 2^{3m} + 1$. We repeat, in some way, the starting argument. Take q the least prime divisor of m, $2^{6m} \equiv 1 \pmod q$, $2^{q-1} \equiv 1 \pmod q$, and $\delta = \gcd(6m, q-1)$. By the definition of q we can have $\delta = 1, 2, 3$ or 6 and we also have $2^\delta \equiv 1 \pmod q$. Thus q can be chosen among prime divisors of the numbers $3, 7, 63$. Since $q > 3$, we can have only $q = 7$. Returning to $m^2 \mid 2^{3m} + 1$, we obtain $49 \mid 2^{3m} + 1$. But we have $2^{3m} + 1 \equiv 2 \pmod 7$, and we get a contradiction.

Thus, $m = 1$ and $n = 3$.

Problem 7.1.16. *Prove that $n \mid 2^{n-1} + 1$ fails for all $n > 1$.*

(Sierpiński)

Solution. Although very short, the proof is tricky. Let $n = \prod_{i=1}^s p_i^{k_i}$, where $p_1 < \cdots < p_s$ are prime numbers. The idea is to look at $v_2(p_i - 1)$. Choose the p_i that minimizes this quantity and write $p_i = 1 + 2^{r_i}m_i$ with m_i odd. Then of course we have $n \equiv 1 \pmod{2^{r_i}}$. Hence we can write $n - 1 = 2^m t$. We have $2^{2^{t_i}t} \equiv -1 \pmod{p_i}$; thus we surely have $-1 \equiv 2^{2^{r_i}tm_i} \equiv 2^{(p_i-1)t} \equiv 1 \pmod{p_i}$ (the last congruence being derived from Fermat's theorem). Thus $p_i = 2$, which is clearly impossible.

Problem 7.1.17. *Prove that for any natural number n, $n!$ is a divisor of*

$$\prod_{k=0}^{n-1}(2^n - 2^k).$$

Solution. Let us take a prime number p. Of course, for the argument to be nontrivial, we take $p \leq n$ (otherwise, it doesn't divide $n!$). First, let us see what happens with $p = 2$. We have

$$e_2(n) = n - S_2(n) \leq n - 1$$

and also

$$v_2\left(\prod_{k=0}^{n-1}(2^n - 2^k)\right) = \sum_{k=0}^{n-1} v_2(2^n - 2^k) \geq n - 1$$

(since $2^n - 2^k$ is even for $k \geq 1$), so we are done with this case. Now let us assume that $p > 2$. We have $p \mid 2^{p-1} - 1$ from Fermat's theorem, so we also have $p \mid 2^{k(p-1)} - 1$ for all $k \geq 1$. Now,

$$\prod_{k=0}^{n-1}(2^n - 2^k) = 2^{\frac{n(n-1)}{2}} \prod_{k=1}^{n}(2^k - 1)$$

and so from the above remarks we infer that

$$v_p\left(\prod_{k=0}^{n-1}(2^n - 2^k)\right) = \sum_{k=1}^{n} v_p(2^k - 1),$$

$$\geq \sum_{1 \leq k(p-1) \leq n} v_p(2^{k(p-1)} - 1) \geq \mathrm{card}\{k \mid 1 \leq k(p-1) \leq n\}.$$

Since

$$\mathrm{card}\{k \mid 1 \leq k(p-1) \leq n\} = \left\lfloor \frac{n}{p-1} \right\rfloor,$$

we have found that

$$v_p\left(\prod_{k=0}^{n-1}(2^n - 2^k)\right) \geq \left\lfloor \frac{n}{p-1} \right\rfloor.$$

But we know that

$$e_p(n) = \frac{n - s_p(n)}{p-1} \leq \frac{n-1}{p-1} < \frac{n}{p-1},$$

and since $e_p(n)$ is an integer, we must have

$$e_p(n) \leq \left\lfloor \frac{n}{p-1} \right\rfloor.$$

From these two inequalities, we conclude that

$$v_p\left(\prod_{k=0}^{n-1}(2^n - 2^k)\right) \geq e_p(n),$$

and now the problem is solved.

7.2 Euler's Theorem

Problem 7.2.5. *Prove that for every positive integer n, there exists a polynomial with integer coefficients whose values at $1, 2, \ldots, n$ are different powers of 2.*

(1999 Hungarian Mathematical Olympiad)

Solution. It suffices to prove the claim when $n \geq 4$, because the same polynomial that works for $n \geq 4$ works for $n \leq 3$. For each $i = 1, 2, \ldots, n$, consider the product $s_i = \prod_{j=1, j \neq i}^{n} (i - j)$. Because $n \geq 4$, one of the terms $i - j$ equals 2, and s_i is even. Thus, we can write $s_i = 2^{q_i} m_i$ for positive integers q_i, m_i with m_i odd. Let $L = \max(q_i)$. For each i there are infinitely many powers of 2 that are congruent to 1 mod m_i. (Specifically, by Euler's theorem, $2^{\phi(m_i)j} \equiv 1 \pmod{m_i}$ for all $j \geq 0$.) Thus there are integers c_i such that $c_i m_i + 1$ is a power of 2. Choose such a c_i so that $c_i m_i + 1$ are distinct powers of 2, and define

$$P(x) = 2^L + \sum_{i=1}^{n} c_i 2^{L-q_i} \prod_{j \neq i} (x - j).$$

For each k, $1 \leq k \leq n$, the term $\prod_{j \neq i} (x - j)$ vanishes at $x = k$ unless $k = i$. Therefore

$$P(k) = 2^L + c_k 2^{L-q_k} \prod_{j \neq k} (k - j) = 2^L (c_k m_k + 1),$$

a power of 2. Moreover, since we choose the $c_i m_i + 1$ to be distinct, they are different powers of 2, as needed.

Problem 7.2.6. *Let $a > 1$ be an odd positive integer. Find the least positive integer n such that 2^{2000} is a divisor of $a^n - 1$.*

(2000 Romanian International Mathematical Olympiad Team Selection Test)

Solution. Since a is odd, $(a, 2^k) = 1$, for any $k \geq 0$. Hence, by Euler's theorem, $a^{\varphi(2^k)} \equiv 1 \pmod{2^k}$. Since $\varphi(2^k) = 2^{k-1}$ and we are looking for the least exponent n such that $a^n \equiv 1 \pmod{2^{2000}}$, it follows that n is a divisor of $2^{1999} = \varphi(2^{2000})$.

If $a \equiv 1 \pmod{2^{2000}}$, it follows that $n = 1$. We shall omit this case.

Consider the decomposition

$$a^{2^m} - 1 = (a - 1)(a + 1)(a^2 + 1)(a^{2^2} + 1) \cdots (a^{2^{m-1}} + 1).$$

Assume $a \equiv 1 \pmod{2^s}$ and $a \not\equiv 1 \pmod{2^{s+1}}$, where $2 \leq s \leq 1999$. That is, $a = 2^s b + 1$, where b is an odd number. Equivalently, a has the binary representation

$$a = 1 \ldots 1 \underbrace{0 0 \ldots 1}_{s \text{ digits}}.$$

It is easy to show that for any integer x, $x^2 + 1$ is not divisible by 4. Then, by the above decomposition, $a^{2^m} - 1$ is divisible by 2^{s+m} and it is not divisible by 2^{s+m+1}. Hence, the required number is 2^{2000-s}.

Assume that $a \equiv -1 \pmod{2^s}$ and $a \not\equiv -1 \pmod{2^{s+1}}$, where $s \geq 2$. Equivalently, a has the binary representation

$$a = 1 \ldots 0 \underbrace{11 \ldots 1}_{s \text{ digits}}.$$

As before, $a-1$ is divisible by 2 and not divisible by 2^2, and $a^{2^k} + 1$ is divisible by 2 and not divisible by 2^2, for all $k \geq 1$. From the above decomposition, $a^{2^m} - 1$ is divisible by 2^{s+m} and not divisible by 2^{s+m+1}. Hence, in this case, the required exponent is $n = 2^{2000-s}$ when $s \leq 1999$, and $n = 2$ when $s \geq 2000$.

Problem 7.2.7. *Let* $n = p_1^{r_1} \cdots p_k^{r_k}$ *be the prime factorization of the positive integer n and let $r \geq 2$ be an integer. Prove that the following are equivalent:*
(a) The equation $x^r \equiv a \pmod{n}$ has a solution for every a.
(b) $r_1 = r_2 = \cdots = r_k = 1$ and $\gcd(p_i - 1, r) = 1$ for every $i \in \{1, 2, \ldots, k\}$.

(1995 UNESCO Mathematical Contest)

Solution. If (b) holds, then $\varphi(n) = (p_1 - 1) \cdots (p_k - 1)$ is coprime to r; thus there exists s with $rs \equiv 1 \pmod{\phi(n)}$, and the unique solution of $x^r \equiv a \pmod{n}$ is $x \equiv a^s \pmod{n}$. Conversely, suppose $x^r \equiv a \pmod{n}$ has a solution for every a; then $x^r \equiv a \pmod{p_i^{r_i}}$ also has a solution for every a. However, if $r_i > 1$ and a is a number divisible by p but not by p^2, then x^r cannot be congruent to a, since it is not divisible by p unless x is divisible by p, in which case it is already divisible by p^2. Hence $r_1 = 1$.

If $d = \gcd(p_i - 1, r)$ and every a is congruent to x^r for some x, then $a^{(p_i-1)/d} \equiv 1 \pmod{p_i}$ for all a. Hence by Lagrange's theorem, the polynomial $P(t) = t^{(p_i-1)/d} - 1$ must have degree $p_i - 1$, that is, $d = 1$.

7.3   The Order of an Element

Problem 7.3.6. *Find all ordered triples of primes (p, q, r) such that*

$$p \mid q^r + 1, \quad q \mid r^p + 1, \quad r \mid p^q + 1.$$

(2003 USA International Mathematical Olympiad Team Selection Test)

Solution. It is quite clear that p, q, r are distinct. Indeed, if, for example, $p = q$, then the relation $p \mid q^r + 1$ is impossible. We will prove that we cannot have $p, q, r > 2$. Suppose this is the case. The first condition $p \mid q^r + 1$ implies

$p \mid q^{2r} - 1$, and so $o_p(q) \mid 2r$. If $o_p(q)$ is odd, it follows that $p \mid q^r - 1$, which combined with $p \mid q^r + 1$ yields $p = 2$, which is impossible. Thus, $o_p(q)$ is either 2 or $2r$. Could we have $o_p(q) = 2r$? No, since this would imply that $2r \mid p - 1$, and so $0 \equiv p^q + 1 \equiv 2 \pmod{r}$, that is, $r = 2$, false. Therefore, the only possibility is $o_p(q) = 2$, and so $p \mid q^2 - 1$. We cannot have $p \mid q - 1$, because $p \mid q^r + 1$ and $p \neq 2$. Thus, $p \mid q + 1$ and in fact $p \mid \frac{q+1}{2}$. In the same way, we find that $q \mid \frac{r+1}{2}$ and $r \mid \frac{p+1}{2}$. This is clearly impossible, just by looking at the largest among p, q, r. So, our assumption was wrong, and indeed one of the three primes must equal 2. Suppose without loss of generality that $p = 2$. Then q is odd, $q \mid r^2 + 1$, and $r \mid 2^q + 1$. Similarly, $o_r(2) \mid 2q$. If $q \mid o_r(2)$, then $q \mid r - 1$, and so $q \mid r^2 + 1 - (r^2 - 1) = 2$, which contradicts the already established result that q is odd. Thus, $o_r(2) \mid 2$ and $r \mid 3$. As a matter of fact, this implies that $r = 3$ and $q = 5$, yielding the triple $(2, 5, 3)$. It is immediate to verify that this triple satisfies all conditions of the problem. Moreover, all solutions are given by cyclic permutations of the components of this triple.

Problem 7.3.7. *Find all primes p, q such that $pq \mid 2^p + 2^q$.*

Solution. Note that $(p, q) = (2, 2), (2, 3), (3, 2)$ satisfy this property and let us show that there are no other such pairs. Assume, by contradiction, that $p \neq 2$ and $q \neq 2$. Write $p - 1 = 2^l n$, $q - 1 = 2^k m$, where m, n are odd positive integers. Because $pq \mid 2^p + 2^q$, using Fermat's little theorem, we obtain $0 \equiv 2^p + 2^q \equiv 2^p + 2 \pmod{q}$. It follows that $2^{p-1} \equiv -1 \pmod{q}$. If we set $x = 2^n$, then we have $x^{2^l} \equiv -1 \pmod{q}$; hence $o(x) = 2^{l+1}$ (since $x^{2^{l+1}} \equiv 1 \pmod{q}$ and $x^{2^l} \not\equiv 1 \pmod{q}$). It follows that $2^{l+1} = o_q(x) \mid \varphi(q) = q - 1 = 2^k m$, i.e., $l + 1 \leq k$.

In a similar way we can prove that $k + 1 \leq l$, and we get $l \leq k - 1 \leq l - 2$, a contradiction. Therefore, it is necessary to have $p = 2$ or $q = 2$. If, for example, $q = 2$, then $p \mid 2^p + 2^q = 2^p + 2^2$, $0 \equiv 2^p + 2^2 \equiv 2 + 2^2 = 6 \pmod{p}$, and we get $p \in \{2, 3\}$.

Problem 7.3.8. *Prove that for any integer $n \geq 2$, $3^n - 2^n$ is not divisible by n.*

Solution. Assume by contradiction that $n \mid 3^n - 2^n$ for some positive integer n. Let us denote by p the smallest prime divisor of n. Since $n \mid 3^n - 2^n$, it follows that $p \geq 5$. Consider a positive integer a such that $2a \equiv 1 \pmod{p}$. From $3^n \equiv 2^n \pmod{p}$ we obtain $(3a)^n \equiv 1 \pmod{p}$. Let $d = o_p(3a)$. It follows that $d \mid p - 1$ and $d \mid n$. But $d < p$ and $d \mid n$ implies $d = 1$, because of the minimality of p. We get $3a \equiv 1 \pmod{p}$ and $2a \equiv 1 \pmod{p}$, i.e., $a \equiv 0 \pmod{p}$, a contradiction to $2a \equiv 1 \pmod{p}$.

Problem 7.3.9. *Find all positive integers m, n such that $n \mid 1 + m^{3^n} + m^{2 \cdot 3^n}$.*

(Bulgarian International Mathematical Olympiad Team Selection Test)

Solution. From $n \mid 1 + m^{3^n} + m^{2 \cdot 3^n}$ it follows that $n \mid m^{3^{n+1}} - 1$; hence $d = o_n(m)$ divides 3^{n+1}, i.e., $d = 3^k$ for some positive integer k. If $k \leq n$, then $d \mid 3^n$ implies

$n \mid m^{3^n} - 1$. Combining with $n \mid 1 + m^{3^n} + m^{2 \cdot 3^n}$ it follows that $n = 3$. If $k \geq n+1$, then $d = 3^{n+1}$ and $d \mid \varphi(n)$ implies $d < n$, impossible, since $3^{n+1} > n$. Therefore $n = 3$ and, consequently, $m \equiv 1 \pmod 3$.

Problem 7.3.10. *Let $a, n > 2$ be positive integers such that $n \mid a^{n-1} - 1$ and n does not divide any of the numbers $a^x - 1$, where $x < n - 1$ and $x \mid n - 1$. Prove that n is a prime number.*

Solution. Set $d = o_n(a)$. Since $n \mid a^{n-1} - 1$, it follows that $d \mid n - 1$. If $d < n - 1$, then we contradict the hypothesis that n does not divide $a^d - 1$. Hence $d \geq n - 1$ and consequently $d = n - 1$.

On the other hand, we have $d \mid \varphi(n)$; hence $n - 1 \mid \varphi(n)$. Taking into account that $\varphi(n) \leq n - 1$, we find that $\varphi(n) = n - 1$, and it follows that n must be a prime number.

Problem 7.3.11. *Find all prime numbers p, q for which the congruence*

$$\alpha^{3pq} \equiv \alpha \pmod{3pq}$$

holds for all integers α.

(1996 Romanian Mathematical Olympiad)

Solution. Without loss of generality assume $p \leq q$; the unique solution will be $(11, 17)$, for which one may check the congruence using the Chinese remainder theorem. We first have $2^{3pq} \equiv 2 \pmod 3$, which means that p and q are odd. In addition, if α is a primitive root mod p, then $\alpha^{3pq-1} \equiv 1 \pmod p$ implies that $p - 1$ divides $3pq - 1$ as well as $3pq - 1 - 3q(p - 1) = 3q - 1$, and conversely that $q - 1$ divides $3p - 1$. If $p = q$, we now deduce $p = q = 3$, but $4^{27} \equiv 1 \pmod{27}$, so this fails. Hence $p < q$.

Since p and q are odd primes, $q \geq p + 2$, so $(3p - 1)/(q - 1) < 3$. Since this quantity is an integer, and it is clearly greater than 1, it must be 2. That is, $2q = 3p + 1$. On the other hand, $p - 1$ divides $3q - 1 = (9p + 1)/2$ as well as $(9p + 1) - (9p - 9) = 10$. Hence $p = 11, q = 17$.

Remark. A composite integer n such that $a^n \equiv a \pmod n$ for all integers a is called a *Carmichael number*. Very recently, W.R. Alford, A. Granville, and C. Pomerance [*Annals Math.*, 139(1994), 703–722] proved that there are infinitely many Carmichael numbers. Using the ideas outlined in the solution of the above problem, one can show that n is a Carmichael number if and only if it is of the form $p_1 p_2 \cdots p_k$, with p_i different prime numbers such that $p_i - 1 \mid n - 1$ for all $i = 1, 2, \ldots, k$ and $k > 1$.

Problem 7.3.12. *Let p be a prime number. Prove that there exists a prime number q such that for every integer n, the number $n^p - p$ is not divisible by q.*

(44th International Mathematical Olympiad)

Solution. Note that $\frac{p^p-1}{p-1} = p^{p-1}+\cdots+p+1$ must have at least one prime factor q that is not congruent to 1 (mod p^2). We will show that this q works. Note that $p^p \equiv 1$ (mod q) and that $q \nmid p-1$. For the latter, we note that if $q \mid p-1$ then $p^{p-1}+\cdots+p+1 \equiv p \equiv 1$ (mod q), a contradiction. Hence $o_q(p) = p$ and $q \equiv 1$ (mod p). Suppose now that $q \mid n^p - p$ for some integer n. Then $n^p \equiv p$ (mod q) and $n^{p^2} \equiv p^p \equiv 1$ (mod q). Hence $o_q(n) = p^2$ and $q \equiv 1$ (mod p^2), contrary to our assumption.

Remark. Taking $q \equiv 1$ (mod p) is natural, because for every other q, n^p takes all possible residues modulo p (including p too). Indeed, if $p \nmid q-1$, then there is an $r \in \mathbb{N}$ satisfying $pr \equiv 1$ (mod $q-1$); hence for any a the congruence $n^p \equiv a$ (mod q) has the solution $n \equiv a^r$ (mod q) (see also the lemma from Problem 7.1.10).

7.4 Wilson's Theorem

Problem 7.4.5. *Let p be an odd prime. Prove that*

$$1^2 \cdot 3^2 \cdots (p-2)^2 \equiv (-1)^{\frac{p+1}{2}} \quad (\text{mod } p)$$

and

$$2^2 \cdot 4^2 \cdots (p-1)^2 \equiv (-1)^{\frac{p+1}{2}} \quad (\text{mod } p).$$

Solution. Using Wilson's theorem, we have $(p-1)! \equiv -1$ (mod p); hence

$$\big(1 \cdot 3 \cdots (p-2)\big)\big(2 \cdot 4 \cdots (p-1)\big) \equiv -1 \quad (\text{mod } p).$$

On the other hand,

$$1 \equiv -(p-1) \quad (\text{mod } p), \quad 3 \equiv -(p-3) \quad (\text{mod } p), \quad \ldots,$$

$$p-2 \equiv -\big(p-(p-2)\big) \quad (\text{mod } p).$$

Therefore

$$1 \cdot 3 \cdots (p-2) \equiv (-1)^{\frac{p-1}{2}}(2 \cdot 4 \cdots (p-1)) \quad (\text{mod } p),$$

and the conclusion follows.

Problem 7.4.6. *Show that there do not exist nonnegative integers k and m such that $k! + 48 = 48(k+1)^m$.*

(1996 Austrian–Polish Mathematics Competition)

Solution. Suppose such k, m exist. We must have $48 \mid k!$, so $k \geq 6$; one checks that $k = 6$ does not yield a solution, so $k \geq 7$. In that case $k!$ is divisible by 32 and by 9, so that $(k! + 48)/48$ is relatively prime to 6, as then is $k + 1$.

If $k+1$ is not prime, it has a prime divisor greater than 3, but this prime divides $k!$ and not $k! + 48$. Hence $k + 1$ is prime, and by Wilson's theorem, $k! + 1$ is a multiple of $k + 1$. Since $k! + 48$ is as well, we find that $k + 1 = 47$, and we need only check that $46!/48 + 1$ is not a power of 47. We check that $46!/48 + 1 \equiv 29$ (mod 53) (by canceling as many terms as possible in $46!$ before multiplying), but that 47 has order 13 modulo 53 and that none of its powers is congruent to 29 modulo 53.

Remark. Another argument for why $(46!/48) + 1$ is not a power of 47 is the following. One has that $(46!/48) + 1 \equiv 329 = 7 \cdot 47 \pmod{47^2}$. The least computational argument I could find was that clearly $(46!/48) + 1$ is congruent to 1 mod 5, 7, and 11 (as well as many other primes) and that $o_5(47) = o_5(2) = 4$, $o_7(47) = o_7(5) = 6$, and $o_{11}(47) = o_{11}(3) = 5$. Thus the least power of 47 that is congruent to 1 mod all three of these primes is $47^{\mathrm{lcm}(4,6,5)} = 47^{60}$. But clearly $(46!/48) + 1 < 47^{46} < 47^{60}$. Thus $(46!/48) + 1$ is not a power of 47.

Problem 7.4.7. *For each positive integer n, find the greatest common divisor of $n! + 1$ and $(n + 1)!$.*

(1996 Irish Mathematical Olympiad)

Solution. Let $f(n) = \gcd(n! + 1, (n+1)!)$. If $n+1$ is composite, then each prime divisor of $(n + 1)!$ is a prime less than n, which also divides $n!$ and so does not divide $n! + 1$. Hence $f(n) = 1$. If $n + 1$ is prime, the same argument shows that $f(n)$ is a power of $n+1$, and in fact $n+1 \mid n!+1$ by Wilson's theorem. However, $(n + 1)^2$ does not divide $(n + 1)!$, and thus $f(n) = n + 1$.

Problem 7.4.8. *Let $p \geq 3$ be a prime and let σ be a permutation of $\{1, 2, \ldots, p - 1\}$. Prove that there are $i \neq j$ such that $p \mid i\sigma(i) - j\sigma(j)$.*

(1986 Romanian International Mathematical Olympiad Team Selection Test)

Solution. Assume by contradiction that p does not divide $i\sigma(i) - j\sigma(j)$ for any $i, j = 1, 2, \ldots, p - 1, i \neq j$. Then, the integers $i\sigma(i), i = 1, 2, \ldots, p - 1$, are all not divisible by p and give distinct residues modulo p. We have

$$\prod_{i=1}^{p-1}(i\sigma(i)) \equiv \prod_{i=1}^{p-1} i = (p - 1)! \equiv -1 \pmod{p}.$$

On the other hand, $\prod_{i=1}^{p-1}(i\sigma(i)) = \prod_{i=1}^{p-1}((p - 1)!)^2 \equiv 1 \pmod{p}$, a contradiction.

8

Diophantine Equations

8.1 Linear Diophantine Equations

Problem 8.1.4. *Solve in integers the equation*

$$(x^2 + 1)(y^2 + 1) + 2(x - y)(1 - xy) = 4(1 + xy).$$

Solution. The equation is equivalent to

$$x^2y^2 - 2xy + 1 + x^2 + y^2 - 2xy + 2(x - y)(1 - xy) = 4,$$

or

$$(xy - 1)^2 + (x - y)^2 + 2(x - y)(1 - xy) = 4.$$

Hence $(1 - xy + x - y)^2 = 4$, and consequently, $|(1 + x)(1 - y)| = 2$.

We have two cases:

(i) $|x + 1| = 1$ and $|y - 1| = 2$, giving $(0, 3)$, $(0, -1)$, $(-2, 3)$ and $(-2, -1)$, and

(ii) $|x + 1| = 2$ and $|y - 1| = 1$, giving $(1, 2)$, $(1, 0)$, $(-3, 2)$ and $(-3, 0)$.

Problem 8.1.5. *Determine the side lengths of a right triangle if they are integers and the product of the legs' lengths equals three times the perimeter.*

(1999 Romanian Mathematical Olympiad)

First solution. Let a, b, c be the lengths of the triangle's sides. We have

$$a^2 = b^2 + c^2$$

and

$$bc = 3(a + b + c).$$

T. Andreescu and D. Andrica, *Number Theory*, DOI: 10.1007/b11856_19,
© Birkhäuser Boston, a part of Springer Science + Business Media, LLC 2009

Let $P = a + b + c$. Then $bc = 3P$ and

$$b^2 + c^2 = (b + c)^2 - 2bc = (P - a)^2 - 6P = P^2 + a^2 - 2aP - 6P.$$

It follows that

$$a^2 = P^2 + a^2 - 2aP - 6P,$$

so

$$P = 2a + 6,$$

that is,

$$a = b + c - 6.$$

We have then

$$b^2 + c^2 = b^2 + c^2 + 2bc - 12b - 12c + 36$$

if and only if

$$bc - 6b - 6c + 18 = 0,$$

that is,

$$(b - 6)(c - 6) = 18.$$

Analyzing the ways in which 18 can be written as a product of integers, we find the following solutions:

$$(a, b, c) \in \{(25, 7, 24), (25, 24, 7), (17, 8, 15), (17, 15, 8), (15, 9, 12), (15, 12, 9)\}.$$

Second solution. From $bc = 3(a + b + c)$, we get $bc - 3b - 3c = 3a$ and square to get $(bc - 3b - 3c)^2 = 9a^2 = 9(b^2 + c^2)$. From there multiplying out gives $bc[(b - 6)(c - 6) - 18] = 0$. Since $bc \neq 0$, we get $(b - 6)(c - 6) = 18$, and enumerating the factors of 18 gives the list of solutions.

Problem 8.1.6. *Let a, b, and c be positive integers, each two of them being relatively prime. Show that $2abc - ab - bc - ca$ is the largest integer that cannot be expressed in the form $xbc + yca + zab$, where x, y, and z are nonnegative integers.*

(24th International Mathematical Olympiad)

Solution. We will solve the problem in two steps.

First step. The number $2abc - ab - bc - ca$ cannot be expressed in the required form. Assume the contrary, that

$$2abc - ab - bc - ca = xbc + yca + zab,$$

where $x, y, z \geq 0$. Then, one obtains the combination

$$2abc = bc(x + 1) + ca(y + 1) + ab(z + 1),$$

where $x+1 > 0$, $y+1 > 0$, $z+1 > 0$. This leads to the divisibility $a \mid bc(x+1)$.

Since a is relatively prime to b and c, a divides $x + 1$ and then $a \leq x + 1$. Using similar arguments, $b \leq y + 1$ and $c \leq z + 1$. Thus, $2abc = bc(x + 1) + ca(y + 1) + ab(z + 1) \geq 3abc$. This is a contradiction.

Second step. Any number N, $N > 2abc - ab - bc - ca$, can be expressed in the form $N = xbc + yca + zab$.

Since $\gcd(ab, bc, ca) = 1$, by Theorem 8.1.1 we can write $N = abx + bcy + caz$ for some integers x, y, z. If (x, y, z) is one solution, then so is $(x \pm c, y \mp a, z)$. Hence we may assume $0 \leq x \leq c - 1$. Similarly, if $(x, y \pm a, z \mp b)$ is a solution, so we may assume $0 \leq y \leq a - 1$. But then

$$z = \frac{1}{ac}[N - abx - bcy] > \frac{1}{ac}[2abc - ab - bc - ca - ab(c-1) - bc(b-1)] = -1.$$

Thus z is again a nonnegative integer and we are done.

Remark. One can prove that if $a_1, a_2, \ldots, a_k \in \mathbb{Z}$ are positive integers such that $\gcd(a_1, \ldots, a_k) = 1$, then any sufficiently large n is a linear combination with nonnegative coefficients of $a_1, \ldots, a_k$. The smallest such n for $k \geq 4$ is unknown. This is the famous problem of Frobenius.

8.2 Quadratic Diophantine Equations

8.2.1 *Pythagorean Equations*

Problem 8.2.3. *Find all Pythagorean triangles whose areas are numerically equal to their perimeters.*

First solution. From (3), the side lengths of such a triangle are

$$k(m^2 - n^2), \quad 2kmn, \quad k(m^2 + n^2).$$

The condition in the problem is equivalent to

$$k^2mn(m^2 - n^2) = 2km(m + n),$$

which reduces to

$$kn(m - n) = 2.$$

A simple case analysis shows that the only possible triples (k, m, n) are $(2, 2, 1)$, $(1, 3, 2)$, $(1, 3, 1)$, yielding the Pythagorean triangles $6 - 8 - 10$ and $5 - 12 - 13$.

Second solution. This solution does not use Theorem 8.2.1, and it is similar to the solution of Problem 8.1.5. Rewrite the equation $a + b + c = ab/2$ as $2c = ab - 2a - 2b$ and square to get $4(a^2 + b^2) = 4c^2 = (ab - 2a - 2b)^2$ and rearrange to get $ab[(a - 4)(b - 4) - 8] = 0$. Then the solutions follow from the factors of 8.

Problem 8.2.4. *Prove that for every positive integer n there is a positive integer k such that k appears in exactly n nontrivial Pythagorean triples.*

(American Mathematical Monthly)

First solution. We will prove by induction that 2^{n+1} appears in exactly n Pythagorean triples. The base case $n = 1$ holds for $(3, 2^2, 5)$, and that is the only such triple. Assume that (x_k, y_k, z_k), where $x_k = u_k^2 - v_k^2$, $y_k = 2u_k v_k$, $z_k = u_k^2 + v_k^2$, $k = 1, \ldots, n$, are the n triples containing 2^{n+1}. Then $(2x_k, 2y_k, 2z_k)$, $k = 1, \ldots, n$, are n imprimitive Pythagorean triples containing 2^{n+2}, and $(2^{2n+2} - 1, 2^{n+2}, 2^{2n+2} + 1)$ is the only such primitive triple.

No other triple with this property exists. Indeed, if $(u^2 - v^2, 2uv, u^2 + v^2)$ were a triple containing 2^{n+2}, then we would have the following cases:

(i) $u^2 + v^2 = 2^{n+2}$. Simplifying by the greatest possible power of 2, we get $a^2 + b^2 = 2^k$, where a and b are not both even. Then the left-hand side is congruent to 1 or 2 (mod 4), while the right-hand side is 0 (mod 4), a contradiction.

(ii) $2uv = 2^{n+2}$. We simplify again by the greatest power of 2 and obtain $ab = 2^s$, where $a > b$ are not both even and $s \geq 1$. It follows that $a = 2^s$ and $b = 1$, yielding the triple generated by $(2^{2s} - 1, 2^{s+1}, 2^{2s} + 1)$ multiplied by a power of 2, which is clearly among the imprimitive triples $(2x_k, 2y_k, 2z_k)$.

(iii) $u^2 - v^2 = 2^{n+2}$. Simplifying again by the greatest power of 2, we arrive at $a^2 - b^2 = 2^t$, where a and b are not both even and $t \geq 3$. If one of a and b is even, then the left-hand side is odd, while the right-hand side is even, a contradiction. If a and b are both odd, then $a - b = 2$ and $a + b = 2^{t-1}$, yielding $a - 2^{t-2}$ and $b = 2^{t-2} - 1$. Again, we get a triple generated by $(2^t, 2(2^{2t-4} - 1), 2(2^{2t-4} + 1))$ multiplied by a power of 2, which is clearly already an imprimitive triple of the form $(2x_k, 2y_k, 2z_k)$.

Second solution. We show that 3^N appears in exactly N Pythagorean triples. Obviously 3^N cannot be the even side. To see that 3^N cannot be the hypotenuse, suppose N is the least power of 3 that occurs as the hypotenuse. Squares are 0 or 1 (mod 3); hence $3 \mid x^2 + y^2$ implies $3 \mid x$ and $3 \mid y$. But then canceling a common factor of 3 gives a Pythagorean triple with 3^{N-1} as the hypotenuse, contradicting the choice of N. Thus the number of Pythagorean triples in which it appears is exactly the number of solutions to $3^N = k(m^2 - n^2)$ with $\gcd(m, n) = 1$. For such a solution $k = 3^r$ for some $0 \leq r \leq N - 1$ and $(m - n)(m + n) = 3^{N-r}$. But if $m + n$ and $m - n$ are both nontrivial powers of 3, then we get $3 \mid m$ and $3 \mid n$, contradicting the fact that $\gcd(m, n) = 1$. Thus $m + n = 3^{N-r}$ and $m - n = 1$. Thus we get exactly N solutions.

Problem 8.2.5. *Find the least perimeter of a right-angled triangle whose sides and altitude are integers.*

(Mathematical Reflections)

Solution. The answer for the least possible perimeter is 60. It holds for a right-angled triangle $(15, 20, 25)$, whose altitude is 12.

Let x, y, z be a Pythagorean triple with z the hypotenuse and let h be the altitude of the triangle. Let $d = \gcd(x, y, z)$ be the greatest common divisor of x, y, z. We have that $x = d \cdot a$, $y = d \cdot b$, and $z = d \cdot c$ for a, b, c a primitive Pythagorean triple. Now calculating the area of the triangle in two ways, we obtain that $h = \frac{xy}{z} = \frac{abd}{c}$. Using the fact that $\gcd(ab, c) = 1$, we get $c \mid d$, which tells us that $c \leq d$, since both are positive integers. Because the perimeter is equal to $d(a + b + c) \geq c(a + b + c)$, we can minimize it by taking $c = d$. Then the altitude of a right-angled triangle having sides ca, cb, c^2 (with c^2 the hypotenuse) is ab, an integer. We get the perimeter

$$p = c(a + b + c). \tag{1}$$

We know that a, b, c is a primitive Pythagorean triple if and only if there exist $m, n \in \mathbb{Z}^+$ such that $\gcd(m, n) = 1$, $m \not\equiv n \pmod{2}$, $m > n > 0$, that satisfy $a = m^2 - n^2$, $b = 2mn$, $c = m^2 + n^2$. Replacing in (1), we notice that all we need to find is the minimum value of

$$p = (m^2 + n^2)(2m^2 + 2mn).$$

Clearly $m > n > 0$; therefore $m \geq 2$ and $n \geq 1$. Thus

$$p \geq (2^2 + 1)(2 \cdot 2^2 + 2 \cdot 2 \cdot 1) = 60.$$

Now the triangle with sides $(15, 20, 25)$ satisfies all the conditions of the original problem, and its perimeter is 60. the problem is solved.

8.2.2 Pell's Equation

Problem 8.2.6. *Let p be a prime number congruent to 3 modulo 4. Consider the equation*
$$(p + 2)x^2 - (p + 1)y^2 + px + (p + 2)y = 1.$$

Prove that this equation has infinitely many solutions in positive integers, and show that if $(x, y) = (x_0, y_0)$ is a solution of the equation in positive integers, then $p \mid x_0$.

(2001 Bulgarian Mathematical Olympiad)

Solution. We show first that $p \mid x$. Substituting $y = z + 1$ and rewriting, we obtain
$$x^2 = (z - x)((p + 1)(z + x) + p).$$

Let $q = \gcd(z - x, (p + 1)(z + x) + p)$. Then $q \mid x$, and therefore $q \mid z$, and also $q \mid p$. On the other hand, $q \neq 1$, because otherwise both factors on the right-hand side must be perfect squares, yet $(p + 1)(z + x) + p \equiv 3 \pmod{4}$. Thus $q = p$ and $p \mid x$ as desired.

Now write $x = px_1$ and $z = pz_1$ to obtain

$$x_1^2 = (z_1 - x_1)((p + 1)(z_1 + x_1) + 1).$$

By what we showed above, the two terms on the right are coprime and thus must be perfect squares. Therefore, for some a, b we have

$$z_1 - x_1 = a^2, \quad (p + 1)(z_1 + x_1) + 1 = b^2, \quad x_1 = ab.$$

The above equality implies

$$b^2 = (p + 1)(a^2 + 2ab) + 1,$$

i.e.,

$$(p + 2)b^2 - (p + 1)(a + b)^2 = 1.$$

Conversely, given a and b satisfying the last equation, there exists a unique pair (x_1, y_1) satisfying the equation above, and hence a unique pair (x, y) satisfying the original equation.

Thus, we reduced the original equation to a "Pell-type" equation. To get some solutions, look at the odd powers of $\sqrt{p + 2} + \sqrt{p + 1}$. It follows easily that

$$(\sqrt{p + 2} + \sqrt{p + 1})^{2k+1} = m_k\sqrt{p + 2} + n_k\sqrt{p + 1}$$

for some positive integers m_k, n_k. Then

$$(\sqrt{p + 2} - \sqrt{p + 1})^{2k+1} = m_k\sqrt{p + 2} - n_k\sqrt{p + 1},$$

and multiplying the left and right sides gives

$$(p + 2)m_k^2 - (p + 1)n_k^2 = 1.$$

Clearly, $n_k > m_k$, so setting $b_k = m_k, a_k = n_k - m_k$ gives a solution for (a, b). Finally, it is easy to see that the sequences $\{m_k\}, \{n_k\}$ are strictly increasing, so we obtain infinitely many solutions this way.

Problem 8.2.7. *Determine all integers a for which the equation*

$$x^2 + axy + y^2 = 1$$

has infinitely many distinct integer solutions (x, y).

(1995 Irish Mathematical Olympiad)

Solution. The equation has infinitely many solutions if and only if $a^2 \geq 4$. Rewrite the given equation in the form

$$(2x + ay)^2 - (a^2 - 4)y^2 = 4.$$

If $a^2 < 4$, the real solutions to this equation form an ellipse, and so only finitely many integer solutions occur. If $a = \pm 2$, there are infinitely many solutions, since the equation becomes $(x \pm y)^2 = 1$. If $a^2 > 4$, then $a^2 - 4$ is not a perfect square, and so the Pell's equation $u^2 - (a^2 - 4)v^2 = 1$ has infinitely many solutions. But setting $x = u - av$, $y = 2v$ gives infinitely many solutions of the given equation.

Problem 8.2.8. *Prove that the equation*

$$x^3 + y^3 + z^3 + t^3 = 1999$$

has infinitely many integral solutions.

(1999 Bulgarian Mathematical Olympiad)

Solution. Observe that $(m - n)^3 + (m + n)^3 = 2m^3 + 6mn^2$. Now suppose we want a general solution of the form

$$(x, y, z, t) = \left(a - b, a + b, \frac{c}{2} - \frac{d}{2}, \frac{c}{2} + \frac{d}{2}\right)$$

for integers a, b and odd integers c, d. One simple solution to the given equation is $(x, y, z, t) = (10, 10, -1, 0)$, so we try setting $a = 10$ and $c = -1$. Then

$$(x, y, z, t) = \left(10 - b, 10 + b, -\frac{1}{2} - \frac{d}{2}, -\frac{1}{2} + \frac{d}{2}\right)$$

is a solution exactly when

$$(2000 + 60b^2) - \frac{1 + 3d^2}{4} = 1999, \quad \text{i.e., } d^2 - 80b^2 = 1.$$

The second equation is a Pell's equation with solution $(d_1, b_1) = (9, 1)$. We can generate infinitely many more solutions by setting

$$(d_{n+1}, b_{n+1}) = (9d_n + 80b_n, 9b_n + d_n) \text{ for } n = 1, 2, 3, \ldots$$

This can be proved by induction, and it follows from a general recursion

$$(p_{n+1}, q_{n+1}) = (p_1 p_n + q_1 q_n D, p_1 q_n + q_1 p_n)$$

for generating solutions to $p^2 - Dq^2 = 1$ given a nontrivial solution (p_1, q_1).

A quick check also shows that each d_n is odd. Thus because there are infinitely many solutions (b_n, d_n) to the Pell's equation (and with each d_n odd), there are infinitely many integral solutions

$$(x_n, y_n, z_n, t_n) = \left(10 - b_n, 10 + b_n, -\frac{1}{2} - \frac{d_n}{2}, -\frac{1}{2} + \frac{d_n}{2}\right)$$

to the original equation.

8.2.3   Other Quadratic Equations

Problem 8.2.11. *Prove that the equation*

$$x^2 + y^2 + z^2 + 3(x + y + z) + 5 = 0$$

has no solutions in rational numbers.

(1997 Bulgarian Mathematical Olympiad)

Solution. Let $u = 2x + 3$, $v = 2y + 3$, $w = 2z + 3$. Then the given equation is equivalent to

$$u^2 + v^2 + w^2 = 7.$$

It is equivalent to show that the equation

$$x^2 + y^2 + z^2 = 7w^2$$

has no nonzero solutions in integers; assume to the contrary that (x, y, z, w) is a nonzero solution with $|w| + |x| + |y| + |z|$ minimal. Modulo 8, we have $x^2 + y^2 + z^2 \equiv 7w^2$, but every perfect square is congruent to 0, 1, or 4 modulo 8. Thus we must have x, y, z, w even, and $(x/2, y/2, z/2, w/2)$ is a smaller solution, contradiction.

Remark. Try to prove the following theorem of Davenport and Cassels: *for $n \in \mathbb{Z}$, the equation $x^2 + y^2 + z^2 = n$ has rational solutions if and only if it has integer solutions.* There is a beautiful elementary geometric proof. Try to find it!

Problem 8.2.12. *Find all integers x, y, z such that $5x^2 - 14y^2 = 11z^2$.*

(2001 Hungarian Mathematical Olympiad)

Solution. The only solution is $(0, 0, 0)$.

Assume, for the sake of contradiction, that there is a triple of integers $(x, y, z) \neq (0, 0, 0)$ satisfying the given equation, and let $(x, y, z) = (x_0, y_0, z_0)$ be a nonzero solution that minimizes $|x| + |y| + |z| > 0$.

Because $5x_0^2 - 14y_0^2 = 11z_0^2$, we have

$$-2x_0^2 \equiv 4z_0^2 \pmod 7,$$

or $x_0^2 \equiv -2z_0^2 \equiv 5z_0^2 \pmod 7$. Therefore, we have $z_0 \equiv 0 \pmod 7$, because otherwise we have

$$5 \equiv (x_0 z_0^{-1})^2 \pmod 7,$$

which is impossible because 5 is not a square modulo 7. (The squares modulo 7 are 0, 1, 2, and 4.)

It follows that x_0 and z_0 are divisible by 7, so that $14y^2 = 5x^2 - 11z^2$ is divisible by 49. Therefore, $7 \mid y_0$. Then $\left(\frac{x_0}{7}, \frac{y_0}{7}, \frac{z_0}{7}\right)$ is also a solution, but $\left|\frac{x_0}{7}\right| + \left|\frac{y_0}{7}\right| + \left|\frac{z_0}{7}\right| < |x_0| + |y_0| + |z_0|$, contradicting the minimality of (x_0, y_0, z_0).

Therefore, our original assumption was false, and the only integer solution is $(0, 0, 0)$.

Remark. A solution mod 8 also works. If x or z is even, then so is the other, and hence y is even. Thus we can cancel a 2 and get a smaller solution. Suppose we have a solution with x and z odd; then we get $5 - (0 \text{ or } 6) \equiv 3 \pmod 8$, which cannot occur.

Problem 8.2.13. *Let n be a nonnegative integer. Find the nonnegative integers a, b, c, d such that*

$$a^2 + b^2 + c^2 + d^2 = 7 \cdot 4^n.$$

(2001 Romanian JBMO Team Selection Test)

Solution. For $n = 0$, we have $2^2 + 1^2 + 1^2 + 1^2 = 7$; hence $(a, b, c, d) = (2, 1, 1, 1)$ and all permutations. If $n \geq 1$, then $a^2 + b^2 + c^2 + d^2 \equiv 0 \pmod 4$; hence the numbers have the same parity. We analyze two cases.

(a) The numbers a, b, c, d are odd. We write $a = 2a' + 1$, etc. We obtain

$$4a'(a' + 1) + 4b'(b' + 1) + 4c'(c' + 1) + 4d'(d' + 1) = 4(7 \cdot 4^{n-1} - 1).$$

The left-hand side of the equality is divisible by 8; hence $7 \cdot 4^{n-1} - 1$ must be even. This happens only for $n = 1$. We obtain $a^2 + b^2 + c^2 + d^2 = 28$, with the solutions $(3, 3, 3, 1)$ and $(1, 1, 1, 5)$.

(b) The numbers a, b, c, d are even. Write $a = 2a'$, etc. We obtain

$$a'^2 + b'^2 + c'^2 + d'^2 = 7 \cdot 4^{n-1},$$

so we proceed recursively.

Finally, we obtain the solutions $(2^{n+1}, 2^n, 2^n, 2^n)$, $(3 \cdot 2^{n-1}, 3 \cdot 2^{n-1}, 3 \cdot 2^{n-1}, 2^{n-1})$, $(2^{n-1}, 2^{n-1}, 2^{n-1}, 5 \cdot 2^{n-1})$, and the respective permutations.

Problem 8.2.14. *Prove that the equation*

$$x^2 + y^2 + z^2 + t^2 = 2^{2004},$$

where $0 \leq x \leq y \leq x \leq t$, has exactly two solutions in the set of integers.

(2004 Romanian Mathematical Olympiad)

Solution. The solutions are $(0, 0, 0, 2^{1002})$ and $(2^{1001}, 2^{1001}, 2^{1001}, 2^{1001})$.

In order to prove the statement, let (x, y, z, t) be a solution. Observe that for odd a we have $a = 4n \pm 1$, and a^2 gives the remainder 1 when divided by 8. Since the right-hand side is 0 (mod 8), the equation has no solution with an odd component.

We thus must have $x = 2x_1$, $y = 2y_1$, $z = 2z_1$, $t = 2t_1$, where $0 \le x_1 \le y_1 \le z_1 \le t_1$ are integers and $x_1^2 + y_1^2 + z_1^2 + t_1^2 = 2^{2002}$. By the same argument, $x_1 = 2x_2$, $y_1 = 2y_2$, $z_1 = 2z_2$, $t_1 = 2t_2$, where $0 \le x_2 \le y_2 \le z_2 \le t_2$ are integers and $x_2^2 + y_2^2 + z_2^2 + t_2^2 = 2^{2000}$.

We can proceed recursively as long as the right-hand side is zero mod 8. Eventually we will arrive at $x = 2^{2001}a$, $y = 2^{2001}b$, $z = 2^{2001}c$, $t = 2^{2001}d$, where $0 \le a \le b \le c \le d$ are integers and $a^2 + b^2 + c^2 + d^2 = 4$. The only solutions to this are $(1, 1, 1, 1)$ and $(0, 0, 0, 2)$ and the conclusion follows.

Problem 8.2.15. *Let n be a positive integer. Prove that the equation*

$$x + y + \frac{1}{x} + \frac{1}{y} = 3n$$

does not have solutions in positive rational numbers.

Solution. Suppose $x = \frac{a}{b}$, $y = \frac{c}{d}$ satisfies the given equation, where $\gcd(a, b) = \gcd(c, d) = 1$. Clearing denominators,

$$(a^2 + b^2)cd + (c^2 + d^2)ab = 3nabcd.$$

Thus, $ab \mid (a^2 + b^2)cd$ and $cd \mid (c^2 + d^2)ab$. Now $\gcd(a, b) = 1$ implies $\gcd(a, a^2 + b^2) = \gcd(a, b^2) = 1$, so $ab \mid cd$; likewise, $cd \mid ab$, and together these give $ab = cd$. Thus,

$$a^2 + b^2 + c^2 + d^2 = 3nab.$$

Now each square on the left is congruent to either 0 or 1 modulo 3. Hence, either all terms are divisible by 3 or exactly one is. The first case is impossible by the assumption $\gcd(a, b) = \gcd(c, d) = 1$, and the second is impossible because $ab = cd$.

8.3 Nonstandard Diophantine Equations

8.3.1 Cubic Equations

Problem 8.3.5. *Find all triples (x, y, z) of natural numbers such that y is a prime number, y and 3 do not divide z, and $x^3 - y^3 = z^2$.*

(1999 Bulgarian Mathematical Olympiad)

Solution. We rewrite the equation in the form

$$(x - y)(x^2 + xy + y^2) = z^2.$$

Any common divisor of $x - y$ and $x^2 + xy + y^2$ also divides both z^2 and $(x^2 + xy + y^2) - (x + 2y)(x - y) = 3y^2$. Because z^2 and $3y^2$ are relatively prime by assumption, $x - y$ and $x^2 + xy + y^2$ must be relatively prime as well. Therefore, both $x - y$ and $x^2 + xy + y^2$ are perfect squares.

Writing $a = \sqrt{x - y}$, we have

$$x^2 + xy + y^2 = (a^2 + y)^2 + (a^2 + y)y + y^2 = a^4 + 3a^2y + 3y^2$$

and

$$4(x^2 + xy + y^2) = (2a^2 + 3y)^2 + 3y^2.$$

Writing $m = 2\sqrt{x^2 + xy + y^2}$ and $n = 2a^2 + 3y$, we have

$$m^2 = n^2 + 3y^2,$$

or

$$(m - n)(m + n) = 3y^2,$$

so $(m - n, m + n) = (1, 3y^2)$, $(y, 3y)$, or $(3, y^2)$.

In the first case, $2n = 3y^2 - 1$ and $4a^2 = 2n - 6y = 3y^2 - 6y - 1$. Hence, $a^2 \equiv 2 \pmod{3}$, which is impossible.

In the second case, $n = y < 2a^2 + 3y = n$, a contradiction.

In the third case, write $4a^2 = 2n - 6y = y^2 - 6y - 3$ as $4a^2 = (y - 3)^2 - 12$, and so $12 = (y - 3)^2 - a^2 = (y - a - 3)(y + a - 3)$. Since these factors are congruent mod 2, we must have $y - a - 3 = 2$ and $y + a + 3 = 6$, so $a = 1$, $y = 7$. This yields the unique solution $(x, y, z) = (8, 7, 13)$.

Problem 8.3.6. *Find all the positive integers a, b, c such that*

$$a^3 + b^3 + c^3 = 2001.$$

(2001 Junior Balkan Mathematical Olympiad)

Solution. Assume without loss of generality that $a \leq b \leq c$.

It is obvious that $1^3 + 10^3 + 10^3 = 2001$. We prove that $(1, 10, 10)$ is the only solution of the equation, except for its permutations.

We start by proving a useful lemma:

Lemma. *Suppose n is an integer. The remainder of n^3 when divided by 9 is 0, 1, or -1.*

Indeed, if $n = 3k$, then $9 \mid n^3$, and if $n = 3k \pm 1$, then $n^3 = 27k^3 \pm 27k^2 + 9k \pm 1 = \mathcal{M}9 \pm 1$.

Since $2001 = 9 \cdot 222 + 3 = \mathcal{M}9 + 3$, then $a^3 + b^3 + c^3 = 2001$ implies $a^3 = \mathcal{M}9 + 1$, $b^3 = \mathcal{M}9 + 1$ and $c^3 = \mathcal{M}9 + 1$; hence a, b, c are numbers of the form $\mathcal{M}3 + 1$. We search for a, b, c in the set $\{1, 4, 7, 10, 13, \dots\}$.

If $c \geq 13$ then $c^3 \geq 2197 > 2001 = a^3 + b^3 + c^3$, which is false. If $c \leq 7$ then $2001 = a^3 + b^3 + c^3 \leq 3 \cdot 343$, which again is false. Hence $c = 10$ and consequently $a^3 + b^3 = 1001$. If $b < c = 10$ then $a \leq b \leq 7$ and $1001 = a^3 + b^3 \leq 2 \cdot 7^3 = 2 \cdot 343$, a contradiction. Thus $b = 10$ and $a = 1$.

Therefore $(a, b, c) \in \{(1, 10, 10), (10, 1, 10), (10, 10, 1)\}$.

Problem 8.3.7. *Determine all ordered pairs (m, n) of positive integers such that*

$$\frac{n^3 + 1}{mn - 1}$$

is an integer.

(35th International Mathematical Olympiad)

First solution. Let $\frac{n^3+1}{mn-1} = k$, k a positive integer.

From $n^3 + 1 = k(mn - 1)$, one obtains $k + 1 = n(km - n^2)$. Thus, n divides $k + 1$ and by noting $km - n^2 = q$ one has $k = nq - 1$. Using this form of k we have

$$n^3 + 1 = (nq - 1)(mn - 1) \Leftrightarrow n(mq - n) = m + q.$$

Since $m + q > 0$, it follows that $x = mq - n > 0$. Thus we have the system

$$\begin{cases} xn = m + q, \\ x + n = mq. \end{cases}$$

By adding these equations we obtain

$$xn + mq = x + n + m + q \Leftrightarrow xn + mq - x - n - m - q + 2 = 2 \Leftrightarrow$$

$$(x - 1)(n - 1) + (m - 1)(q - 1) = 2.$$

The equation

$$(x - 1)(n - 1) + (m - 1)(q - 1) = 2$$

has only a finite number of positive integer solutions. These are listed below:

(1) $x = 1$, $m - 1 = 2$, $q - 1 = 1 \Rightarrow x = 1$, $m = 3$, $q = 2 \Rightarrow m = 3$, $n = 5$.

(2) $x = 1$, $m - 1 = 1$, $q - 1 = 2 \Rightarrow m = 2$, $n = 5$.

(3) $n = 1$, $m - 1 = 2$, $q - 1 = 1 \Rightarrow n = 1$, $m = 3$.

(4) $n = 1$, $m - 1 = 1$, $q - 1 = 2 \Rightarrow n = 1$, $m = 2$.

(5) $m = 1$, $x - 1 = 2$, $n - 1 = 1 \Rightarrow m = 1$, $n = 2$.

(6) $m = 1$, $x - 1 = 1$, $n - 1 = 2 \Rightarrow m = 1$, $n = 3$.

(7) $q = 1$, $x - 1 = 1$, $n - 1 = 2 \Rightarrow n = 3$, $m = 5$.

(8) $q = 1$, $x - 1 = 2$, $n - 1 = 1 \Rightarrow n = 2$, $m = 5$.

(9) $x - 1 = n - 1 = m - 1 = q - 1 = 1 \Rightarrow m = n = 2$.

Thus, we have obtained the following nine pairs (m, n): $(5, 3)$, $(3, 5)$, $(5, 2)$, $(2, 5)$, $(3, 1)$, $(1, 3)$, $(2, 1)$, $(1, 2)$, $(2, 2)$. All pairs are solutions of the problem.

Second solution. Note that (m, n) is a solution if and only if

$$m^2 n^2 + mn + 1 + \frac{n^3 + 1}{mn - 1} = \frac{n^3 (m^3 + 1)}{mn - 1}$$

is an integer. Since $\gcd(n, mn - 1) = 1$, this occurs if and only if $\frac{m^3 + 1}{mn - 1}$ is a solution. That is, (m, n) is a solution if and only if (n, m) is a solution. Suppose (m, n) is a solution and $mn > 1$; then defining q as in the current solution, we see (q, n) is a solution. Further, since $(n^2 - 1)^2 \geq n^3 + 1$ for $n \geq 2$, we see that $q < n$. Similarly, if $m = n > 2$, we get $q < n$. Thus following the steps

(i) If $m < n$, interchange m and n,

(ii) If $m > n > 1$ or $m = n > 2$, replace (m, n) by (q, n),

starting from any solution we can always reduce to a smaller solution until we get to either $m = n = 2$ or $n = 1$. In the latter case we have $m = 2$ or 3. Backtracking gives the chains of solutions $(3, 5(\rightarrow (5, 3) \rightarrow (1, 3) \rightarrow (3, 1)$, $(2, 5) \rightarrow (5, 2) \rightarrow (1, 2) \rightarrow (2, 1)$ and the single solution $(2, 2)$. Thus these nine pairs are all solutions to the problem.

8.3.2 High-Order Polynomial Equations

Problem 8.3.12. *Prove that there are no positive integers x and y such that*

$$x^5 + y^5 + 1 = (x + 2)^5 + (y - 3)^5.$$

Solution. Notice that $z^5 \equiv z \pmod 5$; hence $x + y + 1 \equiv (x + 2) + (y - 3)$ (mod 5), impossible.

Problem 8.3.13. *Prove that the equation $y^2 = x^5 - 4$ has no integer solutions.*

(1998 Balkan Mathematical Olympiad)

Solution. We consider the equation mod 11. Since

$$(x^5)^2 = x^{10} \equiv 0 \text{ or } 1 \pmod{11}$$

for all x, we have $x^5 \equiv -1, 0,$ or $1 \pmod{11}$, so the right-hand side is either 6, 7, or 8 modulo 11. However, all squares are 0, 1, 3, 4, 5, or 9 modulo 11, so the equation $y^2 = x^5 - 4$ has no integer solutions.

Problem 8.3.14. *Let $m, n > 1$ be integers. Solve in positive integers the equation*

$$x^n + y^n = 2^m.$$

(2003 Romanian Mathematical Olympiad)

Solution. Let $d = \gcd(x, y)$ and $x = da$, $y = db$, where $(a, b) = 1$. it is easy to see that a and b are both odd numbers and $a^n + b^n = 2^k$, for some integer k.

Suppose that n is even. Since $a^2 \equiv b^2 \equiv 1 \pmod 8$, we have also $a^n \equiv b^n \equiv 1 \pmod 8$. Since $2^k = a^n + b^n \equiv 2 \pmod 8$, we conclude that $k = 1$ and $a = b = 1$, and thus $x = y = d$. The equation becomes $x^n = 2^{m-1}$, and it has an integer solution if and only if n is a divisor of $m - 1$ and $x = y = 2^{\frac{m-1}{n}}$.

Consider the case that n is odd. From the decomposition

$$a^n + b^n = (a + b)(a^{n-1} - a^{n-2}b + a^{n-3}b^2 - \cdots + b^{n-1}),$$

we easily get $a + b = 2^k = a^n + b^n$, since the second factor above is odd. In this case $a = b = 1$, and the proof goes along the lines of the previous case.

To conclude, the given equations have solutions if and only if $\frac{m-1}{n}$ is an integer, and in this case $x = y = 2^{\frac{m-1}{n}}$.

Problem 8.3.15. *For a given positive integer m, find all pairs (n, x, y) of positive integers such that m, n are relatively prime and $(x^2 + y^2)^m = (xy)^n$, where n, x, y can be represented in terms of m.*

(1995 Korean Mathematical Olympiad)

Solution. If (n, x, y) is a solution, then the AM–GM inequality yields

$$(xy)^n = (x^2 + y^2)^m \geq (2xy)^m > (xy)^m,$$

so $n > m$. Let p be a common prime divisor of x and y and let $p^a \| x$, $p^b \| y$. Then $p^{(a+b)n} \| (xy)^n = (x^2 + y^2)^m$. Suppose $b > a$. Since $p^{2a} \| x^2$, $p^{2b} \| y^2$, we see that $p^{2a} \| x^2 + y^2$ and $p^{2am} \| (x^2 + y^2)^m$. Thus $2am = (a + b)n > 2an$ and $m > n$, a contradiction. Likewise, $a > b$ produces a contradiction, so we must have $a = b$ and $x = y$. This quickly leads to $x = 2^t$ for some integer t, and all solutions are of the form

$$(n, x, y) = (2t + 1, 2^t, 2^t)$$

for nonnegative integers t. Substituting into the equation, we conclude that there is a solution only if m is even, and then $t = \frac{m}{2}$.

8.3.3 *Exponential Diophantine Equations*

Problem 8.3.19. *Determine all triples* (x, k, n) *of positive integers such that*

$$3^k - 1 = x^n.$$

(1999 Italian Mathematical Olympiad)

Solution. All triples of the form $(3^k - 1, k, 1)$ for positive integers k, and $(2, 2, 3)$.

The solutions when $n = 1$ are obvious. Now, n cannot be even because then 3 could not divide $3^k = (x^{\frac{n}{2}})^2 + 1$ (because no square is congruent to 2 modulo 3). Also, we must have $x \neq 1$.

Assume that $n > 1$ is odd and $x \geq 2$. Then $3^k = (x + 1) \sum_{i=0}^{n-1} (-x)^i$, implying that both $x + 1$ and $\sum_{i=0}^{n-1} (-x)^i$ are powers of 3. Because $x + 1 \leq x^2 - x + 1 \leq \sum_{i=0}^{n-1} (-x)^i$, we must have $0 \equiv \sum_{i=0}^{n-1} (-x)^i \equiv n \pmod{x+1}$, so that $x + 1 \mid n$. Specifically, this means that $3 \mid n$.

Write $x' = x^{n/3}$, then we have $3^k - 1 = (x')^3$. Thus repeating the argument of the previous paragraph, now with $n = 3$, shows $x' + 1 \mid 3$. Hence $x' = 2$ and therefore $x = 2$ and $n = 3$.

Remark. In fact, 8 and 9 are the only consecutive powers (other than the trivial 0, 1), as recently proved.

Problem 8.3.20. *Find all pairs of nonnegative integers* x *and* y *that satisfy the equation*

$$p^x - y^p = 1,$$

where p *is a given odd prime.*

(1995 Czech–Slovak Match)

Solution. If (x, y) is a solution, then

$$p^x = y^p + 1 = (y + 1)(y^{p-1} - \cdots + y^2 - y + 1),$$

and so $y + 1 = p^n$ for some n. If $n = 0$, then $x = y = 0$ and p may be arbitrary. Otherwise,

$$p^x = (p^n - 1)^p + 1$$

$$= p^{np} - p \cdot p^{n(p-1)} + \binom{p}{2} p^{n(p-2)} + \cdots - \binom{p}{p-2} p^{2n} + p \cdot p^n.$$

Since p is a prime, all of the binomial coefficients are divisible by p. Hence all terms are divisible by p^{n+1}, and all but the last by p^{n+2}. Therefore the highest power of p dividing the right side is p^{n+1}, and so $x = n + 1$. We also have

$$0 = p^{np} - p \cdot p^{n(p-1)} + \binom{p}{2} p^{n(p-2)} + \cdots - \binom{p}{p-2} p^{2n}.$$

For $p = 3$ this reads $0 = 3^{3n} - 3 \cdot 3^{2n}$, which occurs only for $n = 1$, yielding $x = y = 2$. For $p \geq 5$, the coefficient $\binom{p}{p-2}$ is not divisible by p^2, so every term but the last on the right side is divisible by p^{2n+2}, while the last term is not. Since the terms sum to 0, this is impossible.

Hence the only solutions are $x = y = 0$ for all p and $x = y = 2$ for $p = 3$.

Problem 8.3.21. *Let x, y, z be integers with $z > 1$. Show that*

$$(x + 1)^2 + (x + 2)^2 + \cdots + (x + 99)^2 \neq y^z.$$

(1998 Hungarian Mathematical Olympiad)

Solution. Suppose, to the contrary, that there are integers x, y, z such that $z > 1$, and

$$(x + 1)^2 + (x + 2)^2 + \cdots + (x + 99)^2 = y^z.$$

We notice that

$$\begin{aligned}
y^z &= (x + 1)^2 + (x + 2)^2 + \cdots + (x + 99)^2 \\
&= 99x^2 + 2(1 + 2 + \cdots + 99)x + (1^2 + 2^2 + \cdots + 99^2) \\
&= 99x^2 + \frac{2 \cdot 99 \cdot 100}{2}x + \frac{99 \cdot 100 \cdot 199}{6} \\
&= 33(3x^2 + 300x + 50 \cdot 199),
\end{aligned}$$

which implies that $3 \mid y$. Since $z \geq 2$, $3^2 \mid y^z$, but 3^2 does not divide $33(3x^2 + 300x + 50 \cdot 199)$, we have a contradiction. So our assumption in fact must be false, and the original statement in the problem is correct.

Problem 8.3.22. *Determine all solutions (x, y, z) of positive integers such that*

$$(x + 1)^{y+1} + 1 = (x + 2)^{z+1}.$$

(1999 Taiwanese Mathematical Olympiad)

Solution. Let $a = x + 1$, $b = y + 1$, $c = z + 1$. Then $a, b, c \geq 2$ and

$$a^b + 1 = (a + 1)^c,$$
$$((a + 1) - 1)^b + 1 = (a + 1)^c.$$

Taking the equations mod $(a + 1)$ yields $(-1)^b + 1 \equiv 0$, so b is odd.

Taking the second equation mod $(a + 1)^2$ after applying the binomial expansion yields

$$\binom{b}{1}(a + 1)(-1)^{b-1} + (-1)^b + 1 \equiv 0 \quad (\mathrm{mod}\ (a + 1)^2),$$

so $(a + 1) \mid b$ and a is even. On the other hand, taking the first equation mod a^2 after applying the binomial expansion yields

$$1 \equiv \binom{c}{1}a + 1 \pmod{a^2},$$

so c is divisible by a and is even as well. Write $a = 2a_1$ and $c = 2c_1$. Then

$$2^b a_1^b = a^b = (a + 1)^c - 1 = ((a + 1)^{c_1} - 1)((a + 1)^{c_1} + 1).$$

It follows that $\gcd((a + 1)^{c_1} - 1, (a + 1)^{c_1} + 1) = 2$. Therefore, using the fact that $2a_1$ is a divisor of $(a + 1)^{c_1} - 1$, we may conclude that

$$(a + 1)^{c_1} - 1 = 2a_1^b$$
$$(a + 1)^{c_1} + 1 = 2^{b-1}.$$

We must have $2^{b-1} > 2a_1^b \Rightarrow a_1 = 1$. Then these equations give $c_1 = 1$ and $b = 3$. Therefore the only solution is $(x, y, z) = (1, 2, 1)$.

9

Some Special Problems in Number Theory

9.1 Quadratic Residues; the Legendre Symbol

Problem 9.1.7. *Let* $f, g : Z^+ \to Z^+$ *be functions with the following properties:*
  (i) g is surjective;
  (ii) $2f^2(n) = n^2 + g^2(n)$ *for all positive integers n.*
  If, moreover, $|f(n) - n| \leq 2004\sqrt{n}$ *for all n, prove that f has infinitely many fixed points.*

(2005 Moldavian International Mathematical Olympiad Team Selection Test)

Solution. Let p_n be the sequence of prime numbers of the form $8k + 3$. There are infinitely many such integers. This is a trivial consequence of Dirichlet's theorem, but you can find an elementary proof at the end of solution to Problem 9.1.8. It is obvious that for all n we have

$$\left(\frac{2}{p_n}\right) = (-1)^{\frac{p_n^2-1}{8}} = -1.$$

Using the condition (i) we can find x_n such that $g(x_n) = p_n$ for all n. It follows that $2f^2(x_n) = x_n^2 + p_n^2$, which can be rewritten as $2f^2(x_n) \equiv x_n^2 \pmod{p_n}$. Because $\left(\frac{2}{p_n}\right) = -1$, the last congruence shows that $p_n \mid x_n$ and $p_n \mid f(x_n)$. Thus there exist sequences of positive integers a_n, b_n such that $x_n = a_n p_n$, $f(x_n) = b_n p_n$ for all n. Clearly, (ii) implies the relation $2b_n^2 = a_n^2 + 1$. Finally, using the property $|f(n) - n| \leq 2004\sqrt{n}$, we infer that

$$\frac{2004}{\sqrt{x_n}} \geq \left|\frac{f(x_n)}{x_n} - 1\right| = \left|\frac{b_n}{a_n} - 1\right|,$$

T. Andreescu and D. Andrica, *Number Theory*, DOI: 10.1007/b11856_20,
© Birkhäuser Boston, a part of Springer Science + Business Media, LLC 2009

that is,

$$\lim_{n \to \infty} \frac{\sqrt{a_n^2 + 1}}{a_n} = \sqrt{2}.$$

The last relation immediately implies that $\lim_{n \to \infty} a_n = 1$. Therefore, starting from a certain n, we have $a_n = 1 = b_n$, that is, $f(p_n) = p_n$. The conclusion now follows.

Problem 9.1.8. *Suppose that the positive integer a is not a perfect square. Then $\left(\frac{a}{p}\right) = -1$ for infinitely many primes p.*

Solution. One may assume that a is square-free. Let us write $a = 2^e q_1 q_2 \cdots q_n$, where q_i are distinct odd primes and $e \in \{0, 1\}$. Let us assume first that $n \geq 1$ and consider some odd distinct primes $r_1, \ldots, r_k$ each of them different from $q_1, \ldots, q_n$. We will show that there exists a prime p, different from $r_1, \ldots, r_k$, such that $\left(\frac{a}{p}\right) = -1$. Let s be a quadratic nonresidue modulo q_n.

Using the Chinese remainder theorem, we can find a positive integer b such that

$$\begin{cases} b \equiv 1 & (\mathrm{mod}\ r_i),\ 1 \leq i \leq k, \\ b \equiv 1 & (\mathrm{mod}\ 8), \\ b \equiv 1 & (\mathrm{mod}\ q_i),\ 1 \leq i \leq n-1, \\ b \equiv s & (\mathrm{mod}\ q_n). \end{cases}$$

Now write $b = p_1 \cdots p_m$ with p_i odd primes, not necessarily distinct. Using the quadratic reciprocity law, it follows immediately that

$$\prod_{i=1}^{m} \left(\frac{2}{p_i}\right) = \prod_{i=1}^{m} (-1)^{\frac{p_i^2 - 1}{8}} = (-1)^{\frac{b^2-1}{8}} = 1$$

and

$$\prod_{j=1}^{m} \left(\frac{q_i}{p_j}\right) = \prod_{j=1}^{m} (-1)^{\frac{p_j-1}{2} \cdot \frac{q_i-1}{2}} \left(\frac{p_j}{q_i}\right) = (-1)^{\frac{q_i-1}{2} \cdot \frac{b-1}{2}} \left(\frac{b}{q_i}\right) = \left(\frac{b}{q_i}\right)$$

for all $i \in \{1, 2, \ldots, n\}$. Hence

$$\prod_{i=1}^{m} \left(\frac{a}{p_i}\right) = \left[\prod_{j=1}^{m} \left(\frac{2}{p_j}\right)\right]^e \prod_{i=1}^{n} \prod_{j=1}^{m} \left(\frac{q_i}{p_j}\right)$$

$$= \prod_{i=1}^{n} \left(\frac{b}{q_i}\right) = \left(\frac{b}{q_n}\right) = \left(\frac{s}{q_n}\right) = -1.$$

Thus, there exists $i \in \{1, 2, \ldots, m\}$ such that $\left(\frac{a}{p_i}\right) = -1$. Because $b \equiv 1$ (mod r_i), $1 \leq i \leq k$, we also have $p_i \in \{1, 2, \ldots\} \setminus \{r_1, \ldots, r_k\}$, and the claim is proved.

The only remaining case is $a = 2$. By Theorem 9.1.2, $\left(\frac{2}{p}\right) = -1$ if and only if $\frac{p^2-1}{8}$ is odd, i.e., if and only if $p \equiv \pm 3 \pmod 8$. Thus we need to show there are infinitely many primes congruent to $\pm 3 \pmod 8$. Suppose to the contrary that there are only finitely many such primes $p_1, \ldots, p_n$ with $p_1 = 3$ and consider $N = 8p_2 \cdots p_n + 3$. None of the p_i divide N and neither does 2. Hence every prime divisor of N must be $\pm 1 \pmod 8$. But this is impossible, since $N \equiv 3 \pmod 8$. Thus there must be infinitely many primes of this form.

Problem 9.1.9. *Suppose that $a_1, a_2, \ldots, a_{2004}$ are nonnegative integers such that $a_1^n + a_2^n + \cdots + a_{2004}^n$ is a perfect square for all positive integers n. What is the maximal possible number of nonzero a_i's?*

(2004 Mathlinks Contest)

Solution. Suppose that $a_1, a_2, \ldots, a_k$ are positive integers such that $a_1^n + a_2^n + \cdots + a_k^n$ is a perfect square for all n. We will show that k is a perfect square. In order to prove this, we will use Problem 9.1.8 and show that $\left(\frac{k}{p}\right) = 1$ for all sufficiently large prime p. This is not a difficult task. Indeed, consider a prime p greater than any prime divisor of $a_1 a_2 \cdots a_k$. Using Fermat's little theorem, $a_1^{p-1} + a_2^{p-1} + \cdots + a_k^{p-1} \equiv k \pmod p$, and since $a_1^{p-1} + a_2^{p-1} + \cdots + a_k^{p-1}$ is a perfect square, it follows that $\left(\frac{k}{p}\right) = 1$. Thus k is a perfect square. And now the problem becomes trivial, since we must find the greatest perfect square smaller than 2004. A quick computation shows that this is $44^2 = 1936$, and so the desired minimal number is 68.

Problem 9.1.10. *Find all positive integers n such that $2^n - 1 \mid 3^n - 1$.*

(American Mathematical Monthly)

Solution. We will prove that $n = 1$ is the only solution to the problem. Suppose that $n > 1$ is a solution. Then $2^n - 1$ cannot be a multiple of 3; hence n is odd. Therefore, $2^n \equiv 8 \pmod{12}$. Because any odd prime different from 3 is of one of the forms $12k \pm 1$, $12k \pm 5$, and since $2^n - 1 \equiv 7 \pmod{12}$, it follows that $2^n - 1$ has at least one prime divisor of the form $12k \pm 5$; call it p. Obviously, we must have $(3/p) = 1$, and using the quadratic reciprocity law we finally obtain $(p/3) = (-1)^{(p-1)/2}$. On the other hand $(p/3) = (\pm 2/3) = -(\pm 1)$. Consequently, $-(\pm 1) = (-1)^{(p-1)/2} = \pm 1$, which is the desired contradiction. Therefore the only solution is $n = 1$.

Problem 9.1.11. *Find the smallest prime factor of $12^{2^{15}} + 1$.*

Solution. Let p be this prime number. Because $p \mid 12^{2^{16}} - 1$, we find that $o_p(12) \mid 2^{16}$. Since $12^{2^{15}} \equiv -1 \not\equiv 1 \pmod p$, we find that $o_p(12) = 2^{16}$, and so $2^{16} \mid p - 1$. Therefore $p \geq 1 + 2^{16}$. But it is well known that $2^{16} + 1$ is a prime (and if you do not believe it, you can check; it is not that difficult). So, we might try to see whether this number divides $12^{2^{15}} + 1$. Let $q = 2^{16} + 1$. Then

$$12^{2^{15}} + 1 = 2^{q-1} \cdot 3^{\frac{q-1}{2}} + 1 \equiv 3^{\frac{q-1}{2}} + 1 \pmod{q}.$$

It remains to see whether $(3/q) = -1$. The answer is positive (use the quadratic reciprocity law), so indeed $3^{(q-1)/2} + 1 \equiv 0 \pmod 2$ and $2^{16} + 1$ is the smallest prime factor of the number $12^{2^{15}} + 1$.

9.2   Special Numbers

9.2.1   Fermat Numbers

Problem 9.2.4. *Find all positive integers n such that $2^n - 1$ is a multiple of 3 and $\frac{2^n-1}{3}$ is a divisor of $4m^2 + 1$ for some integer m.*

(1999 Korean Mathematical Olympiad)

Solution. The answer is all $n = 2^k$, where $k = 1, 2, \ldots$.
First observe that $2 \equiv -1 \pmod 3$. Hence $3 \mid 2^n - 1$ if and only if n is even.
Suppose, by way of contradiction, that $l \geq 3$ is a positive odd divisor of n. Then $2^l - 1$ is not divisible by 3 but it is a divisor of $2^n - 1$, so it is a divisor of $4m^2 + 1$ as well. On the other hand, $2^l - 1$ has a prime divisor p of the form $4r + 3$. Then $(2m)^2 \equiv -1 \pmod{4r + 3}$, but we have that a square cannot be congruent to -1 modulo a prime of the form $4r + 3$ (see also Problem 1 in Section 7.1).
Therefore, n is indeed of the form 2^k for $k \geq 1$. For such n, we have

$$\frac{2^n - 1}{3} = (2^{2^1} + 1)(2^{2^2} + 1)(2^{2^3} + 1) \cdots (2^{2^{k-1}} + 1).$$

The factors on the right side are all relatively prime, since they are Fermat numbers. Therefore by the Chinese remainder theorem, there is a positive integer c simultaneously satisfying

$$c \equiv 2^{2^{i-1}} \pmod{2^{2^i} + 1} \text{ for all } i = 1, 2, \ldots, k - 1$$

and $c \equiv 0 \pmod 2$. Putting $c = 2m$, $4m^2 + 1$ is a multiple of $\frac{2^n-1}{3}$, as desired.

Problem 9.2.5. *Prove that the greatest prime factor of f_n, $n \geq 2$, is greater than $2^{n+2}(n + 1)$.*

(2005 Chinese International Mathematical Olympiad Team Selection Test)

Solution. From Problem 9.2.3 we can write

$$f_n = \prod_{i=1}^{s}(1 + 2^{n+2}r_i)^{k_i}, \tag{1}$$

where $p_i = 1 + 2^{n+2} r_i$ are distinct primes and $k_i \geq 1$. Taking relation (1) modulo 4^{n+2}, it follows that

$$0 \equiv \sum_{i=1}^{s} k_i r_i \pmod{2^{n+2}},$$

and hence

$$\sum_{i=1}^{s} k_i r_i \geq 2^{n+2}.$$

From (1) it is clear that

$$f_n \geq (1 + 2^{n+2})^{k_1 + \cdots + k_s};$$

hence

$$k_1 + \cdots + k_s \leq \frac{\lg(1 + 2^{2^n})}{\lg(1 + 2^{n+2})},$$

where $\lg x$ is $\log_2 x$.

It follows that

$$2^{n+2} \leq \left(\max_{1 \leq i \leq s} r_i \right) \sum_{j=1}^{s} k_j \leq \left(\max_{1 \leq i \leq s} r_i \right) \frac{\lg(1 + 2^{2^n})}{\lg(1 + 2^{n+2})}.$$

Assume that $\left(\max_{1 \leq i \leq s} r_i \right) \leq n$. Applying the last inequality, we get

$$2^{n+2} \leq n \frac{\lg(1 + 2^{2^n})}{\lg(1 + 2^{n+2})} < n \frac{\lg(1 + 2^{2^n})}{(n+2) \lg 2},$$

i.e.,

$$\frac{n+2}{n} \cdot 2^{n+2} < \log_2(1 + 2^{2^n}),$$

hence $2^{2^{n+2}} < 1 + 2^{2^n}$, a contradiction. Therefore $\max_{1 \leq i \leq s} r_i \geq n + 1$, and $\max_{1 \leq i \leq s} p_i > 2^{n+2}(n+1)$.

9.2.2 Mersenne Numbers

Problem 9.2.7. *Let P^* denote the set of all odd primes less than 10000, and suppose $p \in P^*$. For each subset $S = \{p_1, p_2, \ldots, p_k\}$ of P^*, with $k \geq 2$ and not including p, there exists a $q \in P^* \setminus S$ such that*

$$(q + 1) \mid (p_1 + 1)(p_2 + 1) \cdots (p_k + 1).$$

Find all such possible values of p.

(1999 Taiwanese Mathematical Olympiad)

Solution. Direct calculation shows that the set T of Mersenne primes less that 10000 is

$$\{M_2, M_3, M_5, M_7, M_{13}\} = \{3, 7, 31, 127, 8191\}.$$

The number $2^{11} - 1$ is not prime; it equals $23 \cdot 89$. We claim that this is the set of all possible values of p.

If some prime p is not in T, then look at the set $S = T$. Then there must be some prime $q \notin S$ less than 10000 such that

$$(q + 1) \mid (M_2 + 1)(M_3 + 1)(M_5 + 1)(M_7 + 1)(M_{13} + 1) = 2^{30}.$$

Thus, $q + 1$ is a power of 2 and q is a Mersenne prime less than 10000, and therefore $q \in T = S$, a contradiction.

On the other hand, suppose p is in T. Suppose we have a set $S = \{p_1, p_2, \ldots, p_k\} \subseteq P^*$ not including p, with $k \geq 2$ and $p_1 < p_2 < \cdots < p_k$. Suppose, by way of contradiction that for all $q \in P^*$ such that $(q + 1) \mid (p_1 + 1) \cdots (p_k + 1)$, we have $q \in S$. Then

$$4 \mid (p_1 + 1)(p_2 + 1) \implies M_2 \in S,$$
$$8 \mid (M_2 + 1)(p_2 + 1) \implies M_3 \in S,$$
$$32 \mid (M_2 + 1)(M_3 + 1) \implies M_5 \in S,$$
$$128 \mid (M_2 + 1)(M_5 + 1) \implies M_7 \in S,$$
$$8192 \mid (M_3 + 1)(M_5 + 1)(M_7 + 1) \implies M_{13} \in S.$$

Then p, a Mersenne prime under 10000, must be in S, a contradiction. Therefore there is some prime $q < 10000$ not in S with $q + 1 \mid (p_1 + 1) \cdots (p_k + 1)$, as desired. This completes the proof.

9.2.3   Perfect Numbers

Problem 9.2.9. *Prove that if n is an even perfect number, then $8n + 1$ is a perfect square.*

Solution. From Problem 1, we have $n = \frac{m(m+1)}{2}$ for some positive integer m; hence

$$8n + 1 = 4m(m + 1) + 1 = (2m + 1)^2.$$

Problem 9.2.10. *Show that if k is an odd positive integer, then $2^{k-1}M_k$ can be written as the sum of the cubes of the first $2^{\frac{k-1}{2}}$ odd positive integers. In particular, any perfect number has this property.*

Solution. Standard summation formulas verify that

$$\sum_{i=1}^{n}(2i - 1)^3 = n^2(2n^2 - 1).$$

With $n = 2^{\frac{k-1}{2}}$, the right-hand side becomes $2^{k-1}(2^k - 1)$; that is, $2^{k-1} M_k$, and we are done.

9.3 Sequences of Integers

9.3.1 Fibonacci and Lucas Sequences

Problem 9.3.5. *Determine the maximum value of $m^2 + n^2$, where m and n are integers satisfying $1 \le m, n \le 1981$ and $(n^2 - mn - m^2)^2 = 1$.*

(22nd International Mathematical Olympiad)

Solution. Let S be the set of pairs (n, m) of positive integers satisfying the equation

$$(x^2 - xy - y^2)^2 = 1. \tag{1}$$

If $n = m$, then $n = m = 1$. Hence $(1, 1) \in S$. It is clear that $(1, 0)$ and $(0, 1)$ are also solutions to equation (1).

We will consider solutions (n, m) with distinct components. Using Fermat's method of infinite descent we obtain the following important result on the set S.

Lemma. *If (n, m) is a positive solution to equation (1) and $n \ne m$, then $n > m > n - m$ and $(m, n - m)$ is also a solution to (1).*

Proof. From $n^2 - nm - m^2 = \pm 1$, we obtain $n^2 = m^2 + nm \pm 1 < m^2$; thus $n > m$. Also from $n^2 - nm - m^2 = \pm 1$, we obtain

$$m^2 - m(n - m) - (n - m)^2 = m^2 + mn - n^2 = \mp 1.$$

Apply the first part to the solution $(m, n - m)$ and obtain $m > n - m$. $\quad\square$

From the lemma we deduce that any pair $(n, m) \in S$ gives rise to a pair $(m, n - m) \in M$, which gives rise to a pair $(a + b, a) \in M$. In this way, by the method of descent $(n, m) \to (m, n - m)$, or by the method of ascent $(a, b) \to (a + b, a)$, we obtain a new solution of the equation. The methods of ascent and descent are the reverse of each other.

By applying the descending method to a pair $(n, m) \in S$ we can only have finitely many steps, because $n - m < m$. Hence, in a finite number of steps we obtain a pair with $n = m$, the pair $(1, 1)$. Thus, all solutions $(n, m) \in S$ are obtained from the pair $(1, 0)$ by applying the ascending method:

$$(1, 0) \to (1, 1) \to (2, 1) \to (3, 2) \to (5, 3) \to \dots$$

The components of all such pairs are Fibonacci numbers F_n. In this way, the ascending transformation is exactly the following:

$$(F_n, F_{n-1}) \to (F_{n+1}, F_n).$$

Thus, to obtain the solution (n, m) with maximum value of $n^2 + m^2$ we consider the members of the Fibonacci sequence, not exceeding 1981:

$$0, 1, 1, 2, 3, 5, 8, 13, 21, 34, 55, 89, 144, 233, 377, 610, 987, 1597.$$

So, the required maximum is $987^2 + 1597^2$.

Remark. Fibonacci numbers F_n have the property

$$F_{n+1}^2 - F_n F_{n+1} - F_n^2 = \pm 1, \text{ for all } n \geq 0.$$

To prove it for $n = 0$ or $n = 1$ it is equivalent to see that $(1, 0) \in S$ and that $(1, 1) \in S$. Further, we can use induction. The relation

$$F_{n+1}^2 - F_n F_{n+1} - F_n^2 = \pm 1$$

implies

$$F_{n+2}^2 - F_{n+1} F_{n+2} - F_{n+1}^2 = (F_{n+1} + F_n)^2 - F_{n+1}(F_{n+1} + F_n) - F_{n+1}^2$$
$$= -(F_{n+1}^2 - F_n F_{n+1} - F_n^2) = \mp 1.$$

So in fact,

$$F_{n+1}^2 - F_n F_{n+1} - F_n^2 = (-1)^n.$$

Another way to prove this relation is to use the matrix form for the Fibonacci numbers and get

$$\begin{pmatrix} 1 & 1 \\ 1 & 0 \end{pmatrix}^{n+1} = \begin{pmatrix} F_{n+2} & F_{n+1} \\ F_{n+1} & F_n \end{pmatrix}.$$

Passing to determinants on both sides yields

$$(-1)^{n+1} = F_{n+2} F_n - F_{n+1}^2 = (F_{n+1} + F_n)F_n - F_{n+1}^2 = F_n^2 + F_n F_{n+1} - F_{n+1}^2.$$

Problem 9.3.6. *Prove that for any integer $n \geq 4$, $F_n + 1$ is not a prime.*

First solution. We have the identity

$$F_n^4 - 1 = F_{n-2} F_{n-1} F_{n+1} F_{n+2}. \tag{2}$$

Assume that $F_n + 1$ is a prime for some positive integer $n \geq 4$. Using (1), it follows that $F_n + 1$ divides at least one of the integers $F_{n-2}, F_{n-1}, F_{n+1}, F_{n+2}$. Since $F_n + 1$ is greater than F_{n-2} and F_{n-1}, it follows that $F_n + 1$ divides F_{n+1} or F_{n+2}. But $F_{n+1} < 2F_n$ and $F_{n+2} < 4F_n$; hence $F_n + 1$ cannot divide F_{n+1} or F_{n+2}, and the desired conclusion follows.

Second solution. Note the four equalities

$$F_{4m+1} + 1 = L_{2n+1} F_{2n-1},$$
$$F_{4n+1} + 1 = L_{2n} F_{2n+1},$$
$$F_{4n+2} + 1 = L_{2n} F_{2n+2},$$
$$F_{4n+3} + 1 = L_{2n+2} F_{2n+1},$$

which follow from the Binet formula or induction.

Problem 9.3.7. *Let k be an integer greater than 1, $a_0 = 4$, $a_1 = a_2 = (k^2 - 2)^2$, and*

$$a_{n+1} = a_n a_{n-1} - 2(a_n + a_{n-1}) - a_{n-2} + 8 \text{ for } n \geq 2.$$

Prove that $2 + \sqrt{a_n}$ is a perfect square for all n.

Solution. The Fibonacci numbers are involved here again, but it is much harder to guess how they are related to the solution.

Let λ, μ be the roots of the equation $t^2 - kt + 1 = 0$. Notice that $\lambda + \mu = k$, $\lambda\mu = 1$. Augmenting the Fibonacci sequence by setting $F_0 = 0$, we claim that

$$a_n = (\lambda^{2F_n} + \mu^{2F_n})^2 \text{ for } n = 0, 1, 2, \ldots$$

This is readily checked for $n = 0, 1, 2$. Assume that it holds for all $k \leq n$. Note that the given recursion can be written as

$$a_{n+1} - 2 = (a_n - 2)(a_{n-1} - 2) - (a_{n-2} - 2),$$

and that $a_k = (\lambda^{2F_k} + \mu^{2F_k})^2$ is equivalent to $a_k - 2 = \lambda^{4F_k} + \mu^{4F_k}$. Using the induction hypothesis for $k = n - 2, n - 1, n$, we obtain

$$a_{n+1} - 2 = (\lambda^{4F_n} + \mu^{4F_n})(\lambda^{4F_{n-1}} + \mu^{4F_{n-1}}) - (\lambda^{4F_{n-2}} + \mu^{4F_{n-2}})$$
$$= \lambda^{4(F_n+F_{n-1})} + \mu^{4(F_n+F_{n-1})} + \lambda^{4(F_{n-1}+F_{n-2})}\mu^{4F_{n-1}}$$
$$\quad + \mu^{4(F_{n-1}+F_{n-2})}\lambda^{4F_{n-1}} - (\lambda^{4F_{n-2}} + \mu^{4F_{n-2}})$$
$$= \lambda^{4F_{n+1}} + \mu^{4F_{n+1}} + (\lambda\mu)^{4F_{n-1}}(\lambda^{4F_{n-2}} + \mu^{4F_{n-2}}) - (\lambda^{4F_{n-2}} + \mu^{4F_{n-2}}).$$

Since $\lambda\mu = 1$, it follows that

$$a_{n+1} = 2 + \lambda^{4F_{n+1}} + \mu^{4F_{n+1}} = (\lambda^{2F_{n+1}} + \mu^{2F_{n+1}})^2,$$

and the induction is complete.

Now

$$2 + \sqrt{a_n} = 2 + \lambda^{2F_n} + \mu^{2F_n} = (\lambda^{F_n} + \mu^{F_n})^2.$$

Since

$$(\lambda^{m-1} + \mu^{m-1})(\lambda + \mu) = (\lambda^m + \mu^m) + \lambda\mu(\lambda^{m-2} + \mu^{m-2}),$$

we have

$$\lambda^m + \mu^m = k(\lambda^{m-1} + \mu^{m-1}) - (\lambda^{m-2} + \mu^{m-2}),$$

leading to an easy proof by induction that $\lambda^m + \mu^m$ is an integer for all nonnegative integers m. The solution is complete.

9.3.2 Problems Involving Linear Recursive Relations

Problem 9.3.12. *Let a, b be integers greater than 1. The sequence $x_1, x_2, \ldots$ is defined by the initial conditions $x_0 = 0$, $x_1 = 1$ and the recursion*

$$x_{2n} = ax_{2n-1} - x_{2n-2}, \qquad x_{2n+1} = bx_{2n} - x_{2n-1}$$

for $n \geq 1$. Prove that for any natural numbers m and n, the product $x_{n+m}x_{n+m-1}$ $\cdots x_{n+1}$ is divisible by $x_m x_{m-1}$.

(2001 St. Petersburg City Mathematical Olympiad)

Solution. We will show that $x_m \mid x_{km}$, and then show that $\gcd(x_m, x_{m-1}) = 1$.

First, consider our sequence modulo x_m for some m. Each x_{k+1} is uniquely determined by x_k, x_{k-1} and the parity of k. Express each x_i as a function $f_i(a, b)$. We have $x_i \equiv f_i(a, b)x_1 \pmod{x_m}$. Suppose $x_r \equiv 0 \pmod{x_m}$ for some r. Since each term is a linear combination of two preceding ones,

$$x_{i+r} \equiv f_i(a, b)x_{r+1} \pmod{x_m} \text{ if } m \text{ is even}, \tag{1}$$

$$x_{i+r} \equiv f_i(b, a)x_{r+1} \pmod{x_m} \text{ if } m \text{ is odd}. \tag{2}$$

Now we need to prove the following statement.

Lemma. *The function $f_i(a, b)$ is symmetric for any odd i.*

Proof. We will prove also that for i even, $f_i(a, b)$ is a symmetric function multiplied by a. Now we are to prove that $f_{2k-1}(a, b)$ is symmetric and $f_{2k-2}(a, b) = ag_{2k-2}(a, b)$, where g_{2k-2} is symmetric too, for any positive integer k. Proceed by induction on k. For $k = 1$ we have $f_1(a, b) = 1$ and $g_0(a, b) = 0$. Suppose that $f_{2k-1}(a, b)$ is symmetric and $f_{2k-2}(a, b) = ag_{2k-2}(a, b)$, where $g_{2k-2}(a, b)$ is symmetric too. Then we can write

$$\begin{aligned}
f_{2k}(a, b) = x_{2k} &= ax_{2k-1} - x_{2k-2} \\
&= a(x_{2k-1} - g_{2k-2}(a, b)) \\
&= a(f_{2k-1}(a, b) - g_{2k-2}(a, b))
\end{aligned}$$

and

$$\begin{aligned}
f_{2k+1}(a, b) = x_{2k+1} &= abx_{2k-1} - bx_{2k-2} - x_{2k-1} \\
&= abx_{2k-1} - abg_{2k-2} - x_{2k-1} \\
&= (ab - 1)f_{2k-1}(a, b) - abg_{2k-2}(a, b).
\end{aligned}$$

Thus f_{2k+1} and g_{2k} are symmetric too, which completes the step of induction. □

Remark. An alternative proof for lemma is the following. Let $y_n = \sqrt{b}x_n$ for n even and $y_n = \sqrt{a}x_n$ for n odd. Then one sees that $y_n = \sqrt{ab}y_{n-1} - y_{n-2}$ for all n. It follows that $f_{2n+1} = y_{2n+1}/\sqrt{a}$ and $g_{2n} = y_{2n}/(a\sqrt{b})$ are symmetric.

Now we are to prove that $x_m \mid x_{km}$. Proceed by induction on k. For $k = 0$ and $k = 1$ this statement is true. Let $x_m \mid x_{km}$. Then from (1) and (2) putting $r = km$ and $i = m$, we obtain the following. If km is even, then

$$x_{m(k+1)} \equiv f_m(a, b)x_{km+1} \equiv x_m x_{km+1} \equiv 0 \pmod{x_m}.$$

For km odd, m is odd too, and $f_m(a, b) = f_m(b, a)$. Hence, we have

$$x_{m(k+1)} \equiv f_m(b, a)x_{km+1} \equiv f_m(a, b)x_{km+1} \equiv x_m x_{km+1} \equiv 0 \pmod{x_m}.$$

So, for each pair of nonnegative integers k, m we have $x_m \mid x_{km}$.

Since the product $x_{n+1}x_{n+2}\cdots x_{n+m}$ has m terms, one term's index is divisible by m and another's index is divisible by $m - 1$. Thus both x_m and x_{m-1} divide the product. If we can show that x_m is relatively prime to x_{m-1}, we will be done. We will prove this by induction. For the base case, x_0 is relatively prime to x_1. Now, $x_{2n} = ax_{2n-1} - x_{2n-2}$. Any prime factor common to x_{2n} and x_{2n-1} must also divide x_{2n-2}, but because x_{2n-2} is relatively prime to x_{2n-1}, there is no such prime factor. A similar argument holds for x_{2n+1} because $x_{2n+1} = bx_{2n} - x_{2n-1}$. Thus $x_m x_{m-1} \mid (x_{n+1}x_{n+2}\cdots x_{n+m})$.

Problem 9.3.13. *Let m be a positive integer. Define the sequence $\{a_n\}_{n\geq0}$ by $a_0 = 0$, $a_1 = m$, and $a_{n+1} = m^2 a_n - a_{n-1}$ for $n \geq 1$. Prove that an ordered pair (a, b) of nonnegative integers, with $a \leq b$, is a solution of the equation*

$$\frac{a^2 + b^2}{ab + 1} = m^2$$

if and only if $(a, b) = (a_n, a_{n+1})$ for some $n \geq 0$.

(1998 Canadian Mathematical Olympiad)

First solution. The "if" direction of the claim is easily proved by induction on n; we prove the "only if" direction by contradiction. Suppose, to the contrary, that there exist pairs satisfying the equation but not of the described form; let (a, b) be such a pair with minimal sum $a + b$. We claim that $(c, a) = (m^2 a - b, a)$ is another such a pair but with smaller sum $c + a$, which leads to a contradiction. Taking cases on the value of a:

(a) $a = 0$. Then $(a, b) = (0, m) = (a_0, a_1)$, a contradiction.

(b) $a = m$. Then $(a, b) = (m, m^3) = (a_1, a_2)$, a contradiction.

(c) $a = 1$. Then $b \geq 1 = 1$ and $(b + 1) \mid (b^2 + 1)$; but $(b + 1) \mid (b^2 - 1)$, thus $(b + 1) \mid [(b^2 + 1) - (b^2 - 1)] = 2$. We have $b = 1$, thus $m = 1$ and $(a, b) = (1, 1) = (a_1, a_2)$, a contradiction.

(d) $2 \le a < m$. Rewrite $(a^2 + b^2)/(ab + 1) = m^2$ as

$$b^2 - m^2ab + a^2 - m^2 = 0.$$

We know that $t = b$ is a root of the quadratic equation

$$t^2 - m^2at + a^2 - m^2 = 0. \tag{1}$$

Thus $m^4a^2 + 4m^2 - 4a^2$, the discriminant of the equation, must be a perfect square. But

$$(m^2a + 1)^2 = m^4a^2 + 2m^2a + 1$$
$$> m^4a^2 + 4m^2 - 4a^2 > (m^2a)^2$$

for $2 \le a < m$. So the discriminant cannot be a perfect square, a contradiction.

(e) $a > m$. Again $t = b$ is a root of (1). It is easy to check that $t = m^2a - b = c$ also satisfies the equation. We have $bc = a^2 - m^2 > 0$; since $b \ge 0$, $c > 0$. Since $a > 0$ and $c > 0$, $ac + 1 > 0$, we have

$$\frac{c^2 + a^2}{ca + 1} = m^2.$$

Since $c > 0$, $b \ge a$, and $bc = a^2 - m^2 < a^2$, we have $c < a$. Thus (c, a) is a valid pair. Also, it cannot be of the form (a_n, a_{n+1}), or else

$$(a, b) = (a_{n+1}, m^2a_{n+1} - a_n) = (a_{n+1}, a_{n+2}).$$

But then, $c + a < a + a \le b + a$, as desired.

From the above, we see that our assumption is false. Therefore every pair satisfying the original equation must be of the described form.

Second solution. Note that if $a = 0$, then necessarily $b = m$, so $(a, b) = (a_0, a_1)$. Also note that there is no solution with a, b both nonzero but of opposite signs. Thus any other solution has $b \ge a > 0$. If (a, b) is a solution, then one easily checks that $(a, m^2a - b)$ is again a solution. Since $a > 0$, this means we must have $m^2a - b \ge 0$. Since $b(m^2a - b) = a^2 - m^2 < a^2$ and $b > a$, we must also have $m^2a - b < a$. Thus we have produced a smaller nonnegative solution. Because we cannot reduce the solution indefinitely, reducing in this way must eventually reach $(0, m)$. Therefore the original solution must have been (a_n, a_{n+1}) for some n.

Problem 9.3.14. *Let b, c be positive integers, and define the sequence $a_1, a_2, \ldots$ by $a_1 = b$, $a_2 = c$, and*

$$a_{n+2} = |3a_{n+1} - 2a_n|$$

for $n \ge 1$. Find all such (b, c) for which the sequence $a_1, a_2, \ldots$ has only a finite number of composite terms.

(2002 Bulgarian Mathematical Olympiad)

Solution. The only solutions are (p, p) for p not composite, $(2p, p)$ for p not composite, and $(7, 4)$.

The sequence $a_1, a_2, \ldots$ cannot be strictly decreasing because each a_n is a positive integer, so there exists a smallest $k \geq 1$ such that $a_{k+1} \geq a_k$. Define a new sequence $b_1, b_2, \ldots$ by $b_n = a_{n+k-1}$, so $b_2 \geq b_1$, $b_{n+2} = |3b_{n+1} - 2b_n|$ for $n \geq 1$, and $b_1, b_2, \ldots$ has only a finite number of composite terms. Now, if $b_{n+1} \geq b_n$,

$$b_{n+2} = |3b_{n+1} - 2b_n| = 3b_{n+1} - 2b_n = b_{n+1} + 2(b_{n+1} - b_n) \geq b_{n+1},$$

so by induction $b_{n+2} = 3b_{n+1} - 2b_n$ for $n \geq 1$.

Using the general theory of linear recursion relations (a simple induction proof also suffices), we have

$$b_n = A \cdot 2^{n-1} + B$$

for $n \geq 1$, where $A = b_2 - b_1$, $B = 2b_1 - b_2$. Suppose (for contradiction) that $A \neq 0$. Then b_n is an increasing sequence, and since it contains only finitely many composite terms, $b_n = p$ for some prime $p > 2$ and some $n \geq 1$. But then $b_{n+l(p-1)}$ would be divisible by p and thus composite for $l \geq 1$, because

$$b_{n+l(p-1)} = A \cdot 2^{n-1} \cdot 2^{l(p-1)} + B \equiv A \cdot 2^{n-1} + B \equiv b_n \equiv 0 \pmod{p}$$

by Fermat's little theorem. This is a contradiction, so $A = 0$ and $b_n = b_1$ for $n \geq 1$. Therefore b_1 is not composite; let $b_1 = p$, where $p = 1$ or p is prime.

We now return to the sequence $a_1, a_2, \ldots$, and consider different possible values of k. If $k = 1$, we have $a_1 = b_1 = b_2 = a_2 = p$, so $b = c = p$ for p not composite. If $k > 1$, consider that $a_{k-1} > a_k$ by the choice of k, but $a_{k+1} = |3a_k - 2a_{k-1}|$, and $a_{k+1} = b_2 = b_1 = a_k$, so $a_{k+1} = 2a_{k-1} - 3a_k$, and thus $a_{k-1} = 2p$. For $k = 2$, this means that $b = 2p$, $c = p$ for p not composite. If $k > 2$, the same approach yields

$$a_{k-2} = \frac{3a_{k-1} \pm a_k}{2} = \tfrac{7}{2}p \text{ or } \tfrac{5}{2}p,$$

so $p = 2$. For $k = 3$, this gives solutions $b = 7$ or $b = 5$, $c = 4$. Because $\frac{3 \cdot 5 \pm 4}{2}$ and $\frac{3 \cdot 7 \pm 4}{2}$ are not integers, there are no solutions for $k > 3$.

Remark. The reader may try to prove the following more general statement: *Let $f \in \mathbb{Z}[X_1, \ldots, X_k]$ be a polynomial and $F(n) = f(n, 2^n, 3^n, \ldots, (k-1)^n)$, $n \geq 1$. If $\lim_{n \to \infty} F(n) = \infty$, then the set of primes dividing the terms of the sequence $(F(n))_{n \geq 1}$ is infinite.*

9.3.3   Nonstandard Sequences of Integers

Problem 9.3.21. *Let* $\{a_n\}$ *be a sequence of integers such that for* $n \geq 1$,

$$(n - 1)a_{n+1} = (n + 1)a_n - 2(n - 1).$$

If 2000 *divides* a_{1999}, *find the smallest* $n \geq 2$ *such that* 2000 *divides* a_n.

(1999 Bulgarian Mathematical Olympiad)

Solution. First, we note that the sequence $a_n = 2n - 2$ works. Then writing $b_n = a_n - (2n - 2)$ gives the recursion

$$(n - 1)b_{n+1} = (n + 1)b_n.$$

For $n \geq 2$, observe that

$$b_n = b_2 \prod_{k=2}^{n-1} \frac{k + 1}{k - 1} = b_2 \frac{\prod_{k=3}^{n} k}{\prod_{k=1}^{n-2} k} = \frac{n(n - 1)}{2} b_2.$$

Thus when $n \geq 2$, the solution to the original equation of the form

$$a_n = 2(n - 1) + \frac{n(n - 1)}{2} c$$

for some constant c. Plugging in $n = 2$ shows that $c = a_2 - 2$ is an integer.

Now, because $2000 \mid a_{1999}$ we have

$$2(1999 - 1) + \frac{1999 \cdot 1998}{2} c \equiv 0 \pmod{2000}.$$

This implies $-4 + 1001c \equiv 0 \pmod{2000}$, hence $c \equiv 4 \pmod{2000}$.

Then $2000 \mid a_n$ exactly when

$$2(n - 1) + 2n(n - 1) \equiv 0 \pmod{2000}$$
$$\Leftrightarrow \quad (n - 1)(n + 1) \equiv 0 \pmod{1000}.$$

The number $(n-1)(n+1)$ is divisible by 8 exactly when n is odd, and it is divisible by 125 exactly when either $n - 1$ or $n + 1$ is divisible by 125. The smallest $n \geq 2$ satisfying these requirements is $n = 249$.

Problem 9.3.22. *The sequence* $(a_n)_{n \geq 0}$ *is defined by* $a_0 = 1$, $a_1 = 3$, *and*

$$a_{n+2} = \begin{cases} a_{n+1} + 9a_n & \text{if } n \text{ is even,} \\ 9a_{n+1} + 5a_n & \text{if } n \text{ is odd.} \end{cases}$$

Prove that

(a) $\sum_{k=1995}^{2000} a_k^2$ *is divisible by* 20,

(b) a_{2n+1} *is not a perfect square for any* $n = 0, 1, 2, \ldots$.

(1995 Vietnamese Mathematical Olympiad)

Solution. (a) We will first prove that the sum is divisible by 4, then by 5. Note that $a_{n+2} \equiv a_{n+1} + a_n \pmod{4}$ whether n is odd or even. The sequence modulo 4 thus proceeds 1, 3, 0, 3, 3, 2, 1, 3, ... in a cycle of 6, so the sum of squares of any six consecutive terms is congruent to $1^2 + 3^2 + 0^2 + 3^2 + 3^2 + 2^2 \equiv 0 \pmod{4}$.

Now let us work modulo 5, in which case $a_{n+2} \equiv a_{n+1} + 4a_n$ if n is even and $a_{n+2} \equiv 4a_{n+1}$ if n is odd. Hence the sequence modulo 5 proceeds 1, 3, 2, 3, 1, 4, 3, 2, 4, 1, 2, 3, ... in a cycle of 8 beginning with a_2. This means that

$$a_{1995}^2 + \cdots + a_{2000}^2 \equiv a_3^2 + \cdots + a_8^2 \equiv 3^2 + 1^2 + 4^2 + 3^2 + 2^2 + 4^2 \equiv 0 \pmod{5}.$$

(b) From part (a) we have $a_{2n+1} \equiv 2$ or 3 $\pmod{4}$, which implies that a_{2n+1} is not a square.

Problem 9.3.23. *Prove that for any natural number $a_1 > 1$, there exists an increasing sequence of natural numbers $a_1, a_2, \ldots$ such that $a_1^2 + a_2^2 + \cdots + a_k^2$ is divisible by $a_1 + a_2 + \cdots + a_k$ for all $k \geq 1$.*

(1995 Russian Mathematical Olympiad)

Solution. We will prove in fact that any finite sequence $a_1, \ldots, a_k$ with the property can be extended by a suitable a_{k+1}. Let $s_k = a_1 + \cdots + a_k$ and $t_k = a_1^2 + \cdots + a_k^2$. Then we are seeking a_{k+1} such that $a_{k+1} + s_k \mid a_{k+1}^2 + t_k$. This is clearly equivalent to $a_{k+1} + s_k \mid s_k^2 + t_k$. Why not, then, choose $a_{k+1} = s_k^2 - s_k + t_k$? Certainly this is greater than a_k and ensures that the desired property is satisfied.

Problem 9.3.24. *The sequence $a_0, a_1, a_2, \ldots$ satisfies*

$$a_{m+n} + a_{m-n} = \tfrac{1}{2}(a_{2m} + a_{2n})$$

for all nonnegative integers m and n with $m \geq n$. If $a_1 = 1$, determine a_n.

(1995 Russian Mathematical Olympiad)

Solution. The relations $a_{2m} + a_{2m} = 2(a_{2m} + a_0) = 4(a_m + a_m)$ imply $a_{2m} = 4a_m$, as well as $a_0 = 0$. Thus we compute $a_2 = 4$, $a_4 = 16$. Also, $a_1 + a_3 = (a_2 + a_4)/2 = 10$ so $a_3 = 9$. At this point we guess that $a_i = i^2$ for all $i \geq 1$.

We prove our guess by induction on i. Suppose that $a_j = j^2$ for $j < i$. Then the given equation with $m = i - 1$, $j = 1$ gives

$$a_i = \tfrac{1}{2}(a_{2i-2} + a_2) - a_{i-2}$$
$$= 2a_{i-1} + 2a_1 - a_{i-2}$$
$$= 2(i^2 - 2i + 1) + 2 - (i^2 - 4i + 4) = i^2.$$

Problem 9.3.25. *The sequence of real numbers $a_1, a_2, a_3, \ldots$ satisfies the initial conditions $a_1 = 2$, $a_2 = 500$, $a_3 = 2000$ as well as the relation*

$$\frac{a_{n+2} + a_{n+1}}{a_{n+1} + a_{n-1}} = \frac{a_{n+1}}{a_{n-1}}$$

for $n = 2, 3, 4, \ldots$ *Prove that all the terms of this sequence are positive integers and that* 2^{2000} *divides the number* a_{2000}.

(1999 Slovenian Mathematical Olympiad)

First solution. From the recursion relation it follows that $a_{n+2}a_{n-1} = a_{n+1}^2$ for $n = 2, 3, \ldots$. No term of our sequence can equal 0, and hence it is possible to write

$$\frac{a_{n+2}}{a_{n+1}a_n} = \frac{a_{n+1}}{a_n a_{n-1}} \tag{1}$$

for $n = 2, 3, \ldots$. It follows by induction that the value of the expression

$$\frac{a_{n+1}}{a_n a_{n-1}}$$

is constant, namely equal to $a_3/a_2 a_1 = 2$. Thus $a_{n+2} = 2a_n a_{n+1}$ and all terms of the sequence are positive integers.

From this new relation, we also know that a_{n+1}/a_n is an even integer for all positive integers n. Write

$$a_{2000} = \frac{a_{2000}}{a_{1999}} \frac{a_{1999}}{a_{1998}} \cdots \frac{a_2}{a_1} a_1.$$

In this product each of the 1999 fractions is divisible by 2, and $a_1 = 2$ is even as well. Thus a_{2000} is indeed divisible by 2^{2000}.

Second solution. Note that $a_n = 2^{F_{n+2}-1} \cdot 5^{3F_n-1}$, proved by induction by using equation (1) in the previous solution, where F_n are the Fibonacci numbers, $n \geq 1$. Hence the divisibility is a consequence of $F_{2002} \geq 2001$.

Problem 9.3.26. *Let k be a fixed positive integer. We define the sequence $a_1, a_2, \ldots$ by $a_1 = k + 1$ and the recursion $a_{n+1} = a_n^2 - ka_n + k$ for $n \geq 1$. Prove that a_m and a_n are relatively prime for distinct positive integers m and n.*

First solution. We claim that

$$a_n = \prod_{i=0}^{n-1} a_i + k, \quad n > 0,$$

assuming that $a_0 = 1$. Since $a_{j+1} - k = a_j(a_j - k)$, we have

$$a_n - k = \prod_{j=1}^{n-1} \frac{a_{j+1} - k}{a_j - k} = \prod_{j=1}^{n-1} a_j,$$

which is what we wanted.

Therefore, we have that $a_n \equiv k \pmod{a_i}$ for $i < n$. Hence, if there exist integers $d > 1$, $x, y \geq 1$ such that $d \mid a_x$ and $d \mid a_y$, d divides k. We now show that for $i > 0$, $a_i \equiv 1 \pmod{k}$ by induction on i. For the base case, $a_1 = k+1 \equiv 1 \pmod{k}$. Now assume that $a_i \equiv 1 \pmod{k}$. Then, $a_{i+1} \equiv a_i^2 - ka_i + k \equiv a_i^2 \equiv 1 \pmod{k}$. Thus, because all common divisors d of a_x and a_y must be divisors of k, we have $a_x \equiv 1 \pmod{d}$ and $a_y \equiv 1 \pmod{d}$. Therefore, no such divisors exist and a_i is relatively prime to a_j for all $i, j > 0$, as desired.

Second solution. First, by induction on n, it follows that $a_n \equiv 1 \pmod{k}$ for all n. Then it follows by induction on m that $a_m \equiv k \pmod{a_n}$ for all $m > n$. Therefore for $m > n$ we have $\gcd(a_m, a_n) = \gcd(k, a_n) = \gcd(k, 1) = 1$.

Problem 9.3.27. *Suppose the sequence of nonnegative integers $a_1, a_2, \ldots, a_{1997}$ satisfies*

$$a_i + a_j \le a_{i+j} \le a_i + a_j + 1$$

for all $i, j \ge 1$ with $i + j \le 1997$. Show that there exists a real number x such that $a_n = \lfloor nx \rfloor$ for all $1 \le n \le 1997$.

(1997 USA Mathematical Olympiad)

Solution. Any x that lies in all of the half-open intervals

$$I_n = \left[\frac{a_n}{n}, \frac{a_n + 1}{n} \right), \quad n = 1, 2, \ldots, 1997,$$

will have the desired property. Let

$$L = \max_{1 \le n \le 1997} \frac{a_n}{n} = \frac{a_p}{p} \quad \text{and} \quad U = \min_{1 \le n \le 1997} \frac{a_n + 1}{n} = \frac{a_q + 1}{q}.$$

We shall prove that

$$\frac{a_n}{n} < \frac{a_m + 1}{m},$$

or equivalently,

$$m a_n < n(a_m + 1) \tag{$*$}$$

for all m, n ranging from 1 to 1997. Then $L < U$, since $L \ge U$ implies that $(*)$ is violated when $n = p$ and $m = q$. Any point x in $[L, U)$ has the desired property.

We prove $(*)$ for all m, n ranging from 1 to 1997 by strong induction. The base case $m = n = 1$ is trivial. The induction step splits into three cases. If $m = n$, then $(*)$ certainly holds. If $m > n$, then the induction hypothesis gives $(m - n)a_n < n(a_{m-n} + 1)$, and adding $n(a_{m-n} + a_n) \le n a_m$ yields $(*)$. If $m < n$, then the induction hypothesis yields $m a_{n-m} < (n - m)(a_m + 1)$, and adding $m a_n \le m(a_m + a_{n-m} + 1)$ gives $(*)$.

Problem 9.3.28. *The sequence $\{a_n\}$ is given by the following relation:*

$$a_{n+1} = \begin{cases} \frac{a_n - 1}{2}, & \text{if } a_n \ge 1, \\ \frac{2a_n}{1 - a_n}, & \text{if } a_n < 1. \end{cases}$$

Given that a_0 is a positive integer, $a_n \neq 2$ for each $n = 1, 2, \ldots, 2001$, and $a_{2002} = 2$, find a_0.

(2002 St. Petersburg City Mathematical Olympiad)

Solution. Answer: $a_0 = 3 \cdot 2^{2002} - 1$.

Suppose we are given a_{n+1}. Then there are exactly two possibilities for a_n. If $a_n \geq 1$, then the first rule gives $a_n = 2a_{n+1} + 1$ (which is at least 1 as required). If $a_n < 1$, then the second rule gives $a_n = \frac{a_{n+1}}{a_{n+1}+2}$ (which is less than 1 as required). Thus the problem amounts to the following: Start with $a_{2002} = 2$ and repeatedly apply one of the two transformations

$$a_n = 2a_{n+1} + 1, \qquad \frac{a_{n+1}}{a_{n+1} + 2}.$$

Suppose you never again get $a_n = 2$ and suppose a_0 is an integer. Then what is a_0?

If we apply the first rule 2002 times, then $a_n + 1$ doubles every step (and in particular $a_n \neq 2$) and we get $a_0 = 3 \cdot 2^{2002} - 1$. We will show that this is the only possibility. Using the two rules, there are two possibilities for a_{2001}, namely 5 or $1/2$. Using the two rules a second time, there are four possibilities for a_{2000}, namely $11, 5/7, 2$, and $1/5$. Since we are not allowed to reuse 2, the third is not actually permitted. Note that the other three are all of the form p/q for p and q odd and relatively prime. We will show by (downward) induction on n that all subsequent a_n's have this form, that the denominator never decreases, and that if we ever use the second rule, then we have $q > 1$. It follows that the only way to get a_0 an integer is always to apply the first rule, as claimed above.

Suppose $a_{n+1} = p/q$ with p and q odd and relatively prime and that we apply the first rule. Then $a_n = 2p/q + 1 = (2p+q)/q$. The numerator and denominator are clearly both odd and $\gcd(2p + q, q) = \gcd(2p, q) = 1$, since q is odd and relatively prime to p. Thus a_n again has the desired form and the denominator was unchanged.

Suppose $a_{n+1} = p/q$ with p and q odd and relatively prime and that we apply the second rule. Then $a_n = p/(2q + p)$. The numerator and denominator are clearly both odd and $\gcd(p, 2q + p) = \gcd(p, 2q) = 1$, since p is odd and relatively prime to q. Thus again a_n has the desired form and the denominator has increased. Thus if we ever apply this rule, the denominator will be greater than 1.

Problem 9.3.29. *Let $x_1 = x_2 = x_3 = 1$ and $x_{n+3} = x_n + x_{n+1}x_{n+2}$ for all positive integers n. Prove that for any positive integer m there is an integer $k > 0$ such that m divides x_k.*

Solution. Observe that setting $x_0 = 0$, the condition is satisfied for $n = 0$.

We prove that there is an integer $k \leq m^3$ such that x_k divides m. Let r_t be the remainder of x_t when divided by m for $t = 0, 1, \ldots, m^3 + 2$. Consider the triples

(r_0, r_1, r_2), (r_1, r_2, r_3), ..., $(r_{m^3}, r_{m^3+1}, r_{m^3+2})$. Since r_t can take m values, it follows by the pigeonhole principle that at least two triples are equal. Let p be the smallest number such that triple (r_p, r_{p+1}, r_{p+2}) is equal to another triple (r_q, r_{q+1}, r_{q+2}), $p < q \le m^3$. We claim that $p = 0$.

Assume by way of contradiction that $p \ge 1$. Using the hypothesis we have

$$r_{p+2} \equiv r_{p-1} + r_p r_{p+1} \pmod{m}$$

and

$$r_{q+2} \equiv r_{q-1} + r_q r_{q+1} \pmod{m}.$$

Since $r_p = r_q$, $r_{p+1} = r_{q+1}$, and $r_{p+2} = r_{q+2}$, it follows that $r_{p-1} = r_{q-1}$ so $(r_{p-1}, r_p, r_{p+1}) = (r_{q-1}, r_q, r_{q+1})$, which is a contradiction to the minimality of p. Hence $p = 0$, so $r_q = r_0 = 0$, and therefore $x_q \equiv 0 \pmod{m}$.

Problem 9.3.30. *Find all infinite bounded sequences $a_1, a_2, \ldots$ of positive integers such that for all $n > 2$,*

$$a_n = \frac{a_{n-1} + a_{n-2}}{\gcd(a_{n-1}, a_{n-2})}.$$

(1999 Russian Mathematical Olympiad)

Solution. The only such sequence is $2, 2, 2, \ldots$, which clearly satisfies the given condition.

Suppose $\gcd(a_{k-1}, a_{k-2}) = 1$ for some k. Then $a_k = a_{k-1} + a_{k-2}$ and hence $\gcd(a_k, a_{k-1}) = \gcd(a_{k-2}, a_{k-1}) = 1$. Hence by induction it follows that $a_n = a_{n-1} + a_{n-2}$ for all $n \ge k$, and the sequence is unbounded.

Therefore we must have $\gcd(a_{n-1}, a_{n-2}) \ge 2$ for all n and

$$a_n = \frac{a_{n-1} + a_{n-2}}{\gcd(a_{n-1}, a_{n-2})} \le \frac{a_{n-1} + a_{n-2}}{2} \le \max(a_{n-1}, a_{n-2})$$

for all n. Thus $\max(a_n, a_{n-1}) \le \max(a_{n-1}, a_{n-2})$. Since this is a decreasing sequence of positive integers, it is eventually constant. If $a_{n-1} < a_{n-2}$, then the argument above gives $a_n < (a_{n-1} + a_{n-2})/2$ and hence $\max(a_n, a_{n-1}) < \max(a_{n-1}, a_{n-2})$. Thus we must eventually have $a_{n-1} = \max(a_{n-1}, a_{n-2})$. Hence the sequence a_n must be eventually constant. But if $a_{n-1} = a_{n-2}$, then we compute $\gcd(a_{n-1}, a_{n-2}) = a_{n-1} = a_{n-2}$ and $a_n = 2$. Thus the sequence must be eventually constant at 2.

Suppose now that $a_{n+1} = a_n = 2$. Then since $\gcd(a_n, a_{n-1}) > 1$, we must have $\gcd(a_n, a_{n-1}) = a_n = 2$ and $2 = a_{n+1} = (a_{n-1} + 2)/2$, implying $a_{n-1} = 2$. Thus the only way the sequence can be eventually constant at the value 2 is if it always has the value 2.

Problem 9.3.31. *Let $a_1, a_2, \ldots$ be a sequence of positive integers satisfying the condition $0 < a_{n+1} - a_n \le 2001$ for all integers $n \ge 1$. Prove that there exists an infinite number of ordered pairs (p, q) of distinct positive integers such that a_p is a divisor of a_q.*

(2001 Vietnamese Mathematical Olympiad)

Solution. Obviously, if $(a_n)_n$ is such a sequence, so is $(a_{n+k})_n$ for all k. Thus it suffices to find $p < q$ such that $a_p \mid a_q$. Observe that from any 2001 consecutive natural numbers, at least one is a term of the sequence. Now, consider the table

$$
\begin{array}{cccc}
a_1 + 1 & a_1 + 2 & \cdots & a_1 + 2001 \\
a_1 + 1 + x_1 & a_1 + 2 + x_1 & \cdots & a_1 + 2001 + x_1 \\
a_1 + 1 + x_1 + x_2 & a_1 + 2 + x_1 + x_2 & \cdots & a_1 + 2001 + x_1 + x_2 \\
\vdots & & &
\end{array}
$$

where

$$
x_1 = \prod_{i=1}^{2001}(a_1 + i), \quad x_2 = \prod_{i=1}^{2001}(a_1 + i + x_1), \quad x_3 = \prod_{i=1}^{2001}(a_1 + x_1 + x_2 + i),
$$

and so on. Observe then that if x, y are in the same column, then $x \mid y$ or $y \mid x$. Now look at the first 2002 lines. We find in this 2002×2001 matrix at least 2002 terms of the sequence (at least one on each line). Thus there are two terms of the sequence in the same column, and one will divide the other.

Problem 9.3.32. *Define the sequence $\{x_n\}_{n \ge 0}$ by $x_0 = 0$ and*

$$
x_n = \begin{cases}
x_{n-1} + \frac{3^{r+1}-1}{2}, & \text{if } n = 3^r(3k+1), \\
x_{n-1} - \frac{3^{r+1}+1}{2}, & \text{if } n = 3^r(3k+2),
\end{cases}
$$

where k and r are nonnegative integers. Prove that every integer appears exactly once in this sequence.

(1999 Iranian Mathematical Olympiad)

First solution. We prove by induction on $t \ge 1$ that

(i) $\{x_0, x_1, \ldots, x_{3^t-2}\} = \left\{ -\frac{3^t-3}{2}, -\frac{3^t-5}{2}, \ldots, \frac{3^t-1}{2} \right\}$;

(ii) $x_{3^t-1} = -\frac{3^t-1}{2}$.

These claims imply the desired result, and they are easily verified for $t = 1$. Now supposing they are true for t, we show they are true for $t + 1$.

For any positive integer m, write $m = 3^r(3k + s)$ for nonnegative integers r, k, s, with $s \in \{1, 2\}$, and define $r_m = r$ and $s_m = s$.

Then for $m < 3^t$, observe that

$$r_m = r_{m+3^t} = r_{m+2\cdot3^t} \quad \text{and} \quad s_m = s_{m+3^t} = s_{m+2\cdot3^t},$$

so that

$$x_m - x_{m-1} = x_{3^t+m} - x_{3^t+m-1} = x_{2\cdot3^t+m} - x_{2\cdot3^t+m-1}.$$

Setting $m = 1, 2, \ldots, k < 3^t$ and adding the resulting equations, we have

$$x_k = x_{3^t+k} - x_{3^t}$$
$$x_k = x_{2\cdot3^t+k} - x_{2\cdot3^t}.$$

Now setting $n = 3^t$ in the recursion and using (ii) from the induction hypothesis, we have $x_{3^t} = 3^t$, and

$$\{x_{3^t}, \ldots, x_{2\cdot3^t-2}\} = \left\{\frac{3^t + 3}{2}, \ldots, \frac{3^{t+1} - 1}{2}\right\},$$

$$x_{2\cdot3^t-1} = \frac{3^t + 1}{2}.$$

Then setting $n = 2 \cdot 3^t$ in the recursion, we have $x_{2\cdot3^t} = -3^t$, giving

$$\{x_{2\cdot3^t}, \ldots, x_{3^{t+1}-2}\} = \left\{-\frac{3^{t+1} - 3}{2}, \ldots, \frac{3^t + 1}{2}\right\}$$

$$x_{2\cdot3^{t+1}-1} = -\frac{3^{t+1} - 1}{2}.$$

Combining this with (i) and (ii) from the induction hypothesis proves the claims for $t + 1$. This completes the proof.

Second solution. For $n_i \in \{-1, 0, 1\}$, let the number

$$[n_m n_{m-1} \cdots n_0]$$

in base 3 equals $\sum_{i=0}^{m} n_i \cdot 3^i$. It is simple to prove by induction on k that the base-3 numbers with at most k digits equal

$$\left\{-\frac{3^k - 1}{2}, -\frac{3^k - 3}{2}, \ldots, \frac{3^k - 1}{2}\right\},$$

which implies that every integer has a unique representation in base 3.

Now we prove by induction on n that if $n = a_m a_{m-1} \cdots a_0$ in base 3, then $x_n = [b_m b_{m-1} \ldots b_0]$ in base 3, where $b_i = -1$ if $a_i = 2$ and $b_i = a_i$ for all other cases.

For the base case, $x_0 = 0 = [0]$. Now assume that the claim is true for $n - 1$. If $n = a_m a_{m-1} \cdots a_{r+1} 1 \underbrace{00 \ldots 0}_{r}$, then

$$
\begin{aligned}
x_n &= x_{n-1} + \frac{3^{r+1} - 1}{2} \\
&= [b_m b_{m-1} \ldots b_i 0 \underbrace{-1 - 1 \cdots - 1}_{r}] + [\underbrace{11 \ldots 1}_{r+1}] \\
&= [b_m b_{m-1} \ldots b_i 1 \underbrace{00 \ldots 0}_{r}].
\end{aligned}
$$

If instead $n = a_m a_{m-1} \cdots a_i 2 \underbrace{00 \ldots 0}_{r}$, then

$$
\begin{aligned}
x_n &= x_{n-1} + \left(-\frac{3^{r+1} + 1}{2} \right) \\
&= [b_m b_{m-1} \ldots b_i 1 \underbrace{-1 - 1 \cdots - 1}_{r}] + [-1 \underbrace{11 \ldots 1}_{r+1}] \\
&= [b_m b_{m-1} \ldots b_i - 1 \underbrace{00 \ldots 0}_{r}].
\end{aligned}
$$

In either case, the claim is true for n, completing the induction.

To finish the proof, note that every integer appears exactly once in base 3. Thus each integer appears exactly once in $\{x_n\}_{n \geq 0}$, as desired.

Problem 9.3.33. *Suppose that $a_1, a_2, \ldots$ is a sequence of natural numbers such that for all natural numbers m and n, $\gcd(a_m, a_n) = a_{\gcd(m,n)}$. Prove that there exists a sequence $b_1, b_2, \ldots$ of natural numbers such that $a_n = \prod_{d \mid n} b_d$ for all integers $n \geq 1$.*

(2001 Iranian Mathematical Olympiad)

First solution. For each n, let rad(n) denote the largest square-free divisor of n (i.e., the product of all distinct prime factors of n). We let b_n equal to the ratio of the following two numbers:

- E_n, the product of all $a_{n/d}$ such that d is square-free, divides n, and has an even number of prime factors.

- O_n, the product of all $a_{n/d}$ such that d is square-free, divides n, and has an odd number of prime factors.

Lemma 1. $\prod_{d \mid a_n} b_d = a_n$.

Proof. Fix n, and observe that $\prod_{d|n} b_n$ equals

$$\frac{\prod_{d|n} E_d}{\prod_{d|n} O_d} = \prod_{d|n} a_{n/d}^{\mu(d)}, \tag{$*$}$$

where μ is the Möbius function.

In the numerator of $(*)$, each E_d is the product of a_m such that $m \mid d$. Also, $d \mid n$, implying that the numerator is the product of various a_m such that $m \mid n$. For fixed m that divides n, how many times does a_m appears in the numerator $\prod_{d|n} E_d$ of $(*)$?

If a_m appears in E_d and $d \mid n$, then let $t = d/m$. By the definition of E_d, we know that (i) t is square-free and (ii) t has an even number of prime factors. Because $d \mid n$ and $t = d/m$, we further know that (iii) t divides n/m.

Conversely, suppose that t is any positive integer satisfying (i), (ii), and (iii), and write $d = tm$. By (iii), d is a divisor of n. Also, t is square-free by (i), is a divisor of d, and has an even number of prime factors by (ii). Thus, a_m appears in E_d.

Suppose that n/m has l distinct prime factors. Then it has $\binom{l}{0} + \binom{l}{2} + \cdots$ factors t satisfying (i), (ii), and (iii), implying that a_m appears in the numerator of $(*)$ exactly

$$\binom{l}{0} + \binom{l}{2} + \cdots$$

times. Similarly, a_m appears in the denominator of $(*)$ exactly

$$\binom{l}{1} + \binom{l}{3} + \cdots$$

times. If $m < n$, then $l \geq 1$, and these expressions are equal, so that the a_m's in the numerator and denominator of $(*)$ cancel each other out. If $m = n$, then $l = 0$, so that a_n appears in the numerator once and in the denominator zero times. Therefore,

$$\prod_{d|n} b_d = \frac{\prod_{d|n} E_d}{\prod_{d|n} O_d} = a_n,$$

as desired. □

Lemma 2. *For any integer α that divides some term in $a_1, a_2, \ldots$, there exists an integer d such that*

$$\alpha | a_n \iff d | n.$$

Proof. Of all the integers n such that $\alpha \mid a_n$, let d be the smallest.

If $\alpha \mid a_n$, then $\alpha \mid \gcd(a_d, a_n) = a_{\gcd(d,n)}$. By the minimality of d, $\gcd(d, n) \geq d$. But $\gcd(d, n) \mid n$ as well, implying $\gcd(d, n) = d$. Hence, $d \mid n$.

If $d \mid n$, then $\gcd(a_d, a_n) = a_{\gcd(d,n)} = a_d$. Thus, $a_d \mid a_n$. Because $\alpha \mid a_d$, it follows that $\alpha \mid a_n$ as well. □

Lemma 3. *For each positive integer n, $b_n = E_n/O_n$ is an integer.*

Proof. Fix n. Call an integer d a top divisor (resp. a bottom divisor) if $d \mid n$, n/d is square-free, and n/d has an even (resp. odd) number of prime factors. By definition, E_d is the product of a_d over all top divisors d, and O_d is the product of a_d over all bottom divisors d.

Fix any prime p. We show that p divides E_n at least as many times as it divides O_n. To do this, it suffices to show the following for any positive integer k:

(1) The number of top divisors d with a_d divisible by p^k is greater than or equal to the number of bottom divisors d with a_d divisible by p^k.

Let k be any positive integer. If p^k divides none of $a_1, a_2, \ldots$, then (1) holds trivially. Otherwise, by the previous lemma, there exists an integer d_0 such that

$$p^k \mid a_m \Leftrightarrow d_0 \mid m.$$

Hence, to show (1) it suffices to show:

(2) The number of top divisors d such that $d_0 \mid d$, is greater than or equal to the number of bottom divisors d such that $d_0 \mid d$.

If $d_0 \nmid n$, then (2) holds because d_0 does not divide d for any divisor d of n, including top or bottom divisors.

Otherwise, $d_0 \mid n$. For which top and bottom divisors d does d_0 divide d? Precisely those for which n/d divides n/d_0. If n/d_0 has $l \geq 1$ distinct prime factors, then there are as many top divisors with this property as there are bottom divisors, namely

$$\binom{l}{0} + \binom{l}{2} + \cdots = 2^{l-1} = \binom{l}{1} + \binom{l}{3} + \cdots.$$

If instead $d_0 = n$ and $l = 0$, then the top divisor 1 is the only value d with $d \mid (n/d_0)$. In either case, there are at least as many top divisors d with $d \mid (n/d_0)$ as there are bottom divisors with the same property. Therefore, (2) holds. This completes the proof.   □

Therefore, $a_n = \prod_{d \mid n} b_d$, and $b_n = E_n/O_n$ is an integer for each n.

Second solution. (Gabriel Dospinescu) Let us define $b_1 = a_1$ and $b_n = a_n / \text{lcm}_{d \mid n, d \neq n} a_d$ for $n > 1$. Of course, if $d \mid n$, then $a_d \mid a_n$ and so $\text{lcm}_{d \mid n, d \neq n} a_d \mid a_n$ and $b_n \in \mathbb{Z}$.

Now comes the hard part, proving that $\prod_{d \mid n} b_d = a_n$, which is the same as

$$\prod_{\substack{d \mid n \\ d \neq n}} b_d = \text{lcm}_{\substack{d \mid n \\ d \neq n}} a_d. \tag{1}$$

We will prove (1) by strong induction. For $n = 1$ it is clear.

Now, for all $d \mid n$, $d \neq n$, by the inductive hypothesis we have

$$a_d = \prod_{d' \mid d} b_{d'} \mid \prod_{\substack{d \mid n \\ d \neq n}} b_d;$$

thus $\prod_{d \mid n, d \neq n} b_d$ is a multiple of $\mathrm{lcm}_{d \mid n, d \neq n} a_d$. It remains to prove that $\prod_{d \mid n, d \neq n} b_d \mid \mathrm{lcm}_{d \mid n, d \neq n} a_d$.

The essential observation is:

Lemma. *If* $\gcd(b_u, b_v) > 1$, *then* $u \mid v$ *or* $v \mid u$.

Proof. We may assume that $u < v$. Assume that u does not divide v. Then

$$b_u = \frac{a_u}{\mathrm{lcm}_{\substack{d \mid u \\ d \neq u}} a_d} \mid \frac{a_u}{a_{\gcd(u,v)}}.$$

Remark. From Problem 9.3.6 (2) we have $\gcd(F_m, F_n) = F_{\gcd(m,n)}$, where F_n is the nth Fibonacci number, so this holds for F_n. We have

$$F_n = \prod_{d \mid n} b_d,$$

where $(b_n)_{n \geq 0}$ is the sequence

$$0, 1, 1, 2, 3, 5, 4, 13, 7, 17, 11, 89, 6, 233, 29, 61, 47, 1597, 152, \ldots .$$

10

Problems Involving Binomial Coefficients

10.1 Binomial Coefficients

Problem 10.1.7. *Show that the sequence*

$$\binom{2002}{2002}, \ \binom{2003}{2002}, \ \binom{2004}{2002}, \ \ldots,$$

considered modulo 2002, is periodic.

(2002 Baltic Mathematical Competition)

Solution. We will show that the sequence, taken modulo 2002, has period $m = 2002 \cdot 2002!$. Indeed,

$$\begin{aligned}
\binom{x+m}{2002} &= \frac{(x+m)(x-1+m)\cdots(x-2001+m)}{2002!} \\
&= \frac{x(x-1)\cdots(x-2001)+km}{2002!} \\
&= \frac{x(x-1)\cdots(x-2001)}{2002!} + 2002k \\
&\equiv \binom{x}{2002} \pmod{2002}.
\end{aligned}$$

Problem 10.1.8. *Prove that*

$$\binom{2p}{p} \equiv 2 \pmod{p^2}$$

for every prime number p.

T. Andreescu and D. Andrica, *Number Theory*, DOI: 10.1007/b11856_21,
© Birkhäuser Boston, a part of Springer Science + Business Media, LLC 2009

Solution. A short solution uses the popular Vandermonde identity

$$\sum_{i=0}^{k}\binom{m}{i}\binom{n}{k-i}=\binom{m+n}{k}.$$

Set $m = n = k = p$ to get

$$\binom{2p}{p}=\binom{p}{0}\binom{p}{p}+\binom{p}{1}\binom{p}{p-1}+\cdots+\binom{p}{p-1}\binom{p}{1}+\binom{p}{p}\binom{p}{0}.$$

The first and the last terms on the right-hand side equal 1. Since p is a prime, it divides each binomial coefficient $\binom{p}{k}$ for $1 \le k \le p-1$. So each of the remaining terms is divisible by p^2, and hence $\binom{2p}{p}$ is congruent to 2 modulo p^2, as required.

Problem 10.1.9. *Let k, m, n be positive integers such that $m + k + 1$ is a prime number greater than $n + 1$. Let us set $C_s = s(s + 1)$. Show that the product*

$$(C_{m+1} - C_k)(C_{m+2} - C_k)\cdots(C_{m+n} - C_k)$$

is divisible by $C_1 C_2 \cdots C_n$.

(18th International Mathematical Olympiad)

Solution. We use the identity

$$C_p - C_q = p(p + 1) - q(q + 1) = (p - q)(p + q + 1),$$

which is valid for all positive integers p and q. Then one has

$$C_{m+i} - C_k = (m - k + i)(m + k + i + 1), \quad \text{for all } i = 1, 2, \ldots, n.$$

For the given products we obtain respectively the formulas

$$(C_{m+1} - C_k)\cdots(C_{m+n} - C_k) = \prod_{i=1}^{n}(m - k + i)\prod_{i=1}^{n}(m + k + 1 + i),$$

$$C_1 C_2 \cdots C_n = n!(n + 1)!.$$

Their quotient is the product of two rational fractions:

$$\frac{\prod_{i=1}^{n}(m - k + i)}{n!} \quad \text{and} \quad \frac{\prod_{i=1}^{n}(m + k + 1 + i)}{(n + 1)!}.$$

It is known that the product of any n consecutive integers is divisible by $n!$ and their quotient is zero or a binomial coefficient, possibly multiplied by -1. In our case we have

$$\frac{1}{n!}\prod_{i=1}^{n}(m - k + i) = \binom{m - k + n}{n}.$$

For the second fraction, a factor is missing in the numerator. We support our argument by using the fact that $m + k + 1$ is a prime number greater than $n + 1$:

$$\frac{1}{(n+1)!} \prod_{i=1}^{n} (m+k+1+i) = \frac{1}{m+k+1} \cdot \frac{1}{(n+1)!} \prod_{i=0}^{n} (m+k+1+i)$$

$$= \frac{1}{m+k+1} \binom{m+k+n+1}{n+1}.$$

Note that

$$\frac{1}{m+k+1} \binom{m+k+n+1}{n+1} = \frac{1}{n+1} \binom{m+k+n+1}{n}.$$

Since the first expression has denominator 1 or the prime $m + k + 1$ and the second expression has denominator at most $n + 1 < m + k + 1$, both must be integers.

Hence the binomial coefficient $\binom{m+k+n+1}{n+1}$ is an integer that is divisible by $m + k + 1$, so our number is integer.

Problem 10.1.10. *Let n, k be arbitrary positive integers. Show that there exist positive integers $a_1 > a_2 > a_3 > a_4 > a_5 > k$ such that*

$$n = \pm \binom{a_1}{3} \pm \binom{a_2}{3} \pm \binom{a_3}{3} \pm \binom{a_4}{3} \pm \binom{a_5}{3}.$$

(2000 Romanian International Mathematical Olympiad Team Selection Test)

Solution. For fixed k, choose $m > k$ such that $n + \binom{m}{3}$ is an odd number. We see that this is possible by considering the parity of n. If n is an odd number, take $m \equiv 0 \pmod 4$, and if n is an even number, take $m \equiv 3 \pmod 4$.

Since $n + \binom{m}{3}$ is an odd number, we express it in the form

$$n + \binom{m}{3} = 2a + 1.$$

Then use the identity

$$2a + 1 = \binom{a}{3} - \binom{a+1}{3} - \binom{a+2}{3} + \binom{a+3}{3}$$

to obtain

$$n = \binom{a}{3} - \binom{a+1}{3} - \binom{a+2}{3} + \binom{a+3}{3} - \binom{m}{3}.$$

Notice that for $m \geq 3$ we may ensure that

$$a = \frac{n - 1 + \binom{m}{3}}{2} > m,$$

yielding the desired representation.

Problem 10.1.11. *Prove that if n and m are integers, and m is odd, then*

$$\frac{1}{3^m n} \sum_{k=0}^{m} \binom{3m}{3k} (3n - 1)^k$$

is an integer.

(2004 Romanian International Mathematical Olympiad Team Selection Test)

Solution. Let $\omega = e^{\frac{2\pi i}{3}}$. Then

$$3 \sum_{k=0}^{m} \binom{3m}{3k} (3n - 1)^k$$
$$= (1 + \sqrt[3]{3n - 1})^{3m} + (1 + \omega\sqrt[3]{3n - 1})^{3m} + (1 + \omega^2 \sqrt[3]{3n - 1})^{3m}. \quad (1)$$

The right side of the above equality is the sum of the $3m$th powers of the roots x_1, x_2, x_3 of the polynomial

$$(X - 1)^3 - (3n - 1) = X^3 - 3X^2 + 3X - 3n.$$

Let $s_k = x_1^k + x_2^k + x_3^k$. Then $s_0 = s_1 = s_2 = 3$ and

$$s_{k+3} = 3s_{k+2} - 3s_{k+1} + 3ns_k. \quad (2)$$

It follows by induction that each s_k is an integer divisible by $3^{\lfloor k/3 \rfloor + 1}$. A repeated application of (2) yields

$$s_{k+7} = 63ns_{k+2} - 9(n^2 - 3n - 3)s_{k+1} + 27n(2n + 1)s_k.$$

Since $s_3 = 9n$, it follows inductively that s_{6k+3} is divisible by $3^{2k+2}n$ for all nonnegative integers k, and the conclusion follows by (1).

Problem 10.1.12. *Show that for every positive integer n the number*

$$\sum_{k=0}^{n} \binom{2n + 1}{2k + 1} 2^{3k}$$

is not divisible by 5.

(16th International Mathematical Olympiad)

Solution. Let us consider the binomial formula:

$$(1 + 2\sqrt{2})^{2n+1} = (1 + 2^{\frac{3}{2}})^{2n+1} = \sum_{i=0}^{2n+1} \binom{2n+1}{i} 2^{\frac{3i}{2}}$$

$$= \sum_{i=0}^{n} \binom{2n+1}{2i} 2^{3i} + \sum_{i=0}^{n} \binom{2n+1}{2i+1} 2^{3i} \cdot 2^{\frac{3}{2}} = a_n + b_n \sqrt{8},$$

where

$$a_n = \sum_{i=0}^{n} \binom{2n+1}{2i} 2^{3i} \quad \text{and} \quad b_n = \sum_{i=0}^{n} \binom{2n+1}{2i+1} 2^{3i}.$$

In a similar way,

$$(1 - 2\sqrt{2})^{2n+1} = a_n - b_n \sqrt{8}.$$

After multiplying these two equalities, we obtain $-7^{2n+1} = a_n^2 - 8b_n^2$. If $b_n \equiv 0 \pmod 5$, the above equality gives $a_n^2 \equiv -2 \pmod 5 \equiv 3 \pmod 5$. Since 3 is not a perfect square modulo 5, we obtain a contradiction.

Problem 10.1.13. *Prove that for a positive integer k there is an integer $n \geq 2$ such that $\binom{n}{1}, \ldots, \binom{n}{n-1}$ are all divisible by k if and only if k is a prime.*

Solution. If k is a prime we take $n = k$, and the property holds (see property 7 in Part I, Section 10.1).

We prove that the set of positive integers k for which the claim holds is exactly the set of primes.

Suppose now that k is not a prime. Then consider two cases:

(a) $k = p^r$, where p is a prime and $r > 1$. We find a value of i for which $k \nmid \binom{n}{i}$.

Suppose, to the contrary, that there is a positive integer n such that for all $1 \leq i \leq n - 1$, $\binom{n}{i}$ is divisible by p^r. Clearly, n is divisible by p^r, and we write $n = p^\alpha \beta$ for some β with $\gcd(\beta, p) = 1$. Take $i = p^{\alpha-1}$. Then

$$\binom{n}{i} = \prod_{j=0}^{p^{\alpha-1}-1} \frac{\beta p^\alpha - j}{p^{\alpha-1} - j}.$$

If $j = 0$, then $\frac{\beta p^\alpha - j}{p^{\alpha-1} - j} = \beta p$. If $\gcd(j, p) = 1$, then both the above numerator and denominator are coprime to p. In all other cases, we write $j = \delta p^\gamma$ for some δ coprime to p and $\gamma \leq \alpha - 2$. Thus,

$$\frac{\beta p^\alpha - j}{p^{\alpha-1} - j} = \frac{\beta p^\alpha - \delta p^\gamma}{p^{\alpha-1} - \delta p^\gamma} = \frac{p^\gamma(\beta p^{\alpha-\gamma} - \delta)}{p^\gamma(p^{\alpha-\gamma-1} - \delta)}.$$

Now, since $\alpha - \gamma - 1 \geq 1$, we have $\beta p^{\alpha-\gamma} - \delta$ and $p^{\alpha-\gamma-1} - \delta$ coprime to p. In this case, the power of p in the above numerator and denominator is γ, and the power of p in the above product of fractions, which is an integer, is 1. This contradicts the assumption that $p^r \mid n$.

(b) k is divisible by at least two distinct primes p, q. Assume by contradiction that there is a positive integer n as required. Then n is divisible by pq and we can write $n = p^\alpha \beta$ where $\gcd(p, \beta) = 1$ and $\beta > 1$ (since q divides β). Take $i = p^\alpha$. Then

$$\binom{n}{i} = \prod_{j=0}^{p^\alpha - 1} \frac{\beta p^\alpha - j}{p^\alpha - j}.$$

When $j = 0$, $\frac{\beta p^\alpha - j}{p^\alpha - j} = \beta$ is coprime to p. In all other cases, both the numerator and the denominator of $\frac{\beta p^\alpha - j}{p^\alpha - j}$ are either coprime to p or are divisible by the same power of p, and therefore the product of those fractions is not divisible by p. But p divides k, and hence $\binom{n}{i}$ is not divisible by k, contrary to our assumption.

Therefore the only positive integers k for which the claim holds are the primes.

10.2   Lucas's and Kummer's Theorems

Problem 10.2.4. *Let p be an odd prime. Find all positive integers n such that $\binom{n}{1}, \binom{n}{2}, \ldots, \binom{n}{n-1}$ are all divisible by p.*

Solution. Express n in base p: $n = n_0 + n_1 p + \cdots + n_m p^m$, where $0 \leq n_0, n_1, \ldots, n_m \leq p - 1$ and $n_m \leq 0$. We also write $k = k_0 + k_1 p + \cdots + k_m p^m$, where $0 \leq k_0, k_1, \ldots, k_m \leq p - 1$, where k_m can be zero. From Lucas's theorem we have

$$\binom{n}{k} \equiv \prod_{j=0}^{m} \binom{n_j}{k_j} \pmod{p}.$$

For $n = p^m$, the property clearly holds. Assume by way of contradiction that $n \neq p^m$. If $n_m > 1$, then letting $k = p^m < n$, we have

$$\binom{n}{k} \equiv n_m \cdot \underbrace{1 \cdot 1 \cdot 1 \cdots 1}_{m-1 \text{ times}} \equiv n_m \neq 0 \pmod{p},$$

a contradiction.

Problem 10.2.5. *Let p be a prime. Prove that p does not divide any of $\binom{n}{1}, \ldots, \binom{n}{n-1}$ if and only if $n = sp^k - 1$ for some positive integer k and some integer s with $1 \leq s \leq p - 1$.*

Solution. If n is of the form $sp^k - 1$, then its representation in base p is

$$n = \overline{(s-1) \underbrace{(p-1)\cdots(p-1)}_{k \text{ times}}}.$$

For $1 \le i \le n-1$, $i = i_0 + i_1 p + \cdots + i_m p^m$, where $0 \le i_h \le p-1$, $h = 1, \ldots, m-1$, and $0 \le i_m \le s-1$. Because p is a prime, it follows that p does not divide either $\binom{p-1}{i_h}$ or $\binom{s-1}{i_m}$. Applying Lucas's theorem, we obtain that p does not divide $\binom{n}{i}$, for all $i = 1, \ldots, n-1$.

Conversely, if n cannot be written in the form $sp^k - 1$, then $n_j < p-1$ for some $0 \le j \le m-1$, where $\overline{n_0 n_1 \cdots n_m}$ is the representation of n in base p. For

$$i = \overline{(p-1) \underbrace{0 \ldots 0}_{j-1 \text{ times}}}$$

in base p, applying again Lucas's theorem, we have

$$\binom{n}{i} \equiv 0 \pmod{p}.$$

11

Miscellaneous Problems

Problem 11.6. *Let* a, b *be positive integers. By integer division of* $a^2 + b^2$ *by* $a + b$ *we obtain the quotient* q *and the remainder* r. *Find all pairs* (a, b) *such that* $q^2 + r = 1977$.

(19th International Mathematical Olympiad)

Solution. There are finitely many possibilities to obtain $1977 = q^2 + r$. Since 1977 is not a perfect square, $0 < r < a + b$. Also, $q \leq \lfloor \sqrt{1977} \rfloor = 44$. From $a^2 + b^2 = q(a + b) + r$, we obtain

$$q = \left\lfloor \frac{a^2 + b^2}{a + b} \right\rfloor \geq \frac{a^2 + b^2}{a + b} - 1 \geq \frac{1}{2}(a + b) - 1 > \frac{r}{2} - 1.$$

Suppose $q \leq 43$. Then $r = 1977 - q^2 \geq 1977 - 43^2 = 128$ and $43 \geq q > \frac{r}{2} - 1 \geq 63$, contradiction.

We have obtained $q = 44$ and $r = 1977 - 44^2 = 41$. To finish, we have to solve in integers the equation

$$a^2 + b^2 = 44(a + b) + 41.$$

Write it in the form

$$(a - 22)^2 + (b - 22)^2 = 1009.$$

It is not difficult to find all pairs of perfect squares having the sum 1009. There exists only the representation $1009 = 28^2 + 15^2$. Then the solutions are $a = 50$, $b = 37$ and $a = 37$, $b = 50$.

Problem 11.7. *Let* m, n *be positive integers. Show that* $25^n - 7^m$ *is divisible by 3 and find the least positive integer of the form* $|25^n - 7^m - 3^m|$, *where* m, n *run over the set of positive integers.*

(2004 Romanian Mathematical Regional Contest)

T. Andreescu and D. Andrica, *Number Theory*, DOI: 10.1007/b11856_22,
© Birkhäuser Boston, a part of Springer Science + Business Media, LLC 2009

Solution. Because $25 \equiv 1 \pmod 3$ and $7 \equiv 1 \pmod 3$, it follows that $25^n - 7^m \equiv 0 \pmod 3$.

For the second part of the problem, we first remark that if m is odd, then any number $a = 25^n - 7^m - 3^m$ is divisible by 15. This follows from the first part together with

$$7^m + 3^m \equiv 2^m + (-2)^m \equiv 0 \pmod 5.$$

Moreover, for $m = n = 1$ one obtains $25 - 7 - 3 = 15$.

Assume now that m is even, say $m = 2k$. Then

$$7^m + 3^m = 7^{2k} + 3^{2k} \equiv ((-3)^{2k} + 3^{2k}) \pmod{10}$$

$$\equiv 2 \cdot 9^k \pmod{10} \equiv \pm 2 \pmod{10} \equiv 2 \text{ or } 8 \pmod{10}.$$

So, the last digit of the number $25^n - 7^m - 3^m$ is either 3 or 7. Because the number $25^n - 7^m - 3^m$ is divisible by 3, the required number cannot be 7. The situation $|25^n - 7^m - 3^m| = 3$ also cannot occur, because $25^n - 7^m - 3^m \equiv -1 \pmod 8$.

Problem 11.8. *Given an integer d, let*

$$S = \{m^2 + dn^2 \mid m, n \in \mathbb{Z}\}.$$

Let $p, q \in S$ be such that p is a prime and $r = \frac{q}{p}$ is an integer. Prove that $r \in S$.

(1999 Hungary–Israel Mathematical Competition)

Solution. Note that

$$(x^2 + dy^2)(u^2 + dv^2) = (xu \pm dyv)^2 + d(xv \mp yu)^2.$$

Write $q = a^2 + db^2$ and $p = x^2 + dy^2$ for integers a, b, x, y. Reversing the above construction yields the desired result. Indeed, solving for u and v after setting $a = xu + dyv, b = xv - yu$, and $a = xu - dyv, b = xv + yu$ gives

$$u_1 = \frac{ax - dby}{p}, \quad v_1 = \frac{ay + bx}{p},$$

$$u_2 = \frac{ax + dby}{p}, \quad v_2 = \frac{ay - bx}{p}.$$

Note that

$$(ay + bx)(ay - bx) = (a^2 + db^2)y^2 - (x^2 + dy^2)b^2 \equiv 0 \pmod p.$$

Hence p divides one of $ay + bx$, $ay - bx$ so that one of v_1, v_2 is an integer. Without loss of generality, assume that v_1 is an integer. Because $r = u_1^2 + dv_1^2$ is an integer and u_1 is rational, u_1 is an integer as well and $r \in S$, as desired.

Problem 11.9. *Prove that every positive rational number can be represented in the form*

$$\frac{a^3 + b^3}{c^3 + d^3},$$

where a, b, c, d are positive integers.

(1999 International Mathematical Olympiad Shortlist)

Solution. We first claim that if m, n are positive integers such that the rational number $r = \frac{m}{n}$ belongs to the interval $(1, 2)$, then r can be represented in the form

$$\frac{a^3 + b^3}{c^3 + d^3}.$$

This can be realized by taking $a^2 - ab + b^2 = a^2 - ad + d^2$, i.e., $b + d = a$ and $a + b = 3m$, $a + d = 2a - b = 3n$; that is, $a = m + n$, $b = 2m - n$, $d = 2n - m$.

We will prove now the required conclusion. If $s > 0$ is a rational number, take positive integers p, q such that $1 < \frac{p^3}{q^3}s < 2$. There exist positive integers a, b, d such that $\frac{p^3}{q^3}s = \frac{a^3+b^3}{a^3+d^3}$, whence $s = \frac{(aq)^3+(bq)^3}{(ap)^3+(dp)^3}$.

Problem 11.10. *Two positive integers are written on the board. The following operation is repeated: if $a < b$ are the numbers on the board, then a is erased and $ab/(b - a)$ is written in its place. At some point the numbers on the board are equal. Prove that again they are positive integers.*

(1998 Russian Mathematical Olympiad)

Solution. Call the original numbers x and y and let $L = \text{lcm}(x, y)$. For each number n on the board consider the quotient L/n; during each operation, the quotients L/b and L/a become L/b and $L/a - L/b$. Thus in terms of the quotients L/a and L/b, the operation is subtracting the smaller quotient from the larger. This is the Euclidean algorithm, so the quantity $\gcd(L/a, L/b)$ is unchanged. Hence the two equal numbers on the board are $L/\gcd(L/x, L/y)$. But $\gcd(L/x, L/y) = 1$, because otherwise x and y would both divide $L/\gcd(L/x, L/y)$ and L would not be a least common multiple. So, the two equal numbers equal $L = \text{lcm}(x, y)$, an integer.

Second solution. Again, let x and y be the original numbers and suppose both numbers eventually equal N. We prove by induction on the number of steps k before we obtain (N, N) that all previous numbers divide N in the sense that N/c is an integer. Specifically, $x \mid N$, so N must be an integer.

The claim is clear for $k = 0$. Now assume that k steps before we obtain (N, N), the numbers on the board are $(c, d) = (N/p, N/q)$ for some integers $p < q$. Then reversing the operation, the number erased in the $(k + 1)$st step must be $cd/(c + d) = N/(p + q)$, completing the inductive step.

Problem 11.11. Let $f(x) + a_0 + a_1 x + \cdots + a_m x^m$, with $m \geq 2$ and $a_m \neq 0$, be a polynomial with integer coefficients. Let n be a positive integer, and suppose that:

   (i) $a_2, a_3, \ldots, a_m$ are divisible by all the prime factors of n;

   (ii) a_1 and n are relatively prime.

   Prove that for every positive integer k, there exists a positive integer c such that $f(c)$ is divisible by n^k.

(2001 Romanian International Mathematical Olympiad Team Selection Test)

Solution. Consider any integers c_1, c_2 such that $c_1 \not\equiv c_2 \pmod{n^k}$. Observe that if $n^k \mid st$ for some integers s, t where t is relatively prime to n, then $n^k \mid s$. In particular, $n^k \nmid (c_1 - c_2)t$ if t is relatively prime to n.

Note that

$$f(c_1) - f(c_2) = \sum_{i=1}^{m} a_i(c_1^i - c_2^i) = (c_1 - c_2)\sum_{i=1}^{m} a_i \sum_{j=0}^{i-1} c_1^j c_2^{i-1-j},$$

since we have

$$c^i - d^i = (c - d)\sum_{j=0}^{i-1} c^j d^{i-1-j}.$$

For any prime p dividing n, p divides $a_2, \ldots, a_m$ but not a_1. Hence, p does not divide the second factor t in the expression above. This implies that t is relatively prime to n, so n^k does not divide the product $(c_1 - c_2)t = f(c_1) - f(c_2)$.

Therefore, $f(0), f(1), \ldots, f(n^k - 1)$ are distinct modulo n^k, and one of them, say $f(c)$, must be congruent to 0 modulo n^k. That is, $n^k \mid f(c)$, as desired.

Problem 11.12. Let x, a, b be positive integers such that $x^{a+b} = a^b b$. Prove that $a = x$ and $b = x^x$.

(1998 Iranian Mathematical Olympiad)

Solution. If $x = 1$, then $a = b = 1$ and we are done. So we may assume $x > 1$. Write $x = \prod_{i=1}^{n} p_i^{\gamma_i}$, where the p_i are the distinct prime factors of x. Since a and b divide x^{a+b}, we have $a = \prod p_i^{\alpha_i}$ and $b = \prod p_i^{\beta_i}$ for some nonnegative integers α_i, β_i.

First suppose that some β_i is zero, that is, p_i does not divide b. Then the given equation implies that $\gamma_i(a + b) = \alpha_i b$, so that $(\alpha_i - \gamma_i)b = a\gamma_i$. Now $p_i^{\alpha_i}$ divides a but is coprime to b, so $p_i^{\alpha_i}$ divides $\alpha_i - \gamma_i$ also. But $p_i^{\alpha_i} > \alpha_i$ for $\alpha_i > 0$, contradiction. We conclude that $\beta_i > 0$.

Now from the fact that

$$\gamma_i(a + b) = \beta_i + b\alpha_i$$

and the fact that p^{β_i} does not divide β_i (again for size reasons), we deduce that p^{β_i} also does not divide a; that is, $\alpha_i < \beta_i$ for all i and so a divides b. Moreover, the equation above implies that a divides β_i, so we may write $b = c^a$ with $c \geq 2$ a positive integer.

Write $x/a = p/q$ in lowest terms (so $\gcd(p, q) = 1$). Then the original equation becomes $x^a p^b = bq^b$. Now p^b must divide b, which can occur only if $p = 1$. That is, x divides a.

If $x \neq a$, then there exists an i with $\alpha_i \geq \gamma_i + 1$, so

$$\gamma_i(a + b) = \beta_i + \alpha_i b \geq (\gamma_i + 1)b$$

and so $\gamma_i a \geq b$. On the other hand, a is divisible by $p_i^{\gamma_i}$, so in particular $a > \gamma_i$. Thus $a^2 > b = c^a$, or $\sqrt{c} < a^{1/a}$; however, $a^{1/a} \leq \sqrt{2}$ for $a \neq 3$, so this can hold only for $c = 2$ and $a = 3$, in which case $b = 8$ is not divisible by a, contrary to our earlier observation.

Thus $x = a$, and from the original equation we get $b = x^x$, as desired.

Problem 11.13. *Let m, n be integers with $1 \leq m < n$. In their decimal representations, the last three digits of 1978^m are equal, respectively, to the last three digits of 1978^n. Find m and n such that $m + n$ is minimal.*

(20th International Mathematical Olympiad)

Solution. Since 1978^n and 1978^m agree in their last three digits, we have

$$1978^n - 1978^m = 1978^m(1978^{n-m} - 1) \equiv 0 \pmod{10^3}.$$

From the decomposition $10^3 = 2^3 \cdot 5^3$ and since $1978^{n-m} - 1$ is odd, we obtain $2^3 \mid 1978^m$. From $1978 = 2 \cdot 989$, it follows that $m \geq 3$.

Let us write $m + n = (n - m) + 2m$. Our strategy is to minimize $m + n$ by taking $m = 3$ and seek the smallest value of $n - m$ such that

$$1978^{n-m} \equiv 1 \pmod{5^3}.$$

Since $(1978, 5) = 1$, the problem is to find the order h of the residue class 1978 (mod 125). It is known that the order h of an invertible residue class modulo m is a divisor of $\varphi(m)$, where φ is the Euler function. In our case,

$$\varphi(125) = 5^2(5 - 1) = 100.$$

Hence, $h \mid 100$. From $1978^h \equiv 1 \pmod{125}$ we also have $1978^h \equiv 1 \pmod 5$. But $1978^h \equiv 3^h \pmod 5$. Since the order of the residue class 3 (mod 5) is 4, it follows that $4 \mid h$. Using the congruence $1978 \equiv -22 \pmod{125}$, we obtain

$$1978^4 \equiv (-22)^4 \equiv 2^4 \cdot 11^4 \equiv 4^2 \cdot 121^2$$
$$\equiv (4 \cdot (-4))^2 \equiv (-1)^2 \equiv 256 \equiv 6 \not\equiv 1 \pmod{125}.$$

So we rule out the case $h = 4$. Because $h \mid 100$, the next possibilities are $h = 20$ and $h = 100$. By a standard computation we have

$$1978^{20} \equiv 6^5 \equiv 2^5 \cdot 3^5 \equiv 32 \cdot (-7) \equiv -224 \equiv 26 \quad (\text{mod } 125) \not\equiv 1 \quad (\text{mod } 125).$$

Hence we necessarily have $h = m - n = 100$ and $n + m = 106$.

Glossary

Arithmetic function
A complex-valued function defined on the positive integers.

Arithmetic–geometric mean inequality (AM–GM)
If n is a positive integer and $a_1, a_2, \ldots, a_n$ are nonnegative real numbers, then

$$\frac{1}{n} \sum_{i=1}^{n} a_i \geq (a_1 a_2 \cdots a_n)^{1/n},$$

with equality if and only if $a_1 = a_2 = \cdots = a_n$. This inequality is a special case of the *power mean inequality*.

Arithmetic–harmonic mean inequality (AM–HM)
If $a_1, a_2, \ldots, a_n$ are n positive numbers, then

$$\frac{1}{n} \sum_{i=1}^{n} a_i \geq \frac{1}{\frac{1}{n} \sum_{i=1}^{n} \frac{1}{a_i}},$$

with equality if and only if $a_1 = a_2 = \cdots = a_n$. This inequality is a special case of the *power mean inequality*.

Base-b representation
Let b be an integer greater than 1. For any integer $n \geq 1$ there is a unique system $(k, a_0, a_1, \ldots, a_k)$ of integers such that $0 \leq a_i < b, i = 0, 1, \ldots, k, a_k \neq 0$, and

$$n = a_k b^k + a_{k-1} b^{k-1} + \cdots + a_1 b + a_0.$$

Beatty's theorem
Let α and β be two positive irrational real numbers such that

$$\frac{1}{\alpha} + \frac{1}{\beta} = 1.$$

The sets $\{\lfloor \alpha \rfloor, \lfloor 2\alpha \rfloor, \lfloor 3\alpha \rfloor, \ldots\}, \{\lfloor \beta \rfloor, \lfloor 2\beta \rfloor, \lfloor 3\beta \rfloor, \ldots\}$ form a partition of the set of positive integers.

Bernoulli's inequality
For $x > -1$ and $a > 1$,

$$(1+x)^a \geq 1 + ax,$$

with equality when $x = 0$.

Bézout's identity
For positive integers m and n, there exist integers x and y such that $mx + by = \gcd(m, n)$.

Binomial coefficient

$$\binom{n}{k} = \frac{n!}{k!(n-k)!},$$

the coefficient of x^k in the expansion of $(x + 1)^n$.

Binomial theorem
The expansion

$$(x+y)^n = \binom{n}{0}x^n + \binom{n}{1}x^{n-1}y + \binom{n}{2}x^{n-2}y + \cdots + \binom{n}{n-1}xy^{n-1} + \binom{n}{n}y^n.$$

Canonical factorization
Any integer $n > 1$ can be written uniquely in the form

$$n = p_1^{\alpha_1} \cdots p_k^{\alpha_k} \text{ and } p_1 < p_2 < \cdots < p_k,$$

where $p_1, \ldots, p_k$ are distinct primes and $\alpha_1, \ldots, \alpha_k$ are positive integers.

Carmichael numbers
The composite integers n satisfying $a^n \equiv a \pmod{n}$ for any integer a.

Ceiling function
The integer $-\lfloor -x \rfloor$ is called the ceiling of x and is denoted by $\lceil x \rceil$.

Complete set of residue classes modulo n
A set S of integers such that for each $0 \leq i < n$ there is an element $s \in S$ with $i \equiv s \pmod{n}$.

Congruence relation
Let a, b, and m be integers. We say that a and b are congruent modulo m if $m \mid a - b$. We denote this by $a \equiv b \pmod{m}$. The relation "$\equiv$" on the set $\mathbb{Z}$ of integers is called the congruence relation.

Convolution product
The arithmetic function defined by

$$(f * g)(n) = \sum_{d|n} f(d)g\left(\frac{n}{d}\right),$$

where f and g are two arithmetic functions.

Division algorithm
For any positive integers a and b there exists a unique pair (q, r) of nonnegative integers such that $b = aq + r$ and $r < a$.

Euclidean algorithm
Repeated application of the division algorithm:

$$m = nq_1 + r_1, \ 1 \leq r_1 < n,$$
$$n = r_1 q_2 + r_2, \ 1 \leq r_2 < r_1,$$
$$\cdots$$
$$r_{k-2} = r_{k-1} q_k + r_k, \ 1 \leq r_k < r_{k-1},$$
$$r_{k-1} = r_k q_{k+1} + r_{k+1}, \ r_{k+1} = 0 \text{ and } r_k = \gcd(m, n).$$

This chain of equalities is finite because $n > r_1 > r_2 > \cdots > r_k > 0$.

Euler's theorem
Let a and m be relatively prime positive integers. Then

$$a^{\varphi(m)} \equiv 1 \pmod{m}.$$

Euler's totient function
The function φ defined by $\varphi(m) =$ the number of all positive integers n less than or equal to m that are relatively prime to m.

Factorial base expansion
Every positive integer k has a unique expansion

$$k = 1! \cdot f_1 + 2! \cdot f_2 + 3! \cdot f_3 + \cdots + m! \cdot f_m,$$

where each f_i is an integer, $0 \leq f_i \leq i$ and $f_m > 0$.

Fermat's little theorem
Let a be any integer and let p be a prime. Then

$$a^p \equiv a \pmod{p}.$$

Fermat numbers
The integers $f_n = 2^{2^n} + 1, n \geq 0$.

Fibonacci sequence
The sequence defined by $F_0 = 0$, $F_1 = 1$, and $F_{n+1} = F_n + F_{n-1}$ for every positive integer n.

Floor function
For a real number x there is a unique integer n such that $n \leq x < n + 1$. We say that n is the greatest integer less than or equal to x or the floor of x and we write $n = \lfloor x \rfloor$.

Fractional part
The difference $x - \lfloor x \rfloor$ is called the fractional part of x and is denoted by $\{x\}$.

Fundamental theorem of arithmetic
Every integer greater than 1 has a unique representation (up to a permutation) as a product of primes.

Hermite's identity
For any real number x and for any positive integer n,

$$\lfloor x \rfloor + \left\lfloor x + \frac{1}{n} \right\rfloor + \left\lfloor x + \frac{2}{n} \right\rfloor + \cdots + \left\lfloor x + \frac{n-1}{n} \right\rfloor = \lfloor nx \rfloor.$$

Lagrange's theorem
If the polynomial $f(x)$ with integer coefficients has degree d modulo p (where p is a prime), then the number of distinct roots of $f(x)$ modulo p is at most p.

Legendre's formula
For any prime p and any positive integer n,

$$e_p(n) = \sum_{i \geq 1} \left\lfloor \frac{n}{p^i} \right\rfloor, \text{ where } n! = \prod_{\substack{p \leq n \\ \text{prime}}} p^{e_p(n)}.$$

Legendre function
Let p be a prime. For any positive integer n, $e_p(n)$ is the exponent of p in the prime factorization of $n!$.

Legendre symbol
Let p be an odd prime and let a be a positive integer not divisible by p. The Legendre symbol of a with respect to p is defined by

$$\left(\frac{a}{p} \right) = \begin{cases} 1 & \text{if } a \text{ is a quadratic residue mod } p, \\ -1 & \text{otherwise.} \end{cases}$$

Linear Diophantine equation
An equation of the form

$$a_1 x_1 + \cdots + a_n x_n = b,$$

where $a_1, a_2, \ldots, a_n, b$ are fixed integers.

Linear recursion of order k
A sequence $x_0, x_1, \ldots, x_2, \ldots$ of complex numbers defined by

$$x_n = a_1 x_{n-1} + a_2 x_{n-2} + \cdots + a_k x_{n-k}, \quad n \geq k,$$

where $a_1, a_2, \ldots, a_k$ are given complex numbers and $x_0 = \alpha_0$, $x_1 = \alpha_1, \ldots,$ $x_{k-1} = \alpha_{k-1}$ are also given.

Lucas's sequence
The sequence defined by $L_0 = 2$, $L_1 = 1$, $L_{n+1} = L_n + L_{n-1}$ for every positive integer n.

Mersenne numbers
The integers $M_n = 2^n - 1, n \geq 1$.

Möbius function
The arithmetic function μ defined by

$$\mu(n) = \begin{cases} 1 & \text{if } n = 1, \\ 0 & \text{if } p^2 \mid n \text{ for some prime } p > 1, \\ (-1)^k & \text{if } n = p_1 \cdots p_k, \text{ where } p_1, \ldots, p_k \text{ are distinct primes.} \end{cases}$$

Möbius inversion formula
Let f be an arithmetic function and let F be its summation function. Then

$$f(n) = \sum_{d \mid n} \mu(d) F\left(\frac{n}{d}\right).$$

Multiplicative function
An arithmetic function $f \neq 0$ with the property that for any relatively prime positive integers m and n,

$$f(mn) = f(m) f(n).$$

Number of divisors
For a positive integer n denote by $\tau(n)$ the number of its divisors. It is clear that

$$\tau(n) = \sum_{d \mid n} 1.$$

Order modulo m
We say that a has order d modulo m, denoted by $o_m(a) = d$, if d is the smallest positive integer such that $a^d \equiv 1 \pmod{m}$. We have $o_n(1) = 1$ and $o_n(0) = 0$.

Pell's equation
The quadratic equation $u^2 - Dv^2 = 1$, where D is a positive integer that is not a perfect square.

Perfect number
An integer n with the property that the sum of its divisors is equal to $2n$.

Power mean inequality
Let $a_1, a_2, \dots, a_n$ be any positive numbers for which $a_1 + a_2 + \dots + a_n = 1$. For positive numbers $x_1, x_2, \dots, x_n$ we define

$$M_{-\infty} = \min\{x_1, x_2, \dots, x_k\},$$
$$M_{\infty} = \max\{x_1, x_2, \dots, x_k\},$$
$$M_0 = x_1^{a_1} x_2^{a_2} \cdots x_n^{a_n}$$
$$M_t = \left(a_1 x_1^t + a_2 x_2^t + \dots + a_k x_k^t\right)^{1/t},$$

where t is a nonzero real number. Then

$$M_{-\infty} \leq M_s \leq M_t \leq M_{\infty}$$

for $s \leq t$.

Prime number theorem
The relation

$$\lim_{n \to \infty} \frac{\pi(n)}{n/\log n} = 1,$$

where $\pi(n)$ denotes the number of primes $\leq n$.

Prime number theorem for arithmetic progressions
Let $\pi_{r,a}^{(n)}$ be the number of primes in the arithmetic progression $a, a+r, a+2r, a+3r, \dots$, less than n, where a and r are relatively prime. Then

$$\lim_{n \to \infty} \frac{\pi_{r,a}(n)}{n/\log n} = \frac{1}{\varphi(r)}.$$

This was conjectured by Legendre and Dirichlet and proved by de la Vallée Poussin.

Pythagorean equation
The Diophantine equation $x^2 + y^2 = z^2$.

Pythagorean triple
A triple of the form (a, b, c) where $a^2 + b^2 = c^2$. All primitive Pythagorean triples are given by $(m^2 - n^2, 2mn, m^2 + n^2)$, where m and n are positive integers.

Quadratic residue mod m
Let a and m be positive integers such that $\gcd(a, m) = 1$. We say that a is a quadratic residue mod m if the congruence $x^2 \equiv a \pmod{m}$ has a solution.

Quadratic reciprocity law of Gauss
If p and q are distinct odd primes, then

$$\left(\frac{q}{p}\right)\left(\frac{p}{q}\right) = (-1)^{\frac{p-1}{2} \cdot \frac{q-1}{2}}.$$

Root mean square–arithmetic mean inequality
For positive numbers $x_1, x_2, \ldots, x_n$,

$$\sqrt{\frac{x_1^2 + x_2^2 + \cdots + x_k^2}{n}} \geq \frac{x_1 + x_2 + \cdots + x_k}{n}.$$

Sum of divisors
For a positive integer n denote by $\sigma(n)$ the sum of its positive divisors including 1 and n itself. It is clear that

$$\sigma(n) = \sum_{d \mid n} d.$$

Summation function
For an arithmetic function f, the function F defined by

$$F(n) = \sum_{d \mid n} f(d).$$

Wilson's theorem
For any prime p, $(p - 1)! \equiv -1 \pmod{p}$. So n is prime if and only if $(n - 1)! \equiv -1 \pmod{n}$.

Bibliography

[1] Andreescu, T.; Feng, Z., *101 Problems in Algebra from the Training of the USA International Mathematical Olympiad Team*, Australian Mathematics Trust, 2001.

[2] Andreescu, T.; Feng, Z., *102 Combinatorial Problems from the Training of the USA International Mathematical Olympiad Team*, Birkhäuser, 2002.

[3] Andreescu, T.; Feng, Z., *103 Trigonometry Problems from the Training of the USA International Mathematical Olympiad Team*, Birkhäuser, 2004.

[4] Feng, Z.; Rousseau, C.; Wood, M., *USA and International Mathematical Olympiads 2005*, Mathematical Association of America, 2006.

[5] Andreescu, T.; Feng, Z.; Loh, P., *USA and International Mathematical Olympiads 2004*, Mathematical Association of America, 2005.

[6] Andreescu, T.; Feng, Z., *USA and International Mathematical Olympiads 2003*, Mathematical Association of America, 2004.

[7] Andreescu, T.; Feng, Z., *USA and International Mathematical Olympiads 2002*, Mathematical Association of America, 2003.

[8] Andreescu, T.; Feng, Z., *USA and International Mathematical Olympiads 2001*, Mathematical Association of America, 2002.

[9] Andreescu, T.; Feng, Z., *USA and International Mathematical Olympiads 2000*, Mathematical Association of America, 2001.

[10] Andreescu, T.; Feng, Z.; Lee, G.; Loh, P., *Mathematical Olympiads: Problems and Solutions from around the World, 2001–2002*, Mathematical Association of America, 2004.

[11] Andreescu, T.; Feng, Z.; Lee, G., *Mathematical Olympiads: Problems and Solutions from around the World, 2000–2001*, Mathematical Association of America, 2003.

[12] Andreescu, T.; Feng, Z., *Mathematical Olympiads: Problems and Solutions from around the World, 1999–2000*, Mathematical Association of America, 2002.

[13] Andreescu, T.; Feng, Z., *Mathematical Olympiads: Problems and Solutions from around the World, 1998–1999*, Mathematical Association of America, 2000.

[14] Andreescu, T.; Kedlaya, K., *Mathematical Contests 1997–1998: Olympiad Problems from around the World, with Solutions*, American Mathematics Competitions, 1999.

[15] Andreescu, T.; Kedlaya, K., *Mathematical Contests 1996–1997: Olympiad Problems from around the World, with Solutions*, American Mathematics Competitions, 1998.

[16] Andreescu, T.; Kedlaya, K.; Zeitz, P., *Mathematical Contests 1995–1996: Olympiad Problems from around the World, with Solutions*, American Mathematics Competitions, 1997.

[17] Andreescu, T.; Enescu, B., *Mathematical Olympiad Treasures*, Birkhäuser, 2003.

[18] Andreescu, T.; Gelca, R., *Mathematical Olympiad Challenges*, Birkhäuser, 2000.

[19] Andreescu, T., Andrica, D., *An Introduction to Diophantine Equations*, GIL Publishing House, 2002.

[20] Andreescu, T.; Andrica, D., *360 Problems for Mathematical Contests*, GIL Publishing House, 2003.

[21] Andreescu, T.; Andrica, D., *On a class of sums involving the floor function*, Mathematical Reflections, **3**(2006), 5 pp.

[22] Andreescu, T.; Andrica, D.; Feng, Z., *104 Number Theory Problems From the Training of the USA International Mathematical Olympiad Team*, Birkhäuser, 2007.

[23] Djuikić, D.; Janković, V.; Matić, I.; Petrović, N., *The International Mathematical Olympiad Compendium*, A Collection of Problems Suggested for the International Mathematical Olympiads: 1959–2004, Springer, 2006.

[24] Doob, M., *The Canadian Mathematical Olympiad 1969–1993*, University of Toronto Press, 1993.

[25] Engel, A., *Problem-Solving Strategies*, Problem Books in Mathematics, Springer, 1998.

[26] Everest, G., Ward, T., *An Introduction to Number Theory*, Springer, 2005.

[27] Fomin, D.; Kirichenko, A., *Leningrad Mathematical Olympiads 1987–1991*, MathPro Press, 1994.

[28] Fomin, D.; Genkin, S.; Itenberg, I., *Mathematical Circles*, American Mathematical Society, 1996.

[29] Graham, R.L.; Knuth, D.E.; Patashnik, O., *Concrete Mathematics*, Addison-Wesley, 1989.

[30] Gillman, R., *A Friendly Mathematics Competition*, The Mathematical Association of America, 2003.

[31] Greitzer, S.L., *International Mathematical Olympiads, 1959–1977*, New Mathematical Library, Vol. 27, Mathematical Association of America, 1978.

[32] Grosswald, E., *Topics from the Theory of Numbers*, Second Edition, Birkhäuser, 1984.

[33] Hardy, G.H.; Wright, E.M., *An Introduction to the Theory of Numbers*, Oxford University Press, 5th Edition, 1980.

[34] Holton, D., *Let's Solve Some Math Problems*, A Canadian Mathematics Competition Publication, 1993.

[35] Kedlaya, K; Poonen, B.; Vakil, R., *The William Lowell Putnam Mathematical Competition 1985–2000*, The Mathematical Association of America, 2002.

[36] Klamkin, M., *International Mathematical Olympiads, 1978–1985*, New Mathematical Library, Vol. 31, Mathematical Association of America, 1986.

[37] Klamkin, M., *USA Mathematical Olympiads, 1972–1986*, New Mathematical Library, Vol. 33, Mathematical Association of America, 1988.

[38] Kürschák, J., *Hungarian Problem Book, volumes I & II*, New Mathematical Library, Vols. 11 & 12, Mathematical Association of America, 1967.

[39] Kuczma, M., *144 Problems of the Austrian–Polish Mathematics Competition 1978–1993*, The Academic Distribution Center, 1994.

[40] Kuczma, M., *International Mathematical Olympiads 1986–1999*, Mathematical Association of America, 2003.

[41] Larson, L.C., *Problem-Solving Through Problems*, Springer-Verlag, 1983.

[42] Lausch, H. *The Asian Pacific Mathematics Olympiad 1989–1993*, Australian Mathematics Trust, 1994.

[43] Liu, A., *Chinese Mathematics Competitions and Olympiads 1981–1993*, Australian Mathematics Trust, 1998.

[44] Liu, A., *Hungarian Problem Book III*, New Mathematical Library, Vol. 42, Mathematical Association of America, 2001.

[45] Lozansky, E.; Rousseau, C. *Winning Solutions*, Springer, 1996.

[46] Mordell, L.J., *Diophantine Equations*, Academic Press, London and New York, 1969.

[47] Niven, I., Zuckerman, H.S., Montgomery, H.L., *An Introduction to the Theory of Numbers*, Fifth Edition, John Wiley & Sons, Inc., New York, Chichester, Brisbane, Toronto, Singapore, 1991.

[48] Savchev, S.; Andreescu, T. *Mathematical Miniatures*, Anneli Lax New Mathematical Library, Vol. 43, Mathematical Association of America, 2002.

[49] Shklarsky, D.O; Chentzov, N.N; Yaglom, I.M., *The USSR Olympiad Problem Book*, Freeman, 1962.

[50] Slinko, A., *USSR Mathematical Olympiads 1989–1992*, Australian Mathematics Trust, 1997.

[51] Szekely, G. J., *Contests in Higher Mathematics*, Springer-Verlag, 1996.

[52] Tattersall, J.J., *Elementary Number Theory in Nine Chapters*, Cambridge University Press, 1999.

[53] Taylor, P.J., *Tournament of Towns 1980–1984*, Australian Mathematics Trust, 1993.

[54] Taylor, P.J., *Tournament of Towns 1984–1989*, Australian Mathematics Trust, 1992.

[55] Taylor, P.J., *Tournament of Towns 1989–1993*, Australian Mathematics Trust, 1994.

[56] Taylor, P.J.; Storozhev, A., *Tournament of Towns 1993–1997*, Australian Mathematics Trust, 1998.

Index of Authors

Subject Index

GPSR Compliance
The European Union's (EU) General Product Safety Regulation (GPSR) is a set
of rules that requires consumer products to be safe and our obligations to
ensure this.

If you have any concerns about our products, you can contact us on

ProductSafety@springernature.com

In case Publisher is established outside the EU, the EU authorized
representative is:

Springer Nature Customer Service Center GmbH
Europaplatz 3
69115 Heidelberg, Germany